CULTURES OF ENERGY

power, practices, technologies

cultures of energy

edited by
sarah strauss
stephanie rupp
thomas love

WALNUT CREEK, CA

Left Coast Press, Inc.
1630 North Main Street, #400
Walnut Creek, CA 94596
http://www.LCoastPress.com

ISBN 978-1-61132-165-4 hardback
ISBN 978-1-61132-166-1 paperback
ISBN 978-1-61132-167-8 institutional eBook
ISBN 978-1-61132-685-7 consumer eBook

Library of Congress Cataloging-in-Publication Data

Cultures of energy : power, practices, technologies / Sarah Strauss, Stephanie Rupp, Thomas Love, editors.
p. cm.
Includes bibliographical references.
ISBN 978-1-61132-165-4 (hbk. : alk. paper) — ISBN 978-1-61132-166-1 (pbk. : alk. paper) — ISBN 978-1-61132-167-8 (institutional eBook) — ISBN 978-1-61132-685-7 (consumer eBook)
1. Energy consumption—Social aspects. 2. Power resources—Social aspects. I. Strauss, Sarah. II. Rupp, Stephanie. III. Love, Thomas F.
HD9502.A2C853 2013
333.79—dc23
2012032963

Printed in the United States of America

∞ ™ The paper used in this publication meets the minimum requirements of American National Standard for Information Sciences—Permanence of Paper for Printed Library Materials, ANSI/NISO Z39.48–1992.

Cover design by Allison Smith
Cover photo: Land of Giants™ ©Choi+Shine Architects

We dedicate this book to all who seek a more sensible

and sustainable energy future for our planet.

CONTENTS

INTRODUCTION

Powerlines: Cultures of Energy in the Twenty-first Century

Sarah Strauss, Stephanie Rupp, and Thomas Love

ENERGY: PARADOXES, PREDICAMENTS, AND PROBLEMS

Our beliefs about energy shape how we use it; our uses of energy simultaneously shape our cultural concepts of and beliefs about energy. People perceive energy as invisible, omnipotent, and dangerous—or as desirable and indispensable, but perhaps unreliable. Through both conceptual essays and case studies, this volume explores cultural conceptions of energy as it is imagined, developed, utilized, and contested in everyday contexts around the globe. The issues of energy use and conservation are crucial for both the United States and the world, yet the degree of social science engagement with this topic—especially in terms of human values, beliefs, and behaviors—has so far been minimal and uneven. Despite the diverse theoretical perspectives of this volume's contributors, we argue in one voice that human use of energy is understood and experienced through cultural frameworks. Given the fast pace of technical innovation, the blockages to responding effectively to the enormous energy challenges facing us all are fundamentally cultural and political rather than technological. This volume examines cultural conceptions, assumptions, and expectations of energy in societies from many parts of the world, enabling readers to think about energy in new ways: as a cultural artifact in particular contexts, and as a broad, cross-cultural concept and concern.

Taken together, these ethnographic cases and theoretical analyses demonstrate that production, distribution, and consumption of energy almost never follow a simple logic of neoclassical economic efficiency; rather, people tend to switch frames of reference among technical, economic, and cultural logics when considering their uses of energy. This tendency to deviate from measures of physical or market efficiency is particularly pronounced in our current era, ironically, as people have rapidly become accustomed to, and indeed increasingly dependent upon, the continued flow of cheap energy to maintain the burgeoning global expectation of continuous economic growth, material accumulation, and "progress."

As the "master resource," energy empowers and transforms the world as it flows in varied forms through natural and social circuitry. While energy originates in natural sources, energy politics comprise both cultural and technological issues. Taking our lead from Appadurai (1990), this volume offers analyses of "energyscapes" at local, national, and transnational levels; contributors consider the problem of energy in motion across social and physical spaces, shifting its cultural, social, economic, and technological values as it flows from one domain to the next.

Three themes unify the sections of the volume, highlighting connections across the disparate examples of energy cultures represented in this collection. First, a foundational pair of ideas—*currents* and *flows*—links chapters throughout the volume. In particular, contributions examine the parallel between physical flows of energy and the circulation of social, economic, and political relationships that emerge in the context of energy systems. The metaphor of *powerlines* expresses the notion of currents and flows and serves to link together chapters that address the relationships between energy and social, political, and economic values—forces that, together, constitute *power*. Second, the theme of *transformation* addresses issues of social change, cultural notions of progress, and the possibility of collapse—economic, political, and social—as a result of human overreliance on finite energy resources. Third, the volume sheds new light on the blurry cultural boundary between *technology* and *magic*, highlighting the multiple and simultaneous interpretations of energy and energy technologies that people in diverse societies hold.

Anthropology has a tremendous role to play in helping to document cultural conceptions of and assumptions about energy uses in all these senses of power—currents/flows, transformations, and technology/magic. In addition, anthropologists are well positioned to facilitate the kinds of conceptual and social change that reducing global dependence on fossil fuels will require, especially by individuals in affluent societies, including the mitigation of social tensions likely to accompany coming energy contraction.

REENERGIZING ANTHROPOLOGY

Energy is an area ripe for anthropological investigation in at least three ways: how people experience and utilize energies of various qualities (types), how we rely on its quantity (continued flow), and how we harness both qualities and quantities of energy to construct socially meaningful worlds.

First, people tend to interact with energy in a variety of forms, not as a monolith. Each form has its own specific qualities such as frequencies, strengths, sources, and potential uses—chemical, mechanical, kinetic, electrical—and availability that ranges from intermittent to constant, institutionally provided or individually generated. These energy sources are sited from deep-sea deposits to windy heights, with properties that endure, decay slowly, or fleet momentarily through circuits. How do such variable properties of energy intersect with people's needs and desires? An anthropology of energy must shuttle back and forth among laws of physics, opportunities and constraints of ecological systems, and processes of culture; furthermore, these layers of reality are necessarily intertwined materially, rhetorically, and metaphorically.

Second, we build our social relationships and cultural understandings to coalesce around the continued flow of energy of familiar qualities in expected quantities. Ensuring access to continued supplies of energy and other resources is one of the central functions of centralized political systems. Shortages of energy—blackouts and queues for gasoline—quickly become political problems and often have political antecedents. However people encounter and experience it, the flow of energy in a place tends to be part of the taken-for-grantedness of unspoken, ordinary social life. Who thinks much about our utter dependence on the sun's regular daily bathing of the lit side of our planet, or about the political structures that people have constructed to harness, manage, and distribute the sun's resources in its many forms? Because of the necessity of institutions to manage energy flows, and because of the necessity of energy flows to individual agency, an anthropology of energy is necessarily political (see Vayda and Walters 1999).

Third, although people realize at some level that "energy" drives our worlds, humans typically think about and experience energy according to what it does and how it enables our goals. As contributors to this volume demonstrate, energy never just "is," existing as some unmediated potentiality; it flows through socionatural systems via the nodes and switches in the social circuitry of power and meaning-making. People make sense of energy in a plethora of ways, from animistic veneration of the sun, wind, or other natural forces to commodity or machine fetishism (Hornborg 2008). An anthropology of energy must therefore also analyze multiple, contested meanings.

Macroscopically, Leslie White (1943, 1949, 2008) reminds us that despite local ebbs and flows, the amount of energy that humans have harnessed per capita has increased steadily over the *longue durée*. Societies that organized themselves to harness more energy per capita outcompeted societies that did not (see Schmookler 1984). Virtually no society in the world today is untouched by fossil-fueled technical, social, political, and economic relations; the ways in which people have constructed, utilized, and contested such energy relations vary across both time and space.

But what happens when the quantity or quality of energy flow changes, and with it the social arrangements and cultural understandings built around a particular energy regime? People tend to think about changes in the quantity and quality of energy in positive terms, emphasizing the potential for social evolution that results from technological ingenuity: the roughly sequential discoveries of whale oil, coal, and petroleum as sources of energy followed by the harnessing of electricity first from steam then fossil fuels, nuclear fission, hydrological sources, and more recently solar, wind, and biofuel sources all contributed to the acceleration of economic and social "progress." The economic intensification and enhanced potential for social prestige that are hallmarks of modernity have, over the unprecedented development of the past century, come to seem not only desirable but almost magical (the American Dream) and unstoppable (the Industrial Revolution). Uncountable "energy slaves"—technical devices that harness cheap energy—have made mass upward socioeconomic mobility not just possible but enthralling and addictive. The global intensity of our era (Tsing 2000) may be taken for granted by those whose supply and consumption of energy is relatively unproblematic, even as the inexorable pressures of globalization are actively resisted through local, renewable energy projects such as those described in this volume. Although the specter of unprecedented long-term reversal of this period of energy intensification increasingly haunts our late-modern world, such an energy contraction seems impossible, paralyzing us by its sheer magnitude.

Even as humanity approaches planetary limits to growth (Meadows, Randers, and Meadows 2004; Catton 1980), it is difficult to see how to halt, or even slow, the widening and accelerating gyre of technological change and socioeconomic intensification. Undergirding the growing malaise in late capitalism is a spreading awareness that unbridled growth cannot continue on a finite planet, that the supplies of fossil fuels we have consumed so prodigiously are in reality a one-time boon of ancient solar energy; the rate of their production seems now to have peaked (Murray and King 2012). *Homo colossus*—a human type whose reliance on technology enables its production and consumption to reach colossal proportions (Catton 1980)—has borrowed heavily against the future. And this future just isn't what "the Future" used to be: it seems unlikely to promise the same economic returns and social progress as our experiences of "futures" have provided over the past century of unprecedented growth, expansion, and intensification. Global limits on the provision of the cheap fuels that power globalization are in view; intensifying conflict over access to energy resources seems inevitable (Klare 2012). Unexpected challenges to the confident worldview of "progress" seem to lie dead ahead (Love 2008). Underlying the contributions to this volume is a shared sense that the time for punditry and half-measures is over. And it is our collective hope that we have a role to play in mitigating the environmental effects and social consequences of our energy choices at individual, communal, national, and global levels.

THE STATE OF ENERGY RESEARCH

Energy has entered the anthropological literature in various guises over the course of the past century. Throughout this literature—some of it technical, and some historical, political-economic, or policy-oriented—energy is presented as a material resource whose presence is evident, whose functions are clear, and whose materiality is obvious and incontrovertible. However, considering energy itself as a phenomenon that merits analysis—energy as a cultural artifact—remains a gap in the large, multidisciplinary literature on energy. In addressing this gap, contributors to this volume seek to highlight the contingent nature of energy: how people describe, discuss, and debate energy as a substance reflects their fundamental values regarding power and agency, and impacts their everyday uses and abuses of energy resources. This volume builds directly on the literature discussed here, guiding the field of energy studies in new directions by highlighting cultural values and relationships through which people perceive, and by means of which people access, energy itself.

Energy and Cultural Evolution: Leslie White's Legacy

Among anthropologists, Leslie White was most prominent in writing about energy, analyzing its relation to the evolution of culture and the development of civilization (1943, 1959). In his earlier work (1943), White advanced the fundamental hypothesis that "other things being equal, the degree of cultural development varies directly as the amount of energy per capita per year harnessed and put to work," a seminal idea that he developed more extensively in his 1959 volume, *The Evolution of Culture*. White struggled with the discipline's rejection of evolutionary theories of development in the first half of the twentieth century, and so his work emphasized an empirical return to analyzing cultural change; he analyzed the role of changing qualities and quantities of energy inputs as a key variable in the development—the step-by-step progression—of civilization. Arguing that "civilization" displaced "barbarism" (cultural existence based on agricultural production) "with the invention of the steam engine as a practical means of harnessing energy" in 1698, White devoted much of his scholarly career to charting the evolution of civilization as a result of the intensified use of energy over the course of the twentieth century (White 1943, 1959, 1973) and spawned related research and writing about energy, society, and change by other authors (Cottrell 1955; Sahlins 1960). White's argument about the evolution of culture based on energy intensification faced robust criticism from anthropologists (in particular the Boasians) who resisted evolutionary schemes and from scholars who debated the fundamental relationship between the utilization of energy and the development of society (Adams 1975; Nader and Beckerman 1978; Rambo 1991). Nevertheless, the application of energy—as *the* universal concept of science (White 1959)—to the study of culture emphasized a concrete materiality in anthropological analyses of human societies and cultural change. Inspired by Morgan, White's basic insight that transformations in technology—including the intensification of use of energy—are coupled with transformations in cultural contexts and social institutions presents a compelling touch point for contemporary research on the impacts of consumption and technology on the natural environment. At the same time, White's narrow focus on technology, the satisfaction of human needs through reliance on material outer forces, such as tools, weapons, and other materials, rather than "inner resources" contained within the human organism, such as myth-making and social associations, seems artificially to separate and unproductively to decouple technology from social values. As the chapters in this volume demonstrate robustly, contemporary uses of and relations to energy are intimately connected with people's social values and images of energy and its associated technologies; how people use energy is related to how people value it; and how people value energy is related to what it enables them to accomplish not only materially but also socially and culturally.

Energy Systems: Linking Science and Society

A considerable body of literature addresses technical understandings of energy systems as they relate to science and society (for example, Odum and Odum 1976; Caldwell 1976; Marchetti 2003; Smil 2008). Odum (1971) and Odum and Odum (1976) produced classic studies of energy flows between nature and society, integrating analysis of the social world into the larger context of energy production and transfer. Building on this work, cultural ecologists focused analytical attention on energy flows in various social contexts, including hunting societies (Kemp 1975), agricultural societies (Rappaport 1975; Finison 1979; Rambo 1980), in rural India (Revelle 1976), and in China (Smil 1978). Concurrently, human ecologists and biologists studied bioenergetics, the flow of energy through living systems, analyzing relations among caloric intake, subsistence practices, and metabolic rates (West 1974; Wen 1976; Norgan 1982; Burnham 1982; Cabrera and Fuster 2002; Grünbühel et al. 2003; Dufour and Piperata 2008). Studies of energy flows, subsistence patterns, and energy resources tended to focus on smaller, more bounded rural communities. Beginning in the 1980s and carrying through current literature, energy studies have examined social patterns of harvesting of energy resources such as firewood (Fergus 1983; Jones 1984; Haile 1989), the importance of the household as the unit of study in rural energy systems (Najib 1993; Mukhopadhyay and Malabika 1996; Hasalkar et al. 2002), the central role of women in conserving energy resources (Klausner 1979; DeFonzo and Warkov 1979; Little et al. 1992; Yahaya et al. 2007), and the access of rural communities to affordable and appropriate energy technologies (Dioffo 1981; Blamont 1989; Eliatamby 1999; Mfune and Boon 2008; Sanchez 2010). In such research on subsistence strategies, rural development, and cultural ecology, energy enters analysis as a quantitative measure of biological sustenance and economic production.

Energy and Social History

Another significant area of research on energy concerns its role in the history of social development. When he passed away in 1975, Leslie White left behind unpublished manuscripts that he had intended as sequels to his classic volume, *The Evolution of Culture*. The second volume in this trilogy was to have been *The Fuel Revolution*, which has been published posthumously as an extensive chapter in the unabridged edition of *Modern Capitalist Culture* (2008), the third volume in the triptych. In *The Fuel Revolution*, White extends his argument about the role of energy intensification and technological change in social advancement by tracing technological and associated social changes in the history of the United States. As White describes,

> Such was the great Fuel Revolution: one of the most dramatic and significant episodes in the entire course of human history. . . . By the beginning of the eighteenth century . . . cultural development had gone just about as far as it could on the basis of animate energy and wind and water; it could not advance appreciably farther without tapping new sources of

> energy. Herein lies the significance of the revolutionary achievement of harnessing the energy of fossil fuels. Vast amounts of energy were locked up in the earth's crust in the form of coal, oil, and gas. The development of steam and internal combustion engines was the means of harnessing and utilizing these energies in ever increasing amounts. And the new technology was extended into all phases of life: into industry, transportation by land and by water, aviation, and into the arts of war as well as those of peace. (White 2008:124)

White was not the only scholar who analyzed energy as a key variable in social change; other historical works pay close attention to the role of energy technologies in the history of the world (Ridgeway 1982; Hughes 1983; Adams 1988; Smil 1994; Black 2000). Within the body of historical literature on energy, scholars also differentiated their focus regionally, with studies on energy in British history (Adams 1982), European history (Schivelbusch 1995), African history (Klieman 2008, 2012, forthcoming), and American history (Nye 1992, 1994, 1998; Williams 1997).

Energy Crises

Periodic energy crises tend to spawn, well, periodic academic research and policy papers on energy issues. Social scientific literature on energy bulged during the period from 1973 through the early 1980s, when America's first energy crisis was triggered by the oil embargo staged by the Organization of Petroleum Exporting Countries (OPEC). Literature on energy ballooned again from 2003 until today, reflecting renewed concern about the industrial nations' overdependence on foreign sources of oil, in particular in the Middle East. Energy crises also lend themselves to literature that offers large-scale analyses of energy systems and their associated socioeconomic forces, representing an effort to ground public audiences in the basic "facts" of energy resources and utilization (for example, Diesendorf 1979; Smil 2003, 2008; Armstrong and Blundell 2007). Energy crises have invited futurist analyses of patterns of energy consumption and supply extended forward (Nader 1980; Smil 2008; Odum and Odum 2001), anticipating significant changes necessary to America's and other nations' patterns of energy use to avoid future, even more serious, crises.

The warnings generated by energy researchers in the years approaching 1980 (for example, Adams 1975) went largely unheeded, as the U.S. restabilized in the 1980s by exploiting new oil resources from Alaska, Mexico, and elsewhere. A return to cheaper oil undermined the nascent development of alternative energy technologies, renewed faith in growth and "progress," and ushered in an era of escalating consumption of energy resources. While in the public mind the energy crisis of the first decade of the twenty-first century peaked during the summer of 2008, when the international price of oil reached a historic $147.27 per barrel, continuing high prices—even during the worst economic downturn since the 1930s—suggest that energy overconsumption is a problem that will not be easy to reverse. In the first decade of the twenty-first century, Americans again sought to quench their thirst for energy to power fast-paced transportation, to control their personal climates, and to consume material objects.

Values and Attitudes Toward Energy

Periods of energy crisis have also tended to foment analyses of social values and attitudes that ordinary people hold regarding energy. Influential opinions regarding American consumption of energy staked out the terrain of public debate (Holdren 1976; Lovins 1976; Gore 2006). Scholars related energy to research on attitudes and values, using social surveys to gauge American attitudes about energy and the environment (Barbour et al. 1982; Conn 1983; Kempton et al. 1995; Erickson 1997; Sovacool and Brown 2007). These collected volumes are important contributions in understanding the social contexts of energy debates, but by treating energy itself as a given, the contributing authors step over the core of the problem. What is it that Americans believe they are conserving or planning when they conserve energy or plan for energy utilization?

Energy Controversies: Risk and Vulnerability

The controversies over nuclear energy, the development of nuclear weapons, and the dangers of nuclear proliferation also have a solid place in the anthropological literature on energy. From historical and anthropological perspectives, scholars have drawn connections between contested identities, national belonging, and nuclear development (Hecht 1998; Topçu 2008). Social science research also examines underlying concepts of risk embedded in nuclear industries (Downey 1986) as well as broader concepts of pollution, health, and identity in the nuclear context (Kirby 2011). Anthropologists have contributed social and symbolic analysis of nuclear laboratories, including the implication of continued use of nuclear energy for peaceful purposes (Gusterson 1996, 2011; Johnston 2011), and have offered ethnographic perspectives on nuclear industrial complexes and their borderlands (Masco 2006). The intractable issues of nuclear waste disposal, storage, and potential contamination have been addressed from environmental, social, and policy perspectives (Holdren 1982; Lenssen 1992; Taubes 2002; Perrin 2006).

Anthropological research has also examined the often simultaneously precarious and strategic position of indigenous communities regarding energy resources. In the context of the United States, a significant body of anthropological literature has focused on Native American communities and their contested relations with the U.S. government and with industrial energy interests (for an overview of preliminary issues, see Owen 1979a, 1979b; Robbins 1984). Studies examined the effects of mining for energy resources on Native American communities (Little 1979; Irvin 1982; Campbell 1987), the potential benefits and costs to Native Americans of drilling for Alaskan fossil fuels (Kruse et al. 1982; Robbins and McNabb 1987), and the still-elusive supply of electricity to remote Native American peoples (Bain et al. 2004). Studies highlight similar concern for indigenous communities impacted by industrial extraction of energy resources in other parts of the world (Topo 1982; Collier 1994; Huertas Castillo 2003).

Energy Sectors

Given the scope of energy's application to technology, society, and the environment, research on energy tends to be compartmentalized according to sectors. For example, significant research has taken up the contested spaces and politics of oil resources. Numerous books—by participants in the oil industry as well as by policy and academic researchers—serve as "primers" on oil, addressing the physical and chemical properties of petroleum products, as well as the technologies, economics, and politics of oil extraction, refining, transportation and distribution, sale, and consumption (Yergin 1991; Mann 2002; Simmons 2005; Shah 2006; Smil 2008, 2010). Analyses of the political economy of oil extraction have also been prominent in recent research, in particular where petro-politics have the potential to destabilize further communities that are already on the margins of socioeconomic development or communities that face environmental collapse as a result of the petroleum industry (Soares de Oliveira 2007; Ghazvinian 2007; Watts 2008; Reyna 2011). Perhaps one of the most useful theoretical insights of this literature is the notion of oil as both artifact and artifice; oil exists as a material substance (an artifact) even as its existence serves to create social, political, and economic structures (artifices) that organize societies for whom petroleum and its derivatives are foundational (Watts 2008). Other scholars address oil addiction and overconsumption (Tamminen 2006), and project visions of economic, political, and social development in a world at the end of oil and beyond (Goodstein 2004; Deffeyes 2005; see also discussions of the energy crisis in blogs such as www.theoildrum.com, www.energybulletin.net, and http://thearchdruidreport.blogspot.com).

Coal is another energy sector featuring extensive literature that is both technical and policy-oriented, as well as ethnographic and historical in its analysis. Policy papers provide large-picture analyses of coal resources, reserves, and management plans projected into the future for major coal-producing nations including China, the U.S., Australia, and India. Analysis provided by geological survey services is particularly robust for leading coal-producing states in the United States, in particular Wyoming, West Virginia, and Kentucky. In both international and regional-U.S. contexts, information about coal and other energy industries is provided by a triangulation of agencies that includes government information services (such as the U.S. Energy Information Agency), nongovernmental organizations (such as the Energy Watch Group), and energy consultancies (such as Energy Ventures Analysis). Studies of coal mining and its history highlight coal's impact on landscapes and livelihoods. Social and historical analyses of coal emphasize its presence both as a material substance and, perhaps more importantly, as a resource that provides an environmental and social context within which individuals, communities, political groups, and industrial interests negotiate—occasionally through violent engagement—issues such as labor rights, environmental values, and gender (Andrews 2008; Connor, Freeman, and Higginbotham 2009; Bell and Braun 2010; Smith 2008, 2010; Lahiri-Dutt 2011). As with oil, coal functions in social life as both object and context (McNeil 2011).

Especially since the turn of the twenty-first century, energy research has shone especially brightly on renewable energy technologies and policy, as scholars, activists, and policy makers alike redouble their efforts to plan for a future premised on clean energy. *Scientific American* published a special issue on the future of energy beyond carbon (2006), and other leading international organizations, such as the United Nations Environment Program (UNEP), have made energy a key focus of their research and policy efforts. Researchers race to design, implement, and analyze clean energy technologies (Inslee and Hendricks 2008) and offer road maps and plans for low-carbon futures (Beck and Martinot 2003; Cipiti 2007; Flavin 2008).

Research on renewable energy resources also highlights the diversity and variation in options, technologies, and policies. Energy generation through hydropower often poses particular problems for rural communities, which often lose traditional lands even as they rarely benefit from the generation of power to facilitate their own daily tasks (Warldram 1984; Coehlo dos Santos 2003; Isaacman 2005). At the same time, untapped hydropower potential still exists to harness energy to provide utilities to people who lack access to predictable flows of electricity, and dams are under construction in fluvially rich nations such as Peru, Ghana, and the Democratic Republic of Congo, often with technical and financial assistance from China (Rupp forthcoming). One of the greatest concerns with hydropower has been the impact of large dams on ecosystems (White 1995; Colombi 2005), as well as the significant displacement of peoples (Horowitz 1991). Recent work on eco-labeling of hydropower systems, however, has helped define the limits of negative impacts in relation to scale of the resource; this policy shift has created new options for hydropower as a viable renewable energy resource, changing perceptions of hydropower as an energy option that creates as many problems as it solves (Bratrich and Truffer 2001) to hydropower as a positively valued alternative energy option, particularly at smaller sites that remain to be developed. Focused research also addresses energy harnessed from wind resources in locations as diverse as the U.S. eastern seaboard (Kempton et al. 2005), Texas and Germany (Drackle and Krauss 2011), and Malawi (Kamkwamba 2009).

Substantial basic research, technology applications, and policy guidelines have been produced concerning energy and climate change. From works that establish the relationship between consumption of energy resources and global climate change (Nordhaus 2001; Weart 2004; Gore 2006) to books that anticipate environmental and social cataclysm as a result of climate change (Kolbert 2006; Romm 2007), extreme positions in research, policy agendas, and ideologies have been demarcated. The anthropology of climate change is a fairly recent venture (Strauss and Orlove 2003; Crate and Nuttall 2009), and the intersection of climate-change impacts with energy resources has been considered only minimally thus far, although some of the earliest work was done by one of this volume's authors (Wilhite 1996). Annette Henning (2005) has emphasized the significance of this critical intersection of energy and climate issues, making an urgent plea for anthropological engagement with energy and climate in terms of cultural context, in particular as a counterbalance to the existing domination of economic studies. A growing body of scholarship addresses the intersection between hydropower and climate change, given the tight integration between water resource issues and climate. Notable research in this area centers on indigenous populations in high mountain areas, where reliance on hydropower is extreme and climate-change impacts are, as in many situations, expected to intensify previously existing challenges rather than generating a completely new type of disaster (Colombi 2005). Linking climate change at the macro level with personal energy behavior at the micro level, Lenora Bohren writes persuasively of the significance of car culture and individual energy choices (Bohren 2009). Other scholars have focused attention on cultural assumptions that underlie larger patterns of consumption, connecting energy use with climate-change impacts (Wilhite et al. 2000; Wilk 2009; Shove 2010).

We have been surprised to discover how little work has been published by anthropologists on solar energy systems of any kind, although it appears that a number of projects are now under way, particularly addressing the use of photovoltaic cells at the household level. For example, Love and Garwood (2011) compare the potential for rural electrification of solar, wind, and small-scale hydro systems in the northern Peruvian Andes. A handful of studies on the use of solar cookers to reduce emissions and improve health outcomes have also been conducted (Webb and Stuart 2007; Austin et al. 2006). Despite the ubiquity of ways through which humans tap solar energy for passive heating, there has been little work on its cultural significance.

Two recent books stand out for their contributions to analyzing energy in cultural contexts: Laura Nader's *The Energy Reader* (2010) and Ingrid Kelley's *Energy in America* (2008). Nader's edited volume presents a broad range of perspectives on energy, combining reprinted materials that draw together understandings of energy in both the physical and social sciences into the same field of view by using an anthropological lens to bring histories and policies of energy in the United States into focus. Kelley's book is a policy-oriented monograph that explains the history of energy use in America and presents policy alternatives.

It is becoming more widely evident that the era of abundant, affordable energy resources is drawing to an end. While the current fuel crisis may seem to subside, the prospects for returning to the earlier status quo are unlikely. Despite a recent—and apparently temporary—glut of natural gas in North America (Berman 2012), oil prices have twice doubled since 2002 yet failed to raise production significantly; global production of oil has remained on a plateau since 2005. It is likely that people throughout the world will face a prolonged period of relatively tight and uncertain supply of increasingly unaffordable energy resources—conditions that will require basic changes in personal attitudes and habits with respect to the consumption and conservation of resources, in particular petroleum (Robinson 1985). This changed energyscape will also require reorientation of public policies.

In the majority of this rapidly growing literature on energy, surprisingly little analysis has been brought to bear on what people think about energy, and how their beliefs and assumptions might shape their behavior. There is a startling paucity of analysis of the everyday life of energy: how people view it, appropriate it, use it, conserve it—and why. If our current predicament of energy overconsumption has any chance of being nudged back in the direction of individual restraint and collective, society-wide conservation, changes will have to be made at the community, household, and individual levels. It would seem apparent that starting with baseline information about how people perceive energy in their everyday lives would be sensible. We offer this volume in the spirit of building such cultural understandings of energy in contemporary life.

INFRASTRUCTURE AND FLOW OF THE VOLUME

Cultures of Energy comprises sixteen ethnographic chapters that are organized into five thematic sections: theorizing energy and culture; technology, meaning, cosmology; electrification and transformation; culture and power; borders and boundaries. The structure of the volume resembles the infrastructure of energy transmission. Taken together, the introduction and the first two theoretical chapters serve as power stations, providing both source (literature and overarching themes) and structure (two different theoretical frameworks) that animate the volume as a whole. The particular ethnographic examples discussed in each of the chapters constitute the multi-stranded, interwoven social and cultural experiences of energy in comparative contexts. In order to integrate and interrogate the multi-stranded themes—or transmission wires—of each section, we have created conceptual pylons that support each section; each section concludes with a conversation in which authors discuss and debate the common themes that emerge across their contributions. The goal of these conversational pylons is both to provide structural support for the conceptual themes that run throughout the volume and to create an opportunity for readers to engage in the conversations that have animated this project from the beginning. It is our hope that such conversations will serve as catalysts for wider discussion of energy issues in classrooms and communities. The conversations also highlight the sense of open discourse that still characterizes this emergent field of scholarship. The afterword provides a final substation in the system that constitutes this volume, from which we hope that the energies of the materials and ideas put forth in this volume will be distributed through and transmitted by the readers in wide-ranging cultural contexts.

Part 1: Theorizing Energy and Culture

The opening theoretical power station of the volume, a section on "Theorizing Energy and Culture," presents two contrasting scales of theoretical engagement: the global setting and individual practice within particular sociocultural contexts. In the first chapter, Hornborg considers technology as total social fact, seeing energy/environment/food/water/climate/financial crises as facets of a single major problem: the fundamental and incongruous relation between modern social institutions and the law of entropy. Put simply, as Yeats expressed nearly a century ago (1919) and as Achebe repeated (1958), *Things Fall Apart*. In Hornborg's analysis, money becomes the condition for technology, enabling the substitution of inorganic for organic energy through the mechanized transformations made possible by the Industrial Revolution. Thus, technology began a process of energy intensification that displaced both animal and human slave labor. Yet now the twin pressures of climate change and peak oil are forcing a radical rethinking of our dependence on cheap fossil fuels, and we may yet come around to "imagining a future in which land requirements and energy requirements will once again coincide . . . bring[ing] renewed competition over scarce land for food, fodder, fibers, and fuel . . . [and giving] way to a radically different understanding of the very phenomenon of technology as a strategy for redistributing embodied labor and resources in world society" (Hornborg, this volume).

In contrast, by examining the ways that largely OECD countries (countries that belong to the Organisation for Economic Co-operation and Development) have viewed energy consumption and reduction strategies, Hal Wilhite develops a new way of understanding energy sustainability through social practice theory. As he points out, "Individuals bring their knowledge and lived experiences to the interaction with things in practice, but the things they interact with can also influence the action" (Wilhite, this volume). Wilhite highlights that our habits with regard to cooking technologies, lighting strategies, or transportation can be valued differently and practiced in multiple and varied ways. The conversation that follows the first section of the book brings in other voices from the volume, considering the notion of "energy slaves" having replaced slave labor provided by human beings, as well as technological and theoretical shifts across the history of energy production, consumption, and distribution. The notion of "mystification" as it applies to energy production, transmission, and infrastructure also comes into this discussion.

Part 2: Culture and Energy: Technology, Meaning, Cosmology

Part 2 engages issues of meaning with respect to varieties of energy resources and technologies. First, Stephanie Rupp considers the ways that New Yorkers think about energy through the familiar anthropological lenses of science, religion, and magic. She suggests that, because people perceive energy as an invisible force that propels their lives, most people rarely consider the theory or mechanics behind the wide range of energy resources that make life in

New York City possible. Instead, New Yorkers rely on "a palette of images" that includes "magical, spiritual, corporeal, social, political, as well as technical models—to explain the forces that enable their everyday tasks and experiences" (Rupp, this volume). Rupp points out that, rather than muddling scientific models of energy, this multiplicity of meanings offers diverse avenues for developing strategies for managing energy consumption and for promoting energy conservation more effectively.

Also engaging the notion of multiple meanings, in this case among Alaskan communities, Chelsea Chapman contributes a study of ontologies of energy to highlight the different "ways of knowing and acting toward energy sources that are, like water, land, and wildlife, all too often considered neutral and static commodities" (Chapman, this volume). Chapman addresses the mythical and cosmological frameworks associated with energy development across the domains of public and private, indigenous and non-Native, scientist and activist, and fossil versus renewable energy sources.

Questions of social capital and the mobilization of communities for or against energy development in specific places or at particular scales are at the heart of Sarah Strauss and Devon Reeser's chapter. The formation of local wind energy cooperatives in Wyoming, designed to maximize landowner social capital in negotiations with outside energy development companies, was paralleled by the creation of a grassroots opposition group with serious concerns about the siting of wind turbines on these private properties as inherently conflicting with the preservation of Wyoming's landscape and heritage. Ironically, members of the cooperatives, who are the friends, family, and neighbors of the opposition group, see the wind energy development as a key "to keep the heritage of ranching alive for their children or grandchildren" (Strauss and Reeser, this volume).

Arthur Mason, on the other hand, focuses his analytical lens on the top of the social hierarchy, "studying up" (Nader 1974) to explore the culture of energy consulting firms engaged in national and international industry futures. The formalization and commodification of energy knowledge by these information brokers has a significant influence on corporate and national policies, and on the assessment of risks associated with energy development. The meanings and values—both symbolic and material—that they assign to various sectors and institutions within the energy industry are packaged and reproduced across these elite communities, thus constituting new subjectivities and strengthened social capital. One result of this elite "clubbiness" is that "emphasis in energy development increasingly is placed on global financial markets, instead of structural positions within national political systems" (Mason, this volume).

The conversation following Part 2 focuses on notions of power in the sense of both agency and control, as well as its more technical definition of the rate at which energy is used or transformed in the material sense. Participants discuss whether energy is relative or absolute in its quantity or quality, and how users interpret and value it.

Part 3: Electrification and Transformation

In this section, we consider the impact of electrification in Africa and South America. In addition to all of the lifestyle changes that electricity brings, accompanying cultural, moral, symbolic, and political shifts need to be understood more completely. Tom Love and Anna Garwood bring us to rural Cajamarca, Peru, to see the scalar disparities at work in a small village adjacent to a multinational gold mine. The mine itself received electrical power from the high-tension lines that passed over the village of Alto Peru, while the village below remained without power. When an NGO-driven, small-scale renewable energy project was initiated to bring electricity to the village itself, the choices made by villagers about what to do with their newfound power were, tellingly, as much about what it meant as what they could make with it: "Alto Peruanos' main uses of their new electricity serve to break down the feeling of being too rural and disconnected . . . To have electricity is to be engaged with the broader world" (Love and Garwood, this volume). Being able to call family, enabling their children to study at night, and following national elections has helped these villagers feel more engaged with the wider national community. Yet paradoxically they also felt that they could retain the ability to stay and work in the local village more easily, instead of migrating to the city.

Tanja Winther shows us the world of "Space, Time, and Sociomaterial Relationships" in rural Zanzibar. Although the Sultan of Zanzibar first utilized electrical lighting in his palace in 1886, it took another century for artificial light to become available to ordinary residents of the island. In this chapter, Winther examines the social and moral aspects of electricity use in Zanzibari homes, highlighting changes in gender and marital relations. She also analyzes the function of electricity in providing improved security in the public sphere, setting the stage for increased engagement in both religious and municipal contexts. Winther notes that, "in contrast with kerosene-lit homes, newly electrified settings provide bright light, allowing for the use of louder voices and laughter. This light- and soundscape invited a freer atmosphere that resembles daytime setting and interactions," permitting shifts in the gendered use of space (Winther, this volume).

In nearby Dar es Salaam, Tanzania, Michael Degani explores the ways in which community members have strategically manipulated limited electrical lines to their own advantage, again demonstrating the complex relationship between social power and access to energy resources. Residents of Dar es Salaam perceive the circulation of electricity through metaphors of cleverness and trickery (*ujanja*); the ability to access power requires machinations. "Faced with shortage, expense, and unreliability, urban residents have improvised clever arrangements to caulk the gaps" in official electric company service delivery (Degani, this volume). Degani analyzes the ad hoc nature of electricity distribution in this postsocialist context as symptomatic of tenuous political and neoliberal reforms, which rather than empowering citizens require them to develop their own "provisional pathways" of electrical current—power—as best they can.

strategies to get access to electricity

Part 3 closes with a conversation among authors that addresses the multiple ways that changes in the technical circuitry of energy result in shifts to the social and moral circuitry of these transformed communities.

Part 4: Energy Contested: Culture and Power

This section addresses conflicts generated by energy risk and opportunity, examining fracture lines that create stresses within and across communities at the individual, social, and political levels. Elizabeth Cartwright brings readers into "the anthropology of the unknown, the invisible, the just beyond the senses" as she delves into the nexus of health, environment, and culture where the impacts of hydraulic fracturing, or fracking, become apparent (Cartwright, this volume). Cartwright's examination of the parallels between individual bodies and the circulatory system of the earth itself uses the example of fracking as a model for understanding how and why this extractive technology is contested, and why American society urgently needs new models for making sense of the potential human health effects of such large-scale environmental engineering and transformation.

Focusing her attention on the coal mines and coal-bed methane fields of Gillette, Wyoming, Jessica Smith Rolston revisits the Gillette Syndrome first identified in 1974 as a type of social pathology of increased alcohol and drug use, divorce, depression, and other disruptive behaviors and consequences that became prevalent in energy boomtowns such as Gillette. Rolston sets out to correct the negative stereotypes that have persisted for the past several decades as a result of the Gillette Syndrome's emergence in academic discourse. She points out that the assumptions in the general media and scientific literature about the "obvious" negative social consequences of energy development are not necessarily always or evenly borne out. She argues that, in order to grasp the social impacts of energy development, it is of primary importance to address "structural factors that create inequalities" (Rolston, this volume) that compel particular groups of people to take up hazardous work in energy industries to earn their livelihoods.

In the next chapter, Derek Newberry discusses conflict surrounding the sustainability of biofuels in Brazil. As in the chapter by Mason, Newberry's research focuses on an organization that produces expert knowledge with the potential to shift policy at national and global levels, as well as to change public perceptions of alternative energy sources such as biofuels. Newberry argues that "denaturalizing the all-too-familiar work practices and professional discourses of these experts is an especially important component of an agenda for resurgent anthropologies of energy," because of the significance of expert opinion in shaping both policy agendas and public perceptions and practices (Newberry, this volume). Newberry relates his work on the everyday practices of energy scientists to the theoretical frameworks suggested by both Wilhite and Hornborg in this volume.

Scott Vandehey brings the distribution of energy into focus by examining the problem of infrastructure and the transmission of energy across distances. In order to bring electricity generated by renewable energy systems into urban Southern California, new transmission lines need to be built; however, local residents contest the infrastructure needed to bring such new energy to market. As with Strauss and Reeser's chapter in this volume, issues of viewshed and lifestyle are brought forward in contesting energy infrastructure development. Power lines, as with the fracking Cartwright describes, are viewed as a health and environmental risk even as they hold the potential to bring renewable energies within the grasp of consumers. The NIMBY—not in my backyard—sentiment expressed by these Southern Californians, Vandehey argues, is neither positive nor negative; he sees it instead as "an attempt to protect a perceived ideal way of life, assert local control over outside forces, and as a refocusing of citizenship on a local level" (Vandehey, this volume). This example of bottom-up resistance to energy infrastructure, notably in a suburban population not known for its radicalism, highlights locally based interests in maintaining the energy status quo, even as, ironically, it heralds the potential for future engagement and dialogue necessary to innovate and nurture alternatives in a post–peak oil world. Issues surrounding the assertion of control over outside forces emerge in the conversation that concludes this section of the volume.

The discussion following these chapters emphasizes the importance of boundaries across scale, from individual bodies to the planet, flowing directly into the fifth and final section of the book. This conversation also addresses the ways that energy itself seems to have become a mediating force between human society and nature.

Part 5: Energy Contested: Borders and Boundaries

Developing the theme of contestation further, the final section of the book addresses a wider arena in which energy politics have significant transboundary consequences, straddling boundaries between human and supernatural forces and between neighboring nations. In the opening chapter of the final section of the volume, Gisa Weszkalnys argues that anthropologists have a special set of tools that can help us understand the "resource curse" that so often emerges in the process of extracting oil resources in underdeveloped nations. Her contribution exemplifies the utility of ethnographic perspectives in "exploring occult practices to explain the association of natural resources with magical power" in São Tomé and Príncipe (STP), a small island nation off the coast of West Africa (Weszkalnys, this volume). Weszkalnys's ethnographic research highlights the distributed nature of oil in STP where there has been no commercial exploration of oil yet. She underscores the fact that while oil may appear to be a resource of abundance—spilling out of the ground as it does in many places, and also promising abundant wealth—the technology and capital required to extract, process, and distribute oil are not accessible to ordinary citizens. The primary way in which Santomeans have experienced the resource is through the myriad activities carried out to prevent a possible resource curse. She argues, therefore, that to properly understand oil's magic, we need to consider the nature and effects of its specific resource materiality, including infrastructures such as the pipeline or the platform.

Lisa Breglia transports readers to another body of water—the Gulf of Mexico—to consider deepwater oil well development and its implications on borders and identities. The fluid boundary between the United States and Mexico is the center of Breglia's analysis; it is a complex arena for oil extraction, as the waters of the Gulf "have not been legally delimited, and sovereign rights over exploration and resource exploitation are often unclear and quickly become matters of dispute or arbitration" (Breglia, this volume). Breglia describes the relationship between oil exploration and gulf fisheries in relation to questions of national patrimony: "All of the Mexican Gulf is a patrimonial sea for Mexicans because the resources it contains—from fisheries to hydrocarbons—are symbolically tied to the nation-state's project of providing for the national welfare" (Breglia, this volume).

Working on the U.S. coastal edge of the Gulf of Mexico, Thomas McGuire and Diane Austin review the aftermath of the catastrophic explosion of British Petroleum's *Deepwater Horizon* oil well off the Louisiana coast in 2010. McGuire and Austin analyze the importance of the oil industry for lifeways on the Gulf Coast, demonstrating that the regional culture provided an interpretive lens for the explosion and its consequences. Contrary to national discourse in the aftermath of the disaster, "in the communities already impacted by the closure of the gulf waters to fishing, shrimping, and oystering, the moratorium on deepwater drilling caused an immediate and negative reaction" (McGuire and Austin, this volume). Although coastal communities in Louisiana clearly recognized the costs of drilling accidents, they also viscerally perceive that there are few other opportunities for financial security available to them, and so the risks seem worth taking.

Transboundary energy issues provide the conceptual pylon in Conversation 5. As Breglia suggests, oil is pretty sneaky stuff, oozing in, around, and across borders and making its transparent and forthright management elusive. Anthropology's qualitative methods enable research to span registers from individuals to communities, regions, nations, and the planet; to straddle borders that demarcate interests and identities; and to consider power relations in economic, political, and occult spheres. Anthropological perspectives offer the chance to view energy holistically in particular and comparative contexts, to analyze the scope of energy's impact in our communities, and to explore the implications of the limits of energy resources.

Finally, the volume benefits beyond measure from the contribution by Laura Nader, a pioneer of energy studies in anthropology who offers the afterword. Nader brings the volume into the future with a final power station that anticipates the significance of the past, thus setting these many and varied impressions of cultures of energy in the twenty-first century into perspective. We may already be beyond the chance for a frank discussion about our energy system and our lack of a coherent energy policy to make a difference, and it may well be too late for "sustainable development" in the energy sector (Grooten 2012). Still, we can try our best to understand how people experience and interpret, (ab)use and (mis)manage energy resources in particular and comparative cultural contexts. We have no crystal balls, and there will surely be Black Swans along the way (always expect the unexpected!), but at least by remembering history and maintaining our focus on the relationships and meanings that an anthropology of energy demands, we might begin to see our way into a positive energy future.

REFERENCES

Achebe, Chinua
1958 *Things fall apart*. London: William Heinemann Ltd.

Adams, Richard Newbold
1975 *Energy and structure: A theory of social power.* Austin: University of Texas Press.
1982 *Paradoxical harvest: Energy and explanation in British History, 1870-1914.* Cambridge: Cambridge University Press.
1988 *The eighth day: Social evolution as the self-organization of energy*. Austin: University of Texas Press.

Andrews, Thomas
2008 *Killing for coal: America's deadliest labor war*. Cambridge, MA: Harvard University Press.

Appadurai, Arjun
1990 Disjuncture and difference in the global cultural economy. *Public Culture* 2(2): 1–24.

Armstrong, Fraser and Katherine Blundell
2007 *Energy... Beyond oil.* Oxford: Oxford University Press.

Bain, Craig, Crystal Ballentine, Anil DeSouza, Lisa Majure, Dean Howard, and Jill Turek
2004 Navajo electrification for sustainable development: The potential economic and social benefits. *American Indian Culture and Research Journal* 28(2): 67–79.

Barbour, I., H. Brooks, S. Lakoff, and J. Opie
1982 *Energy and American values*. National Humanities Center, Research Triangle Park, NC: Praeger Publishers.

Beck, Fred and Eric Martinot
2004 Renewable energy policies and barriers. In *Encyclopedia of energy,* edited by Cutler J. Cleveland, 365–383. Waltham, MA: Academic Press.

Bell, Shannon Elizabeth and Yvonne A. Braun
2010 Coal, identity, and the gendering of environmental justice activism in central Appalachia. *Gender & Society* 24(6): 794–813.

Blamont, D.
1989 Solar energy in the Indian countryside: The ASVIN program of the SNRS. *Techniques et Culture* 14(1989): 203–227.

Bratrich, Christine and Bernhard Truffer
2001 Green electricity certification for hydropower plants: Concept, procedure, criteria. Green Power Publications, Issue 7. ETH/EAWAG, Switzerland.

Bohren, Lenora
2009 Car culture and decision-making: Choice and climate change. In *Anthropology and climate change*, edited by Susan Crate and Mark Nuttall, 370–379. Walnut Creek, CA: Left Coast Press.

Burnham, P.
1982 Energetics and ecological anthropology: Some issues. In *Energy and Effort*, edited by G. A. Harrison. London: Taylor and Francis.

Cabrera, S and V. Fuster
2002 Energy and sociality in human populations. *Social Biology* 49(1/2): 1–12.

Caldwell, Lynton K.
1976 Energy and the structure of social institutions. *Human Ecology* 4(1): 31–45.

Campbell, Gregory
1987 Northern Cheyenne ethnicity, religion, and coal energy development. *Plains Anthropologist* 32(118): 378–388.

Catton, William
1980 *Overshoot: The ecological basis of revolutionary change*. Urbana: University of Illinois Press.

Cipiti, Ben
2007 *The energy construct: Achieving a clean, domestic, and economical energy future*. Booksurge (self-published on Amazon.com).

Colombi, Benedict J.
2005 Dammed in region six: The Nez Perce tribe, agricultural development, and the inequality of scale. *American Indian Quarterly* 29(3): 560–589.

Coelho dos Santos, Silvio
2003 Water generated electricity in southern Brazil and its social impact. *Etnográfica* 7(1): 87–103.

Conn, W. David
1983 *Energy and material resources: Attitudes, values, and public policies*. Boulder, CO: Westview Press.

Connor, Linda, Sonia Freeman, and Nick Higginbotham
2009 Not just a coalmine: Shifting grounds of community opposition to coal mining in Southeastern Australia. *Ethnos* 74 (4): 490–513.

Cottrell, William Frederick
1955 *Energy and society: The relation between energy, social change, and economic development*. New York: McGraw-Hill.

Crate, Susan and Mark Nuttall
2009 *Anthropology and climate change*. Walnut Creek, CA: Left Coast Press.

Deffeyes, Kenneth
2005 *Beyond oil: The view from Hubbert's Peak*. New York: Hill and Wang.

DeFonzo, J. and S. Warkov
1979 Are female-headed households energy efficient: A test of Klausner's hypothesis among Anglo, Spanish-speaking, and black Texas households. *Human Ecology* 7(2): 191–197.

Diesendorf, Mark
1979 *Energy and people: Social implications of different energy futures*. Canberra, Australia: Society for Social Responsibility in Science.

Dioffo, Abdou
1981 The developing world demands a place in the sun. *UNESCO Courier* 34(7): 14–16.

Downey, Gary L.
1986 Risk in culture: the American conflict over nuclear power. *Cultural Anthropology* 1(4): 388–412.

Downey, Morgan
2009 *Oil 101*. New York: Wooden Table Press, LLC.

Drackle, Dorle and Werner Krauss
2011 Ethnographies of wind and power. *Anthropology News* (May): 9.

Dufour, Dama and Barbara Piperata
2008 Energy expenditure among farmers in developing countries: What do we know? *American Journal of Human Biology* 20(3): 249–258.

Eliatamby, Niresh
1999 In a hole over coal. *Himal* 12(10): 34–35.

Erickson, Rita
1997 *Paper or plastic? Energy, environment, and consumerism in Sweden and America.* Westport, CT: Greenwood Publishing Group.

Fergus, Michael
1983 Firewood or Hydropower: A case study of rural energy markets in Tanzania. *Geographical Journal* 149(1): 29–38.

Finison, Karl
1979 *Energy flow on a nineteenth century farm.* Research Reports #18, Department of Anthropology. Amherst: University of Massachusetts.

Flavin, Christopher
2008 Low-carbon energy: A roadmap. *World Watch Report 178:52.* Washington, DC: World Watch Institute. http://www.worldwatch.org/node/5945.

Ghazvinian, John
2007 *Untapped: The scramble for Africa's oil.* New York: Harcourt.

Goodstein, David
2004 *Out of gas: The end of the age of oil.* New York: W.W. Norton.

Gore, Al
2006 *Earth in the balance.* Emmaus, PA: Rodale Books.

Grünbühel, Clemens, Helmut Haberl, Heinz Schandl, and Verena Winiwarter
2003 Socioeconomic metabolism and colonization of natural processes in SanSaeng Village: Material and energy flows, land use, and cultural change in northeast Thailand. *Human Ecology* 31(1): 53–86.

Gusterson, Hugh
1996 *Nuclear rites: A weapons laboratory at the end of the Cold War.* Berkeley, CA: University of California Press.
2011 The lessons of Fukushima. *Bulletin of the Atomic Scientists* (March 16).

Haile, Fekerte
1989 Women fuelwood carriers and the supply of household energy in Addis Ababa. *Canadian Journal of African Studies* 23: 442–451.

Hasalkar, Suma., P. Sumangala, and K. Ashalatha
2002 A micro view of energy consumption in the household ecosystem. *Journal of Human Ecology* 13(6): 437–441.

Hecht, Gabrielle
1998 *The radiance of France: Nuclear power and national identity after World War II.* Cambridge, MA: Massachusetts Institute of Technology Press.

Henning, Annette
2005 Climate change and energy use. *Anthropology Today*. 21(3):8–12.

Holdren, John P.
1976 Technology, environment, and well-being: Some critical choices. In *Growth in America*, edited by C. Cooper. Westport, CT: Greenwood Press.
1982 Energy hazards: What to measure, what to compare. *Technology Review* 85(3): 32–38; 74–75.

Hornborg, Alf
2008 Machine fetishism and the consumer's burden. *Anthropology Today* 24(5): 4–5.

Horowitz, Michael M.
1991 Victims upstream and down. *Journal of Refugee Studies* 4(2): 164–181.

Huertas Castillo, Beatriz
2003 The Camisea Project and indigenous rights. *Indigenous Affairs* 3: 28–35.

Inslee, Jay and Bracken Hendricks
2008 *Apollo's fire: Igniting America's clean energy economy*. Washington, DC: Island Press.
Irvin, Amelia
1982 Energy development and the effects of mining on the Lakota nation. *Journal of Ethnic Studies* 10(1): 89–101.
Isaacman, Allen
2005 Displaced people, displaced energy, and displaced memories: The case of Cahora Bassa, 1970–2004. *International Journal of African Historical Studies* 38(2): 201–238.
Johnston, Barbara Rose
2011 In this nuclear world, what is the meaning of "safe"? *Bulletin of the Atomic Scientists* (March 18). Available at http://thebulletin.org/node/8641.
Jones, John
1984 Central American energy problem: Anthropological perspectives on fuelwood supply and production. *Culture and Agriculture* 22: 6–9.
Kamkwamba, William
2009 *The boy who harnessed the wind: Creating currents of electricity and hope*. New York: Harper Perennial.
Kempton, Willet, James S. Boster, and Jennifer A. Hartley
1995 *Environmental values in American culture*. Cambridge, MA: Massachusetts Institute of Technology Press.
Kempton, Willett, Jeremy Firestone, Jonathan Lilley, Tracy Rouleau, and Phillip Whitaker
2005 The offshore wind power debate: Views from Cape Cod. *Coastal Management* 33: 119–149.
Kelley, Ingrid
2010 *Energy in America: A tour of our fossil fuel culture and beyond*. Burlington, VT: University of Vermont Press.
Kirby, Peter Wynne
2011 *Troubled natures: Waste, environment, Japan*. Honolulu: University of Hawaii Press.
Klare, Michael
2012 *The race for what's left: The global scramble for the world's last resources*. New York: Metropolitan Books.
Klausner, Samuel Z.
1979 Social order and energy consumption in matrifocal households. *Human Ecology* 7(1): 21–39.
Klieman, Kairn
2008 Oil, politics, and development in the formation of a state: The Congolese petroleum wars, 1963–68. *International Journal of African Historical Studies* 41(2): 169–202.
2012 U.S. oil companies, the Nigerian civil war, and the origins of opacity in the Nigerian oil industry, 1964–1971. Special issue of *Journal of American History* on "Oil in America," edited by Brian Black, Karen Merrill, and Tyler Priest.
Forthcoming From kerosene to Avgas: International oil companies and their expansion in sub-Saharan Africa, 1890s to 1945. In *Environment and Economics in Nigeria*, edited by Toyin Falola and Adam Paddock. New York: Routledge.
Kolbert, Elizabeth
2006 *Field notes from a catastrophe: Man, nature, and climate change*. New York: Bloomsbury Publishing.
Kruse, John, Judith Kleinfeld, and Robert Travis
1982 Energy development on Alaska's North Slope: Effects on the Inupiat population. *Human Organisation* 41(2): 95, 97–106.

Lahiri-Dutt, Kuntala
2011 The shifting gender of coal: Feminist musings on women's work in Indian collieries. *South Asia: Journal of South Asian Studies* 35(2): 456–476.

Lenssen, Nicholas
1992 Confronting nuclear waste. In *State of the world 1992*, edited by Lester Brown, 46–65. New York: W.W. Norton.

Little, Michael, Paul Leslie, and Kenneth Campbell
1992 Energy reserves and parity of nomadic and settled Turkana women. *American Journal of Human Biology* 4(6): 729–738.

Little, Ronald
1979 Energy boom towns: Views from within. *Native Americans and Energy Development*. Cambridge, MA: Anthropology Resource Center, 63–85.

Love, Thomas
2008 Anthropology and the fossil fuel era. *Anthropology Today* 24(2): 3–4.

Love, Thomas and Anna Garwood
2011 Wind, sun and water: Complexities of alternative energy development in rural northern Peru. *Rural Society* 20: 294–307.

Lovins, Amory
1976 Energy strategy: The road not taken. *Foreign Affairs* 55(1): 65-96.
1977 *Soft energy paths: Toward a durable peace*. New York: Penguin.

Mann, Charles C.
2002 Getting over oil. *Technology Review* (January/February): 32–38.

Marchetti, Cesare
2003 On energy systems historically and perspectively. *Human Evolution* 19(1/2): 83–95.

Masco, Joseph
2006 *Nuclear borderlands: The Manhattan Project in post-cold war New Mexico*. Princeton, NJ: Princeton University Press.

McNeil, Bryan
2011 *Combating mountaintop removal: New directions in the fight against big coal*. Champaign: University of Illinois Press.

Meadows, Donella, Jørgen Randers, and Dennis L. Meadows
2004 *The limits to growth: The 30-year update*. White River Junction, VT : Chelsea Green Publishing Co.

Mfune, Orleans and Emmanuel Boon
2008 Promoting renewable energy technologies for rural development in Africa. *Journal of Human Ecology* 24(3): 175–189.

Mukhopadhyay, S. and R. Malabika
1996 Domestic energy consumption and domestic environment: A case study. *South Asian Anthropologist* 17(1): 1–6.

Murray, James and David King
2012 Climate policy: Oil's tipping point has passed. *Nature* 481 (January 26): 433–435.

Nader, Laura
1980 *Energy choices in a democratic society*. Supporting Paper No. 7, Study of Nuclear and Alternative Energy Systems. Washington, DC: National Research Council.
2010 *The energy reader*. Oxford: Wiley-Blackwell.

Nader, Laura, and Stephen Beckerman
1978 Energy as it relates to the quality and style of life. *Annual Review of Energy* 3: 1–28.

Najib, Ali B.

1993 Household energy consumption behaviour in a pre-Saharan small town in Morocco. *Newsletter Commission on Nomadic Peoples* 32: 3–32.

Nordhaus, William

2001 Policy forum: Global warming economics. *Science* 294: 1283–1284.

Norgan, N.

1982 *Human energy stores*. London: Taylor and Francis.

Nye, David E.

1992 *Electrifying America: Social meanings of a new technology*. Cambridge, MA: MIT Press.

1994 *American technological sublime*. Cambridge, MA: MIT Press.

1998 *Consuming power: A social history of American energies*. Cambridge, MA: MIT Press.

Odum, Howard T.

1971 *Environment, power and society*. New York: Wiley.

Odum, Howard T. and Elisabeth C. Odum

1976 *Energy basis for man and nature*. New York: McGraw-Hill.

2001 *A prosperous way down: Principles and policies*. Boulder: University Press of Colorado

Owens, Nancy J.

1979a The effects of reservation bordertowns and energy exploitation on American Indian economic development. *Research in Economic Anthropology* 2: 303–337.

1979b Can tribes control energy development? *American Indian Journal* 5(1): 3–17.

Rambo, A. Terry

1980 Fire and the energy efficiency of swidden agriculture. *Asian Perspectives* 23(2): 309–316.

1991 Energy and the evolution of culture: a reassessment of White's law. In *Profiles in cultural evolution*, edited by A. Terry Rambo and Kathleen Gillogly. Ann Arbor: Museum of Anthropology, University of Michigan.

Rappaport, Roy

1975 The flow of energy in agricultural society. In *Biological anthropology*, 371–387. San Francisco: W. H. Freeman.

Revelle, Roger

1976 Energy use in rural India. *Science* 192(4243): 969–975.

Ridgeway, James

1982 *Powering civilization: The complete energy reader*. New York: Pantheon Books.

Robbins, Lynn

1984 Energy developments and the Navajo nation: An update. In *Native Americans and energy development*, edited by Joseph Jorgensen. Boston, MA: Anthropology Resource Center.

Robbins, Lynn and Steven McNabb

1987 Oil developments and community responses in Norton Sound, Alaska. *Human Organization* 46(1): 10–17.

Robinson, I.

1985 Energy and urban form: Relationships between energy conservation, transportation, and spatial structures. In *Energy and cities*. Energy Policy Studies, Vol. 2, edited by J. Byrne and D. Rich. New Brunswick and Oxford: Transaction Books.

Romm, Joseph

2007 *Hell and high water: Global warming—the solution and the politics—and what we should do*. New York: HarperCollins/William Morrow.

Rupp, Stephanie
Forthcoming Powerplay: Ghana, China, and the politics of energy. *African Studies Review*, special issue "Africa and China: Ethnographic and historical perspectives," edited by Jamie Monson and Stephanie Rupp.

Sahlins, Marshall
1960 Evolution: Specific and general. In *Evolution and culture*, edited by M. Sahlins and E. Service, eds., 12–44. Ann Arbor: University of Michigan Press.

Sanchez, Teodoro
2010 *The hidden energy crisis: How policies are failing the world's poor*. Bourton on Dunsmore, Rugby, Warwickshire, UK: Practical Action Publishing.

Schivelbusch, W.
1995 *Disenchanted night: The industrialization of light in the nineteenth century*. Berkeley: University of California Press.

Scientific American
2006 Energy's future beyond carbon: How to power the economy and still fight global warming. Special issue (September).

Shove, Elizabeth
2010 Social theory and climate change: Questions often, sometimes, and not yet asked. *Theory, Culture, and Society* 27(3): 277–288.

Smil, Vaclav
1978 Energy flows in rural China. *Human Ecology* 7(2): 119–133.
1994 *Energy in world history*. Boulder, CO: Westview Press.
2003 *Energy at the crossroads: Global perspectives and uncertainties*. Cambridge, MA: Massachusetts Institute of Technology Press.
2008 *Energy in nature and society: General energetics of complex systems*. Cambridge, MA: Massachusetts Institute of Technology Press.

Smith, Sonia
2010 Gulf anti-moratorium rally draws big crowd. *Platts Oilgram News* 88(142): 9.

Soares de Oliveira, Ricardo
2007 *Oil and politics in the Gulf of Guinea*. New York: Columbia University Press.

Sovacool, Benjamin and Marilyn Brown
2007 *Energy and American society: Thirteen myths*. Dordrecht, The Netherlands: Springer.

Strauss, Sarah and Ben Orlove, eds.
2003 *Weather, Climate, Culture*. Oxford: Berg Publishers, Ltd.

Tamminen, Terry
2006 *Lives per gallon: The true cost of our oil addiction government*. Washington, DC: Island Press.

Taubes, Gary
2002 Whose nuclear waste? *Technology Review* 105(1): 60–67.

Topçu, Sezin
2008 Confronting nuclear risks: Counter-expertise as politics within the French nuclear industry debate. *Nature and Culture* 3(2): 225–45.

Topo, E.
1982 Energy futures and Venezuela's Indians. *Survival International Annual Review* 7(2): 4–17.

Tsing, Anna
2000 The global situation. *Cultural Anthropology* 15(3): 327–360.

Vayda, Andrew P. and Bradley B. Walters
1999 Against political ecology. *Human Ecology* (27)1: 167–179.

Watts, Michael, with Ed Kashi (photographer)
2008 *The curse of the black gold.* New York: Powerhouse Press.

Waldram, James
1984 Hydro-electric development and the process of negotiation in northern Manitoba. *Canadian Journal of Native Studies* 4(2): 205–239.

Weart, Spencer
2004 *The discovery of global warming.* Cambridge, MA: Harvard University Press.

Wen, Dazhong
1986 Seventeenth century organic agriculture in China: Energy flows through an agro ecosystem in Jianxing region. *Human Ecology* 14: 15–28.

West, Q.
1974 Food shortages: A poor man's energy crisis. *Ecology of Food and Nutrition* (3)3: 243–249.

White, Leslie
1943 Energy and the evolution of culture. *American Anthropologist* 45(3): 335–356.
1949 *The science of culture: A study of man and civilization.* New York: Farrar, Straus and Giroux.
1959 *The evolution of culture; The development of civilization to the fall of Rome.* New York: McGraw-Hill.
2007 [1959] *The evolution of culture.* Walnut Creek, CA: Left Coast Press.

White, Richard
1995 *The organic machine.* New York: Hill and Wang.

Wilhite, Harold, Hidetoshi Nakagami, Takashi Masuda, Yukiko Yamaga, and Hiroshi Haneda
1996 A cross-cultural analysis of household energy-use behaviour in Japan and Norway. *Energy Policy* 24(9): 795–803.

Wilhite, Harold, Elizabeth Shove, Loren Lutzenhiser, and Willett Kempton
2000 The legacy of twenty years of demand side management: We know more about individual behavior but next to nothing about demand. In *Society, behaviour and climate change mitigation,* edited by Eberhard Jochem, Jayant A. Stathaye, and Daniel Bouille, 109–126. Dordrecht, the Netherlands: Kluwer Academic Press.

Wilk, Rick
2009 Consuming ourselves to death: The anthropology of consumer culture and climate change. In *Anthropology and climate change: From encounters to actions,* edited by Susan Crate and Mark Nuttall, 265-276. Walnut Creek, CA: Left Coast Press.

Williams, James C.
1997 *Energy and the making of modern California.* Akron, OH: University of Akron Press.

Yahaya, Mohammed K., R. Nabinta, and Bamidele Rasak Olajide
2007 Gender, energy and environment nexus in female farmers household energy management in Gombe State, Nigeria. *The Anthropologist: International Journal of Contemporary and Applied Studies of Man* 9(3): 203–209.

Yergin, Daniel
1991 *The prize: The epic quest for oil, money and power.* New York: Simon & Schuster.

PART 1

Theorizing Energy and Culture

CHAPTER ONE

The Fossil Interlude: Euro-American Power and the Return of the Physiocrats

Alf Hornborg[1]

INTRODUCTION

In this chapter I discuss the phenomenon of modern technology as a total social fact, viewed from the combined perspectives of history, sociology, economics, thermodynamics, ecology, epistemology, and culture theory. Anthropology is uniquely equipped to assemble such transdisciplinary perspectives on material aspects of contemporary societies and to "defamiliarize" lifestyles and social arrangements that have come to appear natural and desirable. It is high time for those of us who enjoy the benefits of modern technology and patterns of energy consumption to recognize our lifestyle as the privilege of a global minority, and technology itself as a strategy for appropriating and redistributing time and space in global society. Such a reconceptualization of deeply rooted notions of "technological progress" and "modernization" would make it easier to grasp the nature of the global crisis that we are currently facing. Rather than fragment our understanding of this crisis into legitimate but separate worries over energy scarcity, environmental degradation, resource depletion, food shortages, climate change, global inequalities, and financial collapse, we need to realize that all these concerns are aspects of a single problem.

This problem is the incongruous relation between modern social institutions and policies, on one hand, and the second law of thermodynamics, on the other. This natural law (also known as the entropy law) observes that any conversion of energy in the universe will entail a net increase in entropy (cf. Georgescu-Roegen 1971). "Entropy" is a measure of disorder or arbitrariness in the distribution of energy and matter. A currently familiar example of increasing entropy is the rising level of carbon dioxide in the atmosphere. The emission of carbon dioxide is largely an unintentional and long-unacknowledged by-product of maintaining a technological infrastructure founded on the combustion of fossil energy. As most of us are now aware, this by-product of growth is undermining the prospects for future generations of human life. As long as our societal pursuit of "economic growth" is based on an expanding combustion of finite stocks of fossil fuels, our cultural understanding of growth and progress is thus completely at odds with what natural law can tell us.

The social arrangements and aspirations that are most fundamentally at odds with the second law of thermodynamics are general-purpose money and beliefs in economic growth and technological progress. Of these illusions, the one that is most difficult to "defamiliarize" is undoubtedly that of technological progress. For this reason, the main objective of this chapter is to suggest a radical reinterpretation of technology.

THE METABOLISM OF SOCIETIES

Energy flows from the sun are the sine qua non of most living systems, including societies. Whereas organisms are programmed to harness such energy in specific and generally predetermined ways, human social systems have been able to devise a number of different strategies for accessing energy and distributing it among its members. As all human societies are organized in terms of more or less collective understandings of their own operation (the domain of cultural meanings), cultural images of energy and energy use constitute a formidable field for comparative study. Such study is inevitably contested and controversial, because access to and distribution of energy is everywhere closely connected to power. In fact, the very concept of "power" can be used to denote energy as well as social dominance (Hornborg 2001).

For the 99 percent of its existence that our species has lived as mobile hunter-gatherers, humans have occupied specific niches in natural food chains, defined by their technologies for extracting food energy from plants and animals. With the development of increasingly complex cognitive and communicative capacities, human societies became more elaborate, populations more concentrated and sedentary, and energy requirements more demanding. Beginning around ten thousand years ago, the domestication of plants and animals provided a more abundant and reliable energy niche for more complex societies in several parts of the world. The new demands on and sources of energy were recursively connected, so that, for instance, ceremonial feasting and chiefly generosity prompted expanded cultivation; more abundant harvests permitted larger settlements; the concentration of population demanded more intensified production; investments in productive facilities (generally farmland) required even greater concentrations of people for defensive purposes; and so on.

No doubt all these premodern societies, from hunter-gatherers to agricultural chiefdoms, had developed their own understandings of the energy flows that sustained them. Many of them seem to have recognized the sun as the source of vital power animating humans and the rest of the world. Agrarian empires were also ultimately dependent on the productivity of solar energy processed by plants, animals, and humans, and they, too, generally acknowledged (and in fact often worshipped) the sun.

From an abstract, comparative perspective, we can observe that these societies relied almost completely on the photosynthesis of various plant species and their conversion into food and fodder as well as the mechanical work of animals and humans. What we have come to call "land" and "labor" were the ultimate energy resources, but they could be invested in "capital" in the form, for instance, of agricultural terraces, irrigation canals, livestock, roads, ships, armies, and temples. *Capital* is here defined as some kind of material infrastructure through which the extraction of energy can be increased.

Preindustrial agricultural cosmologies invariably recognized the productivity of the land as the foundation of human society. This was evident in Europe as recently as among the eighteenth-century Physiocrats, and continues to be a dominant conception among nonindustrial agriculturalists in other areas of the world system today (Gudeman 1986; Gudeman and Rivera 1990). In fact, even the physical energy of labor in these cosmologies is considered secondary to the "strength of the earth" (Gudeman and Rivera 1990). The labor theories of value of the nineteenth century were a product of industrial societies, for which the ultimate dependence on land had become increasingly opaque.

MONEY AS FICTIVE ENERGY

The mercantile empires that often handled long-distance trade between settled, agrarian polities developed a measure of power that had a much more tenuous connection to energy. Their "niche" was not land or labor but exchange value, or purchasing power. This was particularly evident with the sixteenth- and seventeenth-century emergence of transoceanic trading empires like the Portuguese, Dutch, Spanish, and British. If solar energy had been the vital force flowing through agrarian societies, money became the more abstract and elusive "value" that seemed to flow through and empower mercantile societies. The ambiguous relation between energy and money continues to elude us to this day. Purchasing power certainly appears to suffice to empower modern empires, but the particular way in which access to energy is significant for the economy seems to escape economics as a discipline and profession. Nor does a clear understanding of energy seem to be a part of the general public image of the organization of modern society.

But neither, of course, is there a clear understanding of money. The intellectual efforts that have been expended over the past two millennia to grasp the nature of money are impossible to summarize, and the general public today seems as baffled by its logic as ever. Suffice it to say that concepts of energy and money appear to fill similar functions in denoting a vital essence flowing through society.

Like other species we are still, of course, as dependent on solar-derived food energy as ever, but the dominant cultural image of how modern society operates tends to marginalize such concerns in favor of a preoccupation with flows of money. This alienation from the vital flows that animate the biosphere in part derives from the historical experience of merchants and the social institution of money, in part from the nineteenth-century turn to inorganic, fossil energy, which was itself largely a consequence of the mercantile world order.

The concept of energy may seem as abstract and inaccessible as that of money, but it refers to objective material flows that through various intuitive understandings have been part of human consciousness and rationality for hundreds of thousands of years. Its replacement, over the past few millennia, by the concept of monetary value as the standard against which all things are assessed represents a cultural and ideological shift of momentous proportions. Unlike energy, money is fundamentally a fiction. Karl Marx observed that modern Europeans tend to conceive of money as an item that magically grows on its own account, comparable to the premodern worship of idols in West Africa. This was a quintessentially anthropological reflection in that it turned observations of exotic Others back onto the observer's own familiar society. The economic reality in which modern humans are suspended is as culturally constituted and opaque as that of any pre-modern humans. This has been demonstrated by generations of economic anthropologists, but probably no one has done it more convincingly than David Graeber (2011). Money is truly a very peculiar idea and institution. It generates its own varieties of rationality that paradoxically tend to be both imbued with and divorced from morality, as Graeber shows. A fundamental conclusion from his wide-ranging studies, I suggest, is that the historical inclusion of human obligations in the sphere of "goods" exchanged through the medium of general-purpose money has generated pervasive ambiguities about how to draw boundaries between persons and (commodified) things. It is in this context we should understand the phenomenon of slavery. We could add that the impersonal rationality of managing money is decoupled not only from considerations of face-to-face human morality but from the exigencies of living sustainably on planet Earth.

Numerous philosophers, social thinkers, and spiritual leaders have shown very persuasively that money is indeed a fetish. It is a reified representation of social exchange relations that in itself has no substance and no agency except through the ideas that people have about it. As such, it is the ideal tool for controlling people. The premises for rational transactions (e.g., commodity prices, interest rates, currency exchange rates, etc.) can be changed overnight without undermining basic trust in the rationality of money as such. Relative purchasing power can be redistributed in a population through adjustments of this or that regulation in ways so complex that it is impossible for anyone but the high priests of economics to decipher what is being done.

MONEY AS THE CONDITION FOR TECHNOLOGY

As we turn to the phenomenon of modern technology, we should begin by recalling that technology, like energy, is fundamentally a capacity to conduct work. The Industrial Revolution basically boiled down to the unprecedented substitution of organic with inorganic energy in mechanical work. Industrial technology not only replaced much of the work previously conducted with human and animal energy, it made human societies thoroughly dependent on new forms of energy (that is, fossil fuels).

The notion that monetary exchange value is the substance or at least the driving force of society goes much further back than the Industrial Revolution, but it was a condition for it. There would have been no incentive for British textile manufacturers in the nineteenth century to radically intensify their production of cotton cloth if these commodities could not, by means of money, be exchanged for increasing volumes of embodied labor and land (for instance, in the form of imports of cotton fiber). The rationale of mechanization is inextricably intertwined with global differences in the prices of labor and resources. If the African slaves harvesting cotton fiber on the colonial plantations had been paid standard British wages, and the owners of New World soils had demanded standard British land rent, industrialization would simply not have occurred. The existence of modern technology, like the lucrative trade in spices, silver, or beaver pelts, is founded on strategies of conversion between different parts of the world market, where labor and land are very differently priced. This explains why the density of technological infrastructure continues to be very unevenly distributed over the face of the Earth, as can be observed on any global satellite image of nighttime lights.

Thus money came to replace the concern of the eighteenth-century Physiocrats with fertile farmland as the basis of an affluent society. Although Thomas Malthus had worried about the availability of land as a constraint on economic growth, David Ricardo observed that capital and labor could substitute for land, and Karl Marx, too, was optimistic about the prospects of technological progress. These giants of economics appear not to have been very concerned about the fact that industrialization was fundamentally a strategy for Britain to appropriate, in terms of land area, an ecological footprint several times the size of its entire national territory, and, in terms of embodied labor, the toil of a workforce several times larger than its national population (Pomeranz 2000; Hornborg 2006, 2011).

The adoption of technologies propelled by fossil fuels two centuries ago was strongly connected to the emergence of the modern world view, articulated by Ricardo, that I referred to twenty years ago as the "image of unlimited good" (Hornborg 1992). This view sharply contrasted with that of premodern agricultural societies throughout the world, where the existence of absolute constraints was acknowledged, for example in the concern that one person's gain may be another person's loss. The anthropologist George Foster in 1965 presented such beliefs in a Mexican village as an exotic "image of limited good" (Foster 1965). Although realistic from a local perspective, Foster argued, this cultural image posed an obstacle to development. As we today face renewed Malthusian concerns with the limits to growth, not least in relation to problems of peak oil and global warming, it may be worth asking *who* has been living with a cultural illusion, and whether those villagers are in fact now being vindicated, but at a global level?

The turn to fossil fuels as a source of mechanical energy was revolutionary in many ways. Geopolitically, it turned Britain into the most extensive empire the world had ever seen. In part, this was because its textile industry was able to oust its Indian competitors and thrive from the profitable triangular Atlantic trade that converted cotton cloth into African slaves, which were in turn converted into New World plantation produce, including cotton fiber. But fossil fuels also propelled the railways and the steamships with which Britain, frequently using military coercion, organized the metabolic flows of its global empire.

As already mentioned, fossil fuels also revolutionized economics and the public image of the economy. Up until the Industrial Revolution, energy requirements were basically synonymous with land requirements. The work of animals and humans always required land, either for animal fodder or for human food. This meant that there was a fundamental competition over land for production of food versus fodder, which farmers had been familiar with for millennia. Feeding draft animals such as horses and oxen claimed significant proportions of the agricultural landscape in preindustrial Europe. There was thus a limit on the amount of transport energy that was available, and on the distances that bulk goods such as food or fodder could be transported, before the quantity of energy used to move the goods exceeded the energy content of the goods themselves. This posed a limit on rational transport distances because both kinds of energy represented the product of a quantity of land.

Fossil fuels provided a form of energy that did not compete with food production or other uses of the land. This meant that access to land no longer represented the ultimate constraint, as it had to the Physiocrats and to Malthus. Provided that the price of fossil fuels was low enough, it did not matter if the energy expended in transports exceeded the energy content of the cargo. From now on, the same logic applied to production, including agriculture. Relative market prices of various forms of energy, including labor, determined input-output ratios and the feasibility of different kinds of technology. Industrialized production and mechanized transports (of imports as well as exports) went hand in hand. The economic expansion of Britain was determined by the market prices of cotton textiles, slaves, and coal, not by the ability of British crops to harness solar energy.

It is not difficult to imagine how this fundamental transformation of economic rationality must have affected human perceptions of society. Natural constraints were no longer absolute but could be transcended with the help of new technology. If British soils had been exhausted of nutrients, they could be replenished through the import of guano and phosphates from islands in the Pacific. The extent to which this relied on slave-like working conditions on those islands as well as in the British coal mines was made more or less invisible by the impersonal logic of the market, as were the ecological consequences (Clark and Foster 2012). The concept of "technology" from now on signified the almost magic capacity of (some) humans to improve their conditions through sheer ingenuity. Technology was thus perceived as more or less completely a product of inventiveness, rather than of particular kinds of exchange relations on the global market. Moreover, technology was perceived as inevitably progressing toward higher and higher efficiency. Like rationality, of course, "efficiency" is ubiquitously defined by the cultural and societal context. If conceived in terms of an input-output analysis, the parameters for assessing the efficiency of modern technology are related not to expenditures of energy but rather to the input and output of money and of (elite) human time.

It may be helpful, at this point, to add a reflection on the classic conceptualization of unequal exchange by Arghiri Emmanuel (1972). In a nutshell, he argued that, because of international differences in wages, poor nations are obliged to export greater volumes of embodied labor than they would do if wages were uniform. If we exclude Emmanuel's deliberations on labor "value" (see below), this is a perfectly valid observation. International wage differences generate asymmetric flows of embodied labor time, the appropriation of which contributes to underdevelopment in the periphery. But let us also consider this analysis from the converse perspective. If technological progress such as the Industrial Revolution is understood as a process of capital accumulation in the core, at the receiving end of a relation of unequal exchange, it, too, is a product of international differences in wages. It, too, would not occur if wages were uniform. Needless to say, this conclusion should pose certain problems for those orthodox versions of Marxism that celebrate the "inexorable" progress of the productive forces.

TECHNOLOGY AS THE DISPLACEMENT OF SLAVERY

The relation between "modern" (predominantly fossil fuel) technology and slavery is complex. The colonial slave plantations that supplied the British textile industry with cotton fiber were obviously part of the conditions for industrialization in the first place (Inikori 1989). Ironically, however, industrialization appears to have been an important factor in the (official) abolition of slavery (Mouhot 2011). Today it is common to think of the access to modern technology as analogous to having access to its equivalent number of "energy slaves." In fact, this is actually much more than a metaphor.

To begin with, it is significant that the first "protomachines," water mills, appear to have been constructed in the eastern Mediterranean in the early centuries A.D., as replacements for increasingly expensive slaves (Debeir et al. 1991:39). Part of the legacy from ancient Greece and Rome was the delegation of work to other beings who were more or less degraded to things. The idea of externalizing toil from the bodies of free men and delegating it to purportedly mindless agents was fundamental to these civilizations of antiquity. Faced with increasing difficulties in procuring slaves, the choice between doing the work yourself and devising new mechanical contraptions favored the latter option. In order to maintain a traditional lifestyle and identity, landowners tended to replace slaves with machines. The rationality of such "technological progress" then as now hinges on the relative prices of labor and resources. The fifth-century Roman logic of building a water mill instead of purchasing slaves is essentially the same as, fifteen centuries later, affluent middle-class citizens around the world deciding to use a vacuum cleaner or washing machine instead of hiring a housemaid. In both cases, we could add, the owners of technology are able to imagine that technological progress has done away with degrading, low-wage toil. In both cases, however, a closer familiarity with the socioeconomic conditions under which the new technology is manufactured and maintained might have given them a different perspective. To take the example closest at hand, it is far from evident, even today, that the employees of (Chinese) vacuum cleaner manufacturers are better off than American housemaids.

Whether closer familiarity with the conditions of production would actually restrain privileged consumption is doubtful, not least considering the historical inclination of slave owners to defend slavery, but the point here is that geographical dissociation has been a prerequisite for our narrative of technological progress and "sustainable development." Elsewhere referred to as "commodity fetishism" or "consumer blindness," dissociation from the conditions (and consequences) of production is essential to understanding global inequalities of all kinds. This applies not only to technologies but to commodities in general, but its potential ramifications for realizing what much of our "technology" really *is*, viewed through the lens of social science, are particularly disturbing.

From the perspective of privileged sectors of society, investment in new technology is understandably perceived as progress. This conviction has for at least two centuries been fundamental to dominant conceptions of history, development, and modernization (Adas 1989, 2006; Nye 1998; Smith 1998; Marsden and Smith 2005; Friedel 2007; Headrick 2010). But to a large extent, technological progress has been the privilege of affluent elites, and the very existence of the new technology has relied on the appropriation of resources from an increasingly impoverished periphery. The investments in steam technology in nineteenth-century Britain, for instance, were indissolubly connected to the Atlantic slave trade and the cotton plantations in the American South. They relied on a continuous unequal exchange of embodied labor and land between the industrial core and its colonial periphery (Hornborg 2006). Although the complexity of global commodity chains makes it very difficult to demonstrate similar asymmetries in flows of embodied labor and land today, particularly because of the way trade statistics are organized, there are several reasons to believe that the pattern continues. World society remains highly polarized between high-tech core areas with high levels of per capita purchasing power and energy consumption, on one hand, and peripheral areas with much lower levels of purchasing power and energy consumption, on the other. The unequal exchange of embodied labor time in the modern world was demonstrated forty years ago by Arghiri Emmanuel (1972), and the unequal appropriation of embodied land has been amply documented by the voluminous research on ecological footprints (cf. Wackernagel and Rees 1996).

A relevant question for social scientists (including anthropologists) to ask at this point in world history is whether modern technology has really replaced slavery or merely *dis*placed it. If the category "slavery" is defined not primarily in terms of being victims of immediate violence but more fundamentally in terms of being coerced to perform alienating, low-status tasks for the benefit of a privileged elite, a significant part of the world's population would qualify as slaves. Seemingly neutral phenomena such as "technology" and the "world market" organize the transfer of their embodied labor and resources to an affluent minority. From this perspective, the operation of technology represents the deflected agency (the labor energy) of uncounted millions of laborers, harnessed for the service of a global elite. To view technology in terms of a set of "energy slaves" is thus indeed more than a metaphor.

"VALUE" AS MYSTIFICATION

The most serious criticism of mainstream economics and what Aristotle called "chrematistics" (the art of managing money, as opposed to real resources) has come out of Marxist and ecological economics (cf. Martinez-Alier 1987). These two schools of thought have actually converged in their criticism, as both have shown how exchange rates set by market prices can conceal "unequal exchange" of labor or resources that are significant for macro-level processes of development versus underdevelopment (Bunker 1985). It is interesting to note that both schools have in fact been concerned with the way in which monetary prices mystify flows of energy through society. The net flows of embodied labor power emphasized by Marxists are no less a form of energy than the flows of resources (such as "emergy," originally conceived as shorthand for "embodied energy") studied by ecological economists (Odum 1996). Some theorists have explicitly compared these two approaches to unequal exchange (Lonergan 1988). Analytically, the arguments are indeed identical. Unequal exchange is posited to occur when some kind of "value" (labor value or energy value, respectively) is being underpaid.

Predictably, such arguments will not convince mainstream economists, and in this particular respect I must agree with them. The problem is couching the discussion in the idiom of "value." Anthropologists are well aware that any notion of value must be culturally constituted and cannot be derived from Marxist theory or from physics (Baudrillard 1981; Sahlins 1976; Bourdieu 1984). To suggest that Marxist or ecological economists have a better understanding of what is valuable than market actors, and that the latter consistently "underpay" these more essential values, is the wrong way to approach the problem. The major mistake that these theorists make is to use the concept of "value" for some kind of material flow that is not in itself the object of valuation. As we are all well aware, it is not the quantity of embodied labor or energy that determines how much a consumer is willing to pay for a given commodity. The "surplus value" that provides profits for capitalists is not a metaphysical product of labor power (one of several possible sources of energy) but simply the difference between the cost of inputs and the income from selling the output. In addition to human labor, the inputs may include, for instance, the work of draft animals, fuels, and raw materials. It is of course true that labor power can yield more (monetary) "value" than the cost of its maintenance, but this also applies to, for example, animal traction and the use of fossil fuels. There is no fundamental difference, in terms of the production of surplus or added value, between letting humans or animals power a treadmill, work the soil, or transport a burden. To claim that there is a significant difference is to fetishize capitalist profit into something more than mere money.

The paradox, then, is that critics of mainstream economics, in struggling to expose the ideological function of the market in mystifying asymmetric flows, and to identify various forms of energy as the asymmetric flows thus mystified, have resorted to the mercantile notion of "value" to underpin their argument. This notion has for centuries pertained to money (exchange value) and consumer preferences. It belongs to the vocabulary of the market and should not be confused with the objective material flows that both Marxist and ecological economists are concerned with. In order to propose that the world market conceals asymmetric flows of energy that contribute to global inequalities in the distribution of technology, purchasing power, and environmental quality—a proposition that I would wholeheartedly endorse—let us talk about precisely that: asymmetric flows of *energy*.

"VALUE" AND THE SECOND LAW OF THERMODYNAMICS

In thus distinguishing (material) flows of energy from (semiotic) flows of money values, we would be presenting an argument that would not be easy to dismiss. To observe that the accumulation of technological infrastructure in certain areas of the world (visible on satellite images of nighttime lights) would be impossible without a net input or *appropriation* of available energy is simply based on the second law of thermodynamics. Similarly, to observe that energy is dissipated in economic processes, implying that industrial products contain less available energy than the fuels and labor inputs that were used in making them, is also completely in line with the entropy law. Nicholas Georgescu-Roegen (1971:292, 294) reiterated an essentially Physiocratic position when he observed that, viewed from theoretical physics, industry is "completely tributary" to agriculture and mining. From a physical perspective, in other words, production is destruction. The creation of consumer value (utility) is simultaneously the creation of entropy. This is not to deny that thermodynamic efficiency can be increased and the rate of entropy production slowed down, but the fundamental logic of the entropy law cannot be reversed. Finished products must be priced higher than the inputs (labor, fuels, and raw materials) but inexorably represent less available energy. In this sense, money and energy are inversely related.

The dissipation of finite resources is thus blindly and continuously rewarded by the market with more resources to dissipate. This in turn means that increasing quantities of energy use and materials will be concentrated in what Stephen Bunker (1985) called the "productive" sectors of the economy, while the "extractive" sectors are increasingly impoverished. The argument thus accounts well for the polarizing tendencies of the world economy, without resorting to any normative, contestable propositions regarding what is "valuable" or "underpaid." "Unequal" exchange in this sense simply means "asymmetric." Whether it is also "unfair" is a moral conclusion that is up to the reader, rather than an assumption of the argument as such.

But, someone might object, even if embodied labor or energy is not a "value" but a physical measure, wouldn't it be valid to propose that it is being "underpaid"? Couldn't the inequalities in the world be leveled out by adjusting exchange rates? Unfortunately, the basic problem is not simply that prices of embodied labor, energy, and resources are too low (in relation to those of finished goods) but that if they were higher, there would be no incentive to continue the exchange. What would be the point of mechanization if it was not profitable? As argued above, the rationality underlying British industrialization in the nineteenth century was founded on the much lower price of land and labor in the colonies. The foremost rationale of industrial capitalism is to *not* have to pay for the costs of increasing social and ecological disorder in the surrounding world. This logic continues to pervade "technological progress" to this day.

Another objection, among anthropologists, might be that we are not physicists but social scientists. We can only be concerned with cultural constructions, the argument might go, not with putatively objective conditions. But I would counter that it is precisely in its transdisciplinarity that anthropology has its greatest potential, as mediator between social and natural science. In an unforgettable sentence, Roy Rappaport (1993:154) summarized the predicament of our field: "[A]ny radical separation of [the objective and the subjective] is misguided, not only because meanings are often causal and causes are often meaningful but because, more fundamentally, the relationship between them, in all its difficulty, tension, and ambiguity, expresses the condition of a species that lives, and can only live, in terms of meanings it itself must construct in a world devoid of intrinsic meaning but subject to natural law." To "comprehend the fullness of its subject matter's condition," anthropology must not only remind the natural scientists that humans live in terms of meanings but also remind the social scientists (including economists) that societies are subject to natural law.

THE ANTHROPOLOGY OF SUSTAINABILITY

A defining feature of an anthropological perspective is that it acknowledges the importance of *cultural specificity*. As Marshall Sahlins (1976:170) succinctly put it, "no object, no thing, has being or movement in human society except by the significance men can give it." Any discussion of human-environmental relations, patterns of consumption, power structures, or economic world views would be incomplete without considering the particular systems of meaning that organize them. Yet very little mainstream discourse on sustainability, environmental problems, or the economy is concerned with cultural aspects. Natural scientists, technologists, and economists have to a large extent monopolized these discussions, leaving anthropologists with a sense of dismay at finding publicly accepted problems and their solutions so narrowly and simplistically defined.

Anthropological cultural analysis should have crucial things to say about past, present, and future concerns with sustainability, yet it rarely does. Instead, other disciplines are taking up the charge. For example, a few years back the growing public concern about the prospects of socioecological collapse provided an ornithologist (a biogeographer) with the opportunity to produce a bestseller on the turbulent history of human societies (Diamond 2005). To a large extent, the marginalization of anthropological perspectives on sustainability probably derives from the fact that reflexive cultural approaches are intrinsically more difficult to grasp—and above all, to *apply*—than ecological, technological, or economic ones. But there are doubtlessly also ways in which anthropologists could exert themselves a bit more to become relevant. Let me mention two problems that tend to impede anthropological engagements with global sustainability issues.

The first problem derives from the tenacious separation of social and natural science. Those researchers who seem to be the most concerned about the future of the biosphere (the natural scientists) generally have the bluntest analytical tools for understanding the causes of anthropogenic environmental degradation, while those who possess such tools (social scientists, including anthropologists) generally tend to be less concerned with the biosphere as an objective, biophysical reality. In having been trained (indeed required) to approach the environment with quotation marks, as a "contested" or "negotiated" cultural construction, anthropologists often seem at a loss when expected to say something about the *real* environment beyond human perceptions. Certainly, natural scientists need to realize that cultural sign systems such as language, money, and power are components of ecosystems, organizing significant aspects of their flows of matter and energy, but social scientists conversely need to realize that flows of matter and energy are fundamental to social systems and need to be taken into account in any explanation of development, underdevelopment, and collapse. Anthropologists—at least the four-field kind—should be uniquely equipped to think in truly transdisciplinary ways about how material and semiotic processes are intertwined.

The second problem has to do with whether we choose to define our discipline in terms of general outlook or in terms of methodology. I would argue that the capacity to perceive and analyze the cultural dimension of all human thought and activity is the foremost hallmark of anthropology. But this definition of the anthropological approach is often overshadowed by a ubiquitous emphasis on ethnographic fieldwork as the sine qua non of the discipline. These two attempts to define the field (vis-à-vis other social sciences) deserve some reflection, as one does not necessarily coincide with the other. Many indispensable and central contributions to our field over time (for example, volumes by Marcel Mauss, Claude Lévi-Strauss, Marshall Sahlins, Tim Ingold, and David Graeber) have no obvious connection to the author's own field research, while much dedicated fieldwork fails to deliver observations imbued with a sensitivity to meanings. Which is most essential for our identity or purpose as anthropologists: our way of looking at the world or the conditions under which we do so? To seriously raise questions about the anthropological apotheosis of fieldwork would probably be perceived by most anthropologists as a breach of conduct. But even at meetings of the American Anthropological Association, there are occasional voices questioning if the traditional focus on ethnographic fieldwork is really conducive to making anthropological perspectives relevant to contemporary public concerns with sustainability, the economy, and what it means to be human. Could it be that while we and our students are busy immersing ourselves in obscure representations of exotic, local particularities of experience, the wider comparative discourse on the global human condition is all the more easily usurped by ornithologists?

SUBMITTING TO OBJECTS

Anthropologists have long deliberated on the significance of animism, fetishism, and the so-called Cartesian modern inclination toward objectivism. For the purposes of this discussion, we can define fetishism as the attribution to inanimate objects of some properties of living things, objectivism as the (inverse) denial of such properties even in living things, and animism as the (intermediate) acknowledgment of such properties in living things. The abandonment of animism and adoption of objectivism made modernity possible, notably by dismissing or at least suppressing moral qualms about human exploitation of the rest of the biosphere. To objectify Nature was thus to gain power over it, because "objects" in the Cartesian scheme are morally neutral. In obsessively transforming Nature into commodities for consumption (that is, destruction), modern societies appear to consummate their power over the material world.

Paradoxically, however, the same Cartesian scheme that inspired this modern aspiration to achieve power over objects has generated an unprecedented human *submission* to objects. Premodern forms of fetishism, attributing personhood to inanimate objects, have been replaced by the specifically modern forms identified by Karl Marx. Rather than full personhood, money and commodities are attributed with autonomous agency (as in the notion of "interest" on bank accounts). Although not apparent to Marx, technological objects are also fetishes in the sense that they are inanimate objects attributed with autonomous agency and productivity, obscuring their foundation in asymmetric global relations of exchange. Technology is a modern form of magic that needs to be exposed no less than the financial practices on Wall Street. But capital accumulation, once congealed into technology, seems automatically exempt from political critique.

If the commodities we consume (that is, metaphorically eat) are really em*bodi*ments of other people's life energy, capitalism is a transformation not only of slavery, as Graeber (2007:85–112) has argued, but of cannibalism. The defining feature of capitalism is its specific social and cultural organization of the appropriation of geographically remote labor and land. Modern forms of market exchange, technology, and consumption represent net transfers of embodied (human) time and (natural) space extracted from some social groups for the disposal of others. Rather than directly controlling the labor of other human bodies in the vicinity, as in slavery, this is achieved by controlling the products of labor. Rather than shipping commodified labor (in chains) across the oceans, modern ocean liners thus ship the commodified embodiments of labor. Ever since the first textile factories emerged in early industrial Britain, machines have assumed an illusory dissociation from the social relations of exchange through which their raw materials are extracted, appropriated, transformed, and redistributed. This illusion rests on the cultural assumption that material objects are politically innocent and immune to moral critique. The same, ultimately Cartesian illusion liberates consumers to continue devouring distantly derived objects without any significant moral qualms about the social or ecological implications of consumption.

CONCLUSIONS

For about two hundred years, Euro-American expansion has been largely based on fossil-fuel technologies for production, transports, and warfare. As we consider the future implications of so-called peak oil and climate-related restraints on the combustion of fossil fuels, the ideological mystifications of the age of fossil-fuel capitalism begin to protrude with greater clarity. Until the Industrial Revolution, energy requirements and land requirements were one and the same, converging in the production of food for human labor and fodder for draft animals. For the past two centuries, however, the age of fossil fuels has kept land requirements and energy requirements conceptually distinct, justifying a pervasive trust in limitless economic growth and technological progress. As modern industrial societies are currently contemplating the reality that peak oil and climate change may prompt us to rely increasingly on biofuels, we are in fact imagining a future in which land requirements and energy requirements will once again coincide. It appears as though it will again be possible to calculate the costs of transport distances in terms of eco-productive space, as during the preindustrial era. If such a post-petroleum future will bring renewed competition over scarce land for food, fodder, fibers, and fuel, concepts of ecologically unequal exchange and environmental load displacement will emerge as tangible conditions of human existence. Most importantly, the pervasive trust in new technologies would give way to a radically different understanding of the very phenomenon of technology as a strategy for redistributing embodied labor and resources in world society. Their materiality would no longer grant technological objects immunity from political critique. In terms of economic understandings of history, such a reversion to equivalences between land, labor, and energy would imply the bankruptcy of Ricardian and Marxian concepts of labor value and perhaps inspire the reinstatement of a cosmology of value akin to preindustrial Physiocracy. Contrary to our current illusions, the vital essence that flows through human societies is energy, not money. The sooner we begin to grasp the peculiar incongruity between our material conditions and cultural constructions, the sooner we shall be able to think creatively about how to organize human economies in ways that truly deserve to be called sustainable.

NOTE

1. I gratefully acknowledge financial support from The Swedish Research Council for Environment, Agriculture and Spatial Planning (FORMAS) for supporting the Lund University Centre of Excellence for the Integration of Social and Natural Dimensions of Sustainability (LUCID), which has provided me with the opportunity to write this paper.

REFERENCES

Adas, Michael

1989 *Machines as the measure of men: Science, technology, and ideologies of western dominance*. Ithaca, NY: Cornell University Press.

2006 *Dominance by design: Technological imperatives and America's civilizing mission*. Cambridge, MA: Massachusetts Institute of Technology Press.

Baudrillard, Jean

1981 *For a critique of the political economy of the sign*. St. Louis: Telos Press.

Clark, Brett and John Bellamy Foster

2012 Guano: The global metabolic rift and the fertilizer trade. In *Ecology and power: Struggles over land and material resources in the past, present and future*, edited by Alf Hornborg, Brett Clark, and Kenneth Hermele, 68–82. London: Routledge.

Bourdieu, Pierre

1984 *Distinction: A social critique of the judgment of taste*. London: Routledge.

Bunker, Stephen G.

1985 *Underdeveloping the Amazon: Extraction, unequal exchange, and the failure of the modern state*. Chicago: University of Chicago Press.

Debeir, Jean-Claude, Jean-Paul Deléage, and Daniel Hémery

1991 *In the servitude of power: Energy and civilization through the ages*. London: Zed Books.

Diamond, Jared

2005 *Collapse: How societies choose to fail or succeed*. New York: Penguin.

Emmanuel, Arghiri

1972 *Unequal exchange: A study of the imperialism of trade*. New York: Monthly Review Press.

Foster, George

1965 Peasant society and the image of limited good. *American Anthropologist* 67 (2): 293–315.

Friedel, Robert

2007 *A culture of improvement: Technology and the western millennium*. Cambridge, MA: Massachusetts Institute of Technology Press.

Georgescu-Roegen, Nicholas
1971 *The entropy law and the economic process*. Cambridge, MA: Harvard University Press.

Graeber, David
2007 *Possibilities: Essays on hierarchy, rebellion, and desire*. Oakland: AK Press.
2011 *Debt: The first 5,000 years*. Brooklyn: Melville House.

Gudeman, Stephen
1986 *Economics as culture: Models and metaphors of livelihood*. London: Routledge and Kegan Paul.

Gudeman, Stephen and Alberto Rivera
1990 *Conversations in Colombia: The domestic economy in life and text*. Cambridge: Cambridge University Press.

Headrick, Daniel R.
2010 *Power over peoples: Technology, environments, and Western imperialism, 1400 to the present*. Princeton, NJ: Princeton University Press.

Hornborg, Alf
1992 Machine fetishism, value, and the image of unlimited good: Toward a thermodynamics of imperialism. *Man* (N.S.) 27: 1–18.
2001 *The power of the machine: Global inequalities of economy, technology, and environment*. Walnut Creek, CA: AltaMira Press.
2006 Footprints in the cotton fields: The industrial revolution as time-space appropriation and environmental load displacement. *Ecological Economics* 59 (1): 74–81.
2011 *Global ecology and unequal exchange: Fetishism in a zero-sum world*. London: Routledge.

Inikori, Joseph E.
1989 Slavery and the revolution in cotton textile production in England. *Social Science History* 13 (4): 343–379.

Lonergan, Stephen C.
1988 Theory and measurement of unequal exchange: A comparison between a Marxist approach and an energy theory of value. *Ecological Modeling* 41: 127–145.

Marsden, Ben and Crosbie Smith
200 *Engineering empires: A cultural history of technology in nineteenth-century Britain*. Houndmills: Palgrave Macmillan.

Martinez-Alier, Juan
1987 *Ecological economics: Energy, environment and society*. Oxford: Blackwell.

Mouhot, Jean-François
2011 Past connections and present similarities in slave ownership and fossil fuel usage. *Climatic Change* 105: 329–355.

Nye, David E.
1998 *Consuming power: A social history of American energies*. Cambridge, MA: Massachusetts Institute of Technology Press.

Odum, Howard T.
1996 *Environmental accounting: Emergy and environmental decision-making*. New York: John Wiley and Sons.

Pomeranz, Kenneth
2000 *The great divergence: China, Europe, and the making of the modern world economy*. Princeton, NJ: Princeton University Press.

Rappaport, Roy
1993 "Humanity's evolution and anthropology's future." In *Assessing cultural anthropology*, edited by Robert Borofsky, 153–166. New York: McGraw-Hill.

Sahlins, Marshall
1976 *Culture and practical reason*. Chicago: University of Chicago Press.

Smith, Crosbie
1998 *The science of energy: A cultural history of energy physics in Victorian Britain*. Chicago: University of Chicago Press.

Wackernagel, Mathis and William E. Rees
1996 *Our ecological footprint: Reducing human impact on the Earth*. Gabriola Island: New Society Publishers.

CHAPTER TWO

Energy Consumption as Cultural Practice: Implications for the Theory and Policy of Sustainable Energy Use

Harold Wilhite

INTRODUCTION

Contending with the specters of climate change and resource depletion will require deep reductions in the global use of fossil fuel–based energy over the coming decades. There are essentially three strategies for accomplishing this: changing production from fossil fuels (coal, oil, natural gas) to non-fossil fuels such as renewable and nuclear energy, carbon capture and storage, and reduction in energy consumption. It is becoming increasingly apparent that renewable energies will not be phased in fast enough to make a dent in rising global carbon emissions. Nuclear energy provides a carbon-free alternative, but it is expensive and bears with it risks of severe accidents and problematic environmental side effects. Carbon capture is expensive, the technology is immature, and there is growing political resistance in Europe to carbon storage. Germany has recently decided to close the door on carbon storage. Added to these difficulties is a desperate need for energy among the 1.4 billion people globally who do not have access to electricity. For example, only 50 percent of India's rural population and only 24 percent of sub-Saharan Africa's population have access to electricity. These and other parts of the developing South and East will need energy use for basic services such as light, health care, and schooling, as well as for developing their economies. Energy will also be needed for "intermediate development" in India, China, and Brazil, as well as other Asian and South American countries with rapidly growing economies. These countries will need massive amounts of energy for industrial development and to accommodate the demands of growing middle classes for household energy appliances and automobiles.

The growing need for energy in these non-OECD countries will add to the high levels of energy use in the OECD. To illustrate the imbalance in energy use between OECD and non-OECD consumption: the per capita energy consumption in the United States is six times that of China and fifteen times that of Tanzania. Electricity consumption per capita is 13,400 kilowatt-hours (kWh) in the United States, 1,600 kWh in China, and a mere 55 kWh in Tanzania. High rates of energy use and carbon emissions were essential to sustained economic growth in the United States, Europe, and Japan over a half century. From an ethical point of view, it would be highly questionable to demand reductions or even tight restrictions on energy intensity in non-OECD countries. The bottom line from a climate perspective is that if the consequences of climate change are to be limited, then global energy use will have to be reduced, and the major share of the reduction must take place in the rich OECD countries. The prospects for accomplishing this are grim. From the origins of the energy-conservation effort following on the oil shocks of the 1970s, governments have been reluctant to enact robust energy-reduction policies due to concerns about the consequences for employment, profits, and economic growth. Policy efforts have been weak and fragmented. Policy has been based on a reductive research platform that has failed to capture the dynamics of consumption or to provide a robust theory of change. In this chapter, I will briefly review and critique mainstream theory and policy frameworks for energy consumption and reduction, then flesh out a new approach grounded in social practice theory, which I claim has the potential to reinvigorate research and policy agendas on energy sustainability.

To begin with the energy-savings track record, the efforts to reduce dependence on energy and reduce the environmental consequences of energy use now have a forty-year history, which began with the "oil shocks" created by the Middle Eastern oil boycotts of the 1970s. Energy was to be conserved in North America and Europe in order to reduce dependence on imports and thereby increase "energy security." Energy analysts such as Amory Lovins demonstrated how greater energy efficiency in both energy production and consumption made economic sense, both for society and for energy users (Lovins 1977). The 1980s brought greater awareness of the environmental effects of energy use, including acid rain, nuclear waste, oil spills, and climate change. The point is that the benefits and rationale for energy savings were well understood thirty years ago: energy security, economic efficiency, and environmental amelioration. A field of research and policy was established. The primary concerns were with the manufacture and diffusion of efficient technologies. Research agendas were fairly blank on how to incentivize the "end user," the term used to represent businesses, households, and other actors on the demand side. As an understanding grew that technologies would not make their way into homes and businesses on their own, economists were recruited to explain how energy is used and how energy use could be made more efficient. In the reconstituted theoretical framework, the end user was characterized as economically rational and purposive. Critiques of this characterization began early as other social sciences turned their attention to energy (see Lutzenhiser 1993), but alternative perspectives have not had much impact on research agendas or policy discourses until recently. The failure of energy policies to deliver reductions has called into question policies directed at a socially disembodied consumer. Over the past decade, increasing consensus has grown up around a new overarching framework for energy consumption: social practice theory. In the remainder of the chapter, I will explore this new theoretical framework and its implications for reframing energy-conservation policy.

SOCIAL PRACTICE THEORY

Pierre Bourdieu, the founder of one of the important strands of practice theory, was interested in what he called practical reason and put forward the outline of a theory that explained "practice, action, interaction, activity, experience, and performance" (1977:xx; 1998). Bourdieu proposed that human actions have sociomaterial histories and that this had not been adequately addressed in theories of action. Moving and acting in sociomaterial space carves out predispositions for subsequent actions that are embedded in bodies, practices, and material settings. These embedded predispositions are important to understanding how and why practices stabilize and change. Practice theory was more or less abandoned in the postmodernist and antimaterialist moves of the 1980s, which disembodied and dematerialized consumption (see Baumann 1988; Lash and Urry 1994; Brown and Turley 1997). As Alan Warde wrote about this period, "A decade or more of analysis, founded in political economy and developing a materialist perspective on social life, seemed suddenly to be abandoned for a mode of studying culture which operated on wholly antithetical assumptions, according signs, discourses and mental constructs an exclusive role in understanding social activity" (1997:1). In the ensuing years, a number of social scientists from differing academic disciplines have contributed to the development and application of practice theory to an understanding of energy consumption (Shove 2003; Warde 2005; Wilhite 2008a; Røpke 2009). These efforts draw on newer refinements in the practice theory, such as those proposed by Ortner (2006a, 2006b) and Reckwitz (2002). Reckwitz defines a practice as "a routinized type of behavior which consists of several elements, interconnected to one another: forms of bodily activities, forms of mental activities, 'things' and their use, a background knowledge in the form of understanding, know-how, states of emotion and motivational knowledge" (2002:249, cited in Warde 2005). From a practice perspective:

> Individuals . . . are no longer either passive dupes beholden to broader social structures, or free and sovereign agents revealing their preferences through market decisions, but instead become knowledgeable and skilled 'carriers' of practice who at once follow the rules, norms and regulations that hold practice together, but also, through their active and always localised performance of practices, improvise and creatively reproduce and transform them. (Seyfang et al. 2010:12)

Practice theory implies a shift from agent to agency, defined as "the capacity to influence acts" (Ortner 1999). This agency (capacity) is not a property of individuals but is rather shared by individuals, the things with which they interact, and the social contexts for the action. As Ahearn expressed it, "agency can be considered the socioculturally mediated capacity to act, while praxis (or practice) can be considered the action itself" (2001:118). The practice perspective sets up an entirely new agenda for study of energy consumption and opens a new conceptual basis for policies that aim to reduce the intensity of energy use in everyday practices. It expands the focus of research and policy from motivating individuals (reified as rational and purposive) and implementing technical efficiency to encompass the contributions to consumption of embodied knowledge, habit, and artifacts.

SITES OF EMBEDDED KNOWLEDGE

When individuals perform an action regularly and frequently, predispositions for subsequent performances are embodied. Much of this embodiment takes place due to immersion in a certain cultural field: actions such as walking and eating are embodied early in life, as people are exposed to the ways they are performed by others in their family and social networks. Mauss (1935) called this form for the embedding of knowledge "enculturation." Another source of embodiment takes place as practices become routinized, such as in the frequent repetition of a routine such as mowing the lawn, walking the dog, or taking a shower. Yet another source of embodiment is purposive training such as that associated with sports or the learning of crafts and manual skills. Embodiment is strongly agentive in activities in which the material backdrop is stable (Wilhite 2012), such as in tasks like typing or swimming (keyboard and pool). In performances associated with team sports and dance, there is an added degree of complexity in the form of coordination with other bodies. The need for verbal communication and physical coordination of movements draws on a mix of practical and cognitive knowledge. Nonetheless, one of the purposes of training is to develop embodied instincts that can contribute to skilled actions and reactions to other players and situations.

The degree of complexity of the material space and artifacts involved in the interaction has an effect on the strength of embedded knowledge. In my estimation, neither Bourdieu, Reckwitz, nor other practice theorists have been specific about the ways that things and material contexts contribute to the embedding of knowledge. As I have written about extensively in recent publications (Wilhite 2008a, 2008b), we can look to the research domains of material culture (Miller 1998, 2001; Appadurai 1986) and science and technology studies (STS) (Bijker and Law 1992; Latour 2000; Akrich 2000) for insights on material agency. It is important to note that material agency is not equivalent to technological determinism: "the capacity to turn humans into a cog in a machine" (Veerbeek 2005:3). In fact, users "domesticate" or "appropriate" (terms popular in energy research, see for example Lie and Sørensen 1996) technologies in surprising ways. For example, people use room thermostats like on-off switches, override movement-sensitive or natural-light-sensitive lighting systems by manually manipulating lighting, open windows in thermostatically controlled buildings to regulate heat, and so on. However, once in place and running in a home, household technologies such as refrigerators, cooking appliances, washing machines, and air conditioners bear with them the potential to reshape practices. Individuals bring their knowledge and lived experiences to the interaction with things in practice, but the things they interact with can also influence the action.

Researchers from the STS tradition have explored the ways that things affect practices. Madeleine Akrich defined the potential of things to influence actions as a "technology script" (2000). I can exemplify how technology scripting affects practices with two examples from my research in South India (Wilhite 2008a). Both have to do with refrigeration. The refrigerator/freezer has been agentive in a gradual but important change in cooking and eating preferences. There is a long-standing food ideology in South India that regards cooking, storing, and then reheating foods as unhealthy. Foods should be eaten soon after they are prepared. There is a widely shared belief that reheating and eating stored foods causes laziness and stupidity. This idea explains the lack of initial enthusiasm for the refrigerator when it became available in India in the 1960s. People's initial interest in the refrigerator was mainly because of its potential to store and prolong the life of raw, uncooked foods (not problematic according to local thinking) and to reduce the number of trips to the market. However, as increasing numbers of women have entered the work force, they have had less time to prepare every meal from scratch. The refrigerator's "script" for storing cooked foods has persistently exercised agency over a course of three generations. Young women today routinely make food in bulk and reheat portions for forthcoming meals. The refrigerator has also opened for the storage of frozen and other ready-made foods, the consumption of which is increasing rapidly. The microwave oven completes the convenience-food technology regime. In 2004, the microwave was the fastest-growing household appliance in India.

It is ironic that the STS perspective represented by Akrich's theory has been given so little attention in a research domain that has had an overwhelming interest in the diffusion of efficient technologies (Wilhite and Nørgaard 2004). The embedded potentials for changing energy-consuming routines and for encouraging purchases of new, complementary energy-using devices have been largely ignored in energy research and policy. The fact that new technologies are energy efficient is negated by the encouragement of new practices that demand more energy. This technology-driven increase in energy contributes to what energy analysts call the "rebound effect." The usual explanation is economic: money acquired from energy savings associated with a more efficient technology is used to pay for an increase in comfort, for example increasing living room temperatures from 21 to 22 degrees Celsius, or to buy other energy-using things or services, such as air travel, thus negating the net reductions in energy use from the original efficiency purchase (see Herring 1999; Moezzi 1998; Wilhite and Nørgaard 2004). The transition in practices behind the "rebound" has yet to be explored and deserves more attention in research and policy.[1]

Another refrigeration technology, the air conditioner, is strongly agentive in a range of household practices. In Kerala, India, the roots of the air-conditioning transformation now taking place extend back to the beginning of the twentieth century. Until then, building construction was done mainly by caste-based craftsmen whose building principles took account of Kerala's hot and humid climate. They used wood, mud, unfired bricks, bamboo, straw, and leaves as building materials—all porous materials that allow natural ventilation. Houses had canted roofs, screen porches, and other design principles aimed at reducing heat gain and encouraging drafts. A commercial building industry grew up in India in the mid-twentieth century, using unskilled labor and least-cost building principles. The skills associated with home building in a hot climate disappeared. Cement became the building material of choice and building designs began to incorporate large windows, south-facing views, and flat concrete roofs, all of which tend to contribute to heat gain in hot climates (Wilhite 2008a). Thus, a building "script" was created from the 1970s onwards that begged for air-conditioning. However, air conditioners were classified as luxury goods, fees were placed on imports, and there were very few Indian manufacturers. This situation changed dramatically in the early 1990s, after the "opening of India" to transnational capitalism. Import fees were reduced, foreign-made air conditioners flooded Indian markets, and transnational corporations set up local production. In Kerala, only about 3 percent of homes had air conditioners in 1995. By 2001, this had increased to 15 percent. Air-conditioning has affected many home practices, such as when and where people sleep and eat. Front porches, formerly magnets for casual conversation with neighbors and passersby, are hardly used by households that have installed air-conditioning. Air-conditioning use is straining electricity peak loads and increasing the frequency of blackouts. This has disrupted food preparation, washing, and entertainment practices for all households, not just those with air conditioners. These changes in household practices are consistent with changes in other parts of the world where air-conditioning is pervasive (Cooper 2008, 1998; Strengers 2008).

Things are bearers of predispositions for consumption. Replacing a tool with an energy-using technology can lead to significant changes in consumption practices and in energy use, for example a broom with a vacuum cleaner, a washbasin with a washing machine, or a mortar and pestle with a mixmaster (electric mixer). At the level of physical infrastructures, building highways and a network of petrol stations will obviously favor automobile-based transport habits. On the other hand, low-energy transport infrastructures for public transportation systems, bicycles, and walking can lead to dramatic changes in mobility practices. In Copenhagen, Denmark, after decades of work on making bicycling safe and convenient, over 50 percent of commuters in Copenhagen today commute by bicycle. These perspectives on technology, practice, and energy use need to be brought to center stage on the energy policy agenda.

MOVING PEOPLE

Moving technologies into practices is one important change mechanism. Another is the movement of people into or through a new sociocultural context. Another example from my India research illustrates how movement can affect practices. Many Malayalee (inhabitants of Kerala) work outside of India and spend their working lives moving between two countries, India and their place of work. In Kerala's capital Trivandrum, 40 percent of all families have at least one family member working abroad. The majority of work migrants work in the Oman Gulf countries (such as Kuwait and the United Arab Emirates), but some find work as far afield as Saudi Arabia, Singapore, Europe, and Australia. Most migrants spend more time in their work home than they do with their family in Kerala. They have a lifestyle that involves a dual residence. In their places of work abroad, migrants develop routines that in some cases are very different from typical Kerala routines. The confrontation with new practices and technologies lifts embodied knowledge into what Wilk calls the "discursive sphere of heterodoxy . . . where, eventually, through the exercise of power, they can become re-established as orthodoxy, and eventually sink back into the accepted daily practice" (1999:10). The dual residence of Kerala work migrants works this way, disrupting embedded preferences and leading to a reorganizing of consumption practices such as transport, food preparation, and home cooling. Movement need not be to a radically different sociocultural context in order to initiate changes in practice. In a study Rick Wilk and I conducted in the early 1980s in Northern California, we found that a move into a new home engendered a period of intense reflection by the family over household practices and often initiated a flurry of home projects and changes in practice. Further disruptions occurred in conjunction with the birth of children, as well as when children moved out of the home (Wilk and Wilhite 1985).

TOWARD A MORE ROBUST AGENDA FOR ENERGY REDUCTION

The conceptual move that I am proposing—from viewing energy consumption as something performed by individuals and individual devices to something that is a result of the interaction between things, people, knowledge, and social contexts—has subtle but important implications for policies aimed at reducing energy consumption. From a practice-grounded perspective, individual consumers, their material worlds, and their sociocultural contexts are viewed as agentive. This displaces the starting point for a policy framework from individuals and individual devices to clusters of energy practices, such as those associated with heating, lighting, cooling, preparing food, and so on. For example, concerning personal mobility, efforts to reduce automobility begin with people's transport needs and practices and work through the ways in which public transport systems, bicycles, walking, and automobiles can contribute to satisfying them using less energy. This implies not only the promotion of more fuel-efficient cars but also the consideration of investments in fast and convenient alternatives in the form of coordinated and comprehensive public transport systems, fast intercity trains, and walking/biking infrastructures. Concerning home cooling, in the many parts of the temperate world where passive cooling still dominates (southern Europe, North Africa, India, China), a sustainable-practice perspective would encourage efforts to reinforce existing passive building practices, focus on urban planning and the retention of green spaces, and provide economic incentives that give an advantage to climate-friendly building materials.

Social learning theory is grounded in the same conceptual framework as practice theory. It offers insights on how practices change and can be changed. According to anthropologist Jean Lave, the conventional view of learning aims at filling the cognitive vessel (mind). From a social learning perspective, new knowledge is acquired and deployed through a combination of cognitive and bodily processes (Lave 1993). Learning through participation in practices such as sporting activities (discussed above) is an example of social learning. The learning of a sport requires learning the rules but also participating in exercises and rehearsals in order to build up tacit skills. The practice of apprenticeship deploys social learning. It entails exposure to and participation in practices along with guidance and feedback. When making major purchase decisions, there is evidence that people deploy social networks. Potential buyers are more confident in the experiences of their colleagues, friends, and family than they are in product information or sales pitches. This social knowledge resource constitutes what Lave defined as a "community of practice" (1993). The outline of such a community is emerging from a study that is just getting under way in Norway, one of the objectives of which is to examine how and why people decide to buy and install heat pumps in their homes. Initial findings show that people use members of their extended family, neighbors, and colleagues as sources of information when comparing prices, exploring the choice of entrepreneur, and assessing the quality of performance. Early findings seem to indicate that this form of learning is more important to the buyer's decision-making than the advice of experts, such as Enova, the Norwegian Energy Directorate (Winther and Bouly De Lesdain 2012). Promising policy approaches based on a social learning perspective are the use of demonstration projects and in situ experiments that highlight low-energy-intensive practices, for example life in low-energy homes or in apartment blocks offering the residents shared services such as car use and laundry (see Attali and Wilhite 2001; Jelsma and Knot 2002). Another social learning–based approach would be a better circulation of information on successful energy practice transformations, such as the efforts in many European cities to encourage bicycle commuting.

From a practice perspective, the old distinctions between "upstream" and "downstream" energy policy instruments dissolve. Since the former is intended to move the consumption choices and the latter the choosers, both are about consumption. From this perspective, standards, regulations, and bans on energy-wasting technologies (such as the recent European ban on incandescent light bulbs) are energy-saving policies on a par with motivational instruments such as information and pricing. Information remains an important policy instrument, in the form of prices, labels, incentives, and so on, but as we have seen, its effect is blunted by the momentum of habits and the technologies that support them.

I have argued that agency in consumption is distributed among bodies, technologies, and social contexts. I would argue that responsibility for change is also distributed. Substantial energy savings will not happen through voluntarism or public-policy dictates alone. A transformation to a low-energy society will require aggressive public policies, active consumers, and a shared understanding that sustainable ways of consuming energy does not imply regression or sacrifice but rather readjustments that also offer opportunities to improve quality of life. Redefining a sustainable policy frame in this way faces daunting obstacles. People everywhere are faced with massive flows of information that associate energy-intensive consumption with a good life. This discourse emanates from multiple sources, of which product advertising is one. In 2010, an estimated U.S. $503 billion was spent globally on product advertising. In Norway in 2011, the equivalent of U.S. $4 billion was spent on commercial advertising for this country with a population of five million. By comparison, the total spending for Norway's socialized health system, including hospitals, retirement homes, and care for disabled, was the equivalent of about U.S. $20 billion. The language of politics is also saturated with messages associating greater consumption with progress and economic well-being. Growth in consumption is used as a political benchmark for the health of the economy. The recent global economic crisis has been permeated with worrisome analyses of stagnating consumption and ideas about how to stimulate growth.

The point is that any effort to reduce consumption must do so in the face of an overwhelming volume of countermessages associating more consumption with better lives. Like the Russian doll, there are many layers of entrenched discourse and practice (political, commercial, and household) inhibiting change in the direction of more sustainable consumption. At heart lies a weak theory of consumption and change. It may take a catastrophe, such as a collapse in the planet's thermostatic controls, to break through to this inner core and open for new perspectives. Let us hope that it does not come to that. Energy-savings theory and policy must be renewed and revamped, and anthropology can play an important role in this effort.

NOTE

1. The rebound effect will be researched from a practice perspective at CREE, the Oslo Centre for Research on Environmentally Friendly Energy (http://www.frisch.uio.no/cree/). The work package on rebound consists of a multidisciplinary team of economists and social anthropologists.

REFERENCES

Ahern, Laura M.
2001 Language and agency. *Annual Review of Anthropology* 30: 109–137.

Akrich, Madeleine
2000 The de-scription of technical objects. In *Shaping technology/building society. Studies in sociotechnical change*, edited by Wiebe E. Bijker and John Law, 205–224. Cambridge, MA: Massachusetts Institute of Technology Press.

Appadurai, Arjun
1986 Introduction: Commodities and the politics of value. In *The social life of things: Commodities in a cultural perspective,* edited by Arjun Appadurai, 3–63. Cambridge: Cambridge University Press.

Attali, Sophie and Harold Wilhite
2001 Assessing variables supporting and impeding the development of car sharing. *Proceedings of the ECEEE 2001 Summer Study*. Paris: European Council for an Energy Efficient Economy.

Bauman, Zygmunt
1988 *Freedom.* Milton Keynes, UK: Open University Press.

Bijker, Wiebe and John Law
1992 *Shaping technology—Building society: Studies in sociotechnical change.* Cambridge, MA: Massachusetts Institute of Technology Press.

Bourdieu, Pierre
1977 *Outline of a theory of practice* (R. Nice, Trans.). Cambridge: Cambridge University Press.
1998 *Practical reason*. Cambridge: Polity Press.

Brown, Steven and Darach Turley
1997 *Consumer research: Postcards from the edge*. London: Routledge.

Cooper, Gail
1998 *Air conditioning America: Engineers and the controlled environment, 1900–1960.* Baltimore: The John Hopkins University Press.
2008 Escaping the house: Comfort and the California garden. *Building Research & Information* 36 (4): 373–380.

Herring, Horace
1999 Does energy efficiency save energy? The debate and its consequences. *Applied Energy* 63: 209–226.

Jelsma, Jaap and Marjolijn Knot
2002 Designing environmentally efficient services: A "script" approach. *The Journal of Sustainable Product Design* 2: 119–130.

Lash, Scott and John Urry
1994 *Economies of signs and space*. London: Sage.

Lave, Jean
1991 Situated learning in communities of practice. In *Perspectives on socially shared cognition*, edited by Lauren Resnick, John M. Levine, and Stephanie Teasley, 62–82. Washington, DC: American Psychological Association.
1993 The practice of learning. In *Understanding practice: Perspectives on activity and context*, edited by Seth Chaiklin and Jean Lave, 3–35. Cambridge: Cambridge University Press.

Latour, Bruno
2000 Where are the missing masses? The sociology of a few mundane artifacts. In *Shaping technology/Building society*, edited by Wiebe Bijker and John Law, 225–258. Cambridge, MA: Massachusetts Institute of Technology Press.

Lie, Marianne and Knut H. Sørensen
1996 *Making technology our own? Domesticating technology into everyday life*. Oslo: Scandinavian University Press.

Lovins, Amory
1977 *Soft energy paths: Toward a durable peace*. New York: Penguin.

Lutzenhiser, Loren
1993 Social and behavioral aspects of energy use. *Annual Review of Energy and Environment* 18: 247–289.

Mauss, Marcel
1973 [1935] Techniques of the body (Ben Brewster, trans.). *Economy and Society* 2: 70–88.

Miller, Daniel
1998 Why some things matter. In *Material cultures: Why some things matter*, edited by Daniel Miller, 3–21. Chicago: The University of Chicago Press.
2001 *The dialectics of shopping*. Chicago and London: The University of Chicago Press.

Moezzi, Mitzra
1998 *The predicament of energy efficiency*. Proceedings of the 1998 ACEEE Summer Study: Energy Efficiency in Buildings. Washington, DC: ACEEE.

Ortner, Sherry B.
1999 Thick resistance: Death and the cultural construction of agency in Himalaya mountaineering. In *The fate of "culture": Geertz and beyond*, edited by Sherry B. Ortner, 136–165. Berkeley: University of California Press.
2006a Updating practice theory. In *Anthropology and social theory: Culture, power, and the acting subject*, edited by Sherry B. Ortner. Durham and London: Duke University Press.
2006b Power and projects: Reflections on agency. In *Anthropology and social theory: Culture, power and the acting subject*, edited by Sherry B. Ortner. Durham and London: Duke University Press.

Reckwitz, Andreas
2002 Toward a theory of social practices: A development of culturist theorizing. *European Journal of Social Theory* 5: 243–263.

Røpke, Inge
2009 Theories of practice: New inspiration for ecological economic studies on consumption. *Ecological Economics* 68 (10): 2490–2497.

Shove, Elizabeth
2003 *Comfort, cleanliness, and convenience: The social organization of normality*. Oxford, New York: Berg.

Seyfang, Gill, Alex Haxeltine, Tom Hargreaves, and Noel Longhurst
2010 Energy and communities in transition: Towards a new research agenda on agency and civil society in sustainability transitions. CSERGE working paper EDM 10-13.

Strengers, Yolande
2008 Comfort expectations: The impact of demand management strategies in Australia. *Building Research & Information* 36(4): 381–391.

Verbeek, Peter-Paul
2005 *What things do: Philosophical reflections on technology, agency and design.* University Park: The Pennsylvania State University Press.

Warde, Alan
1997 Afterword: The future of the sociology of consumption. In *Consumption Matters,* edited by S. Edgell, K. Hetherington, and A. Warde. Oxford: Blackwell Publishers.

2005 Consumption and theories of practice. *Journal of Consumer Culture* 5: 131–153.

Wilhite, Harold
2008a *Consumption and the transformation of everyday life: A view from South India.* Basingstoke and New York: Palgrave Macmillan.
2008b New thinking on the agentive relationship between end-use technologies and energy-using practices. *Journal of Energy Efficiency* 1(2): 121–130.
2012 Towards a better accounting of the roles of body, things and habits in consumption. In *The habits of consumption,* edited by Alan Warde and Dale Southerton, COLLEGIUM: Studies across Disciplines in the Humanities and Social Sciences, Volume 12. Helsinki: Helsinki Collegium for Advanced Studies.

Wilhite, Harold and Jørgen Nørgaard
2004 Equating efficiency with reduction: A self-deception in energy policy. *Energy and Environment* 15 (6): 991–1009.

Wilk, Rick
1999 *Towards a useful multigenic theory of consumption.* Proceedings of the 1999 ECEEE Summer Study. Paris: European Council for an Energy Efficient Economy.

Wilk, Rick and Harold Wilhite
1985 Why don't people weatherize their homes? An ethnographic solution. *Energy: The International Journal* 10 (5): 621–630.

Winther, Tanja and Sophie Bouly De Lesdain
2012 Electricity as a cultural concept implicated in everyday practices: A comparison of French and Norwegian responses to policy appeals for sustainable energy. Paper presented at EASA conference, Paris, July 2012.

CONVERSATION 1
Theorizing Energy and Culture

Michael Degani, Alf Hornborg, Thomas Love, Sarah Strauss, Harold Wilhite

The two chapters in this first section set the stage for the remainder of the volume by presenting a pair of broader theoretical frameworks that might be useful in considering how energy and culture are mutually constituted at two key anthropological scales: Hal Wilhite's individual to household level practice, and Alf Hornborg's wide-angle historical and spatial analysis of the general incompatibility between modern social institutions and the law of entropy. Hornborg argues that the "social arrangements and aspirations that are most fundamentally at odds with the second law of thermodynamics are general-purpose money and beliefs in economic growth and technological progress." Since energy flows from the sun have made possible both life on Earth and its transformations through culture, we need to understand the ways that access to and control over solar energy have come to be synonymous with power in its various manifestations. Energy, of course, is the ability to do work; Hornborg takes this concept a step further, to see money as "fictive energy" that mystifies purchasing power as the vital essence of society. On a related front, but very different order of magnitude, Wilhite uses social practice theory to consider our everyday habits as they relate to energy consumption and the problem of sustainability. For both of these theorists, then, anthropologists must engage the fundamental problem of inequality in relation to *power* in all its meanings, including, most importantly, the ability to do work, as it is manifest through both the variety of energy resources exploited by human populations over the course of our history and the second-order implementation of power through human agency. In pursuing some of these themes in conversation, one really resonated with both of these scholars, as well as other contributors: what is the relationship between slavery, technology, and energy? This topic moves us from the internal energetic connections between individual consumption and production, as well as the relationship between human and animal labor and the use of extrasomatic fossil fuels. In addition, the concept of mystification further contributes to our understanding of how hidden these relationships often are, and how frameworks for transportation, power transmission, and other infrastructural elements further distance consumers from the sources of their energy; without greater transparency, it is hard to motivate people to shift their energy-related behaviors to more sustainable patterns, but this is precisely the challenge, as these authors note, that we need to address.

TOM LOVE: Thanks for raising the comparison with ancient slavery. I find the concept of "energy slaves" useful in bringing home our utter dependence, as industrial humans, on prodigious amounts of extrasomatic energy. In a recent article by Richard Douthwaite in the *Energy Bulletin* (November 2011), we learn that a reasonably fit man has a total energy output of about three kilowatt-hours per forty-hour work week, while a liter of gasoline yields nearly nine kilowatt-hours, or the rough equivalent of three weeks of manual labor. There are 150 liters of oil in a barrel, so even $100 (and up) per barrel is very little to pay for 450 weeks of human work—a very cheap slave indeed! Unlike in classical antiquity, though, our slaves are invisible to us. The cultural implications are enormous. One would seem to be our unwarranted belief in our prowess, our demigod sense of self, our indifference toward "stuff" that all seems to just magically be replaceable, our usually unexamined, axiomatic expectation of what Anna Tsing called "the inevitability of global newness."

ALF HORNBORG: Yes, and precisely this invisibility of our energy slaves—the way our lifestyles are subsidized by global, unequal exchange of embodied labor and resources—is fundamental to the specific way in which power is mystified in our fossil-fueled civilization. And the invisibility is made possible by the increasing spatial separation of extraction, production, and consumption, which relies on the massive combustion of fossil fuels for transport energy. So the global scale of economic interdependencies is part and parcel of the mystification of power.

MIKE DEGANI: I do think there is a part of us that inherits the arrogance of the Greek aristocrat, especially in the unreflective habits of day-to-day life. But doesn't that come paired—like the slave owner's occasional paranoia about revolt—with anxious flashes of our own vulnerability? And in these anxious moments, when we become aware of our "slaves," don't *they* suddenly seem more like demigods who have grown unstable and capricious? It's almost as if electric cars or recycled shopping bags (beyond being status symbols) are talismans to appease forces we sense but can't directly see, like peak oil or climate change. More generally I think you could trace this vein of cultural anxiety about energy from the nuclear horror of "I am become Shiva, Destroyer of Worlds" to *Akira* to Hollywood disaster movies and so forth.

TOM LOVE: It is deeply important for us to analyze the mystification that so often accompanies people's understanding and use of energy. Alf alerts us to ways fossil fuels have temporarily severed the synonymy of land and energy that prevailed in all preindustrial social formations, dependent as they (and we, if we could see/theorize appropriately) were/are on the sun's radiation. Animated by invisible energy (fossil sunlight or, more recently, nuclear energy), apparently disconnected from the daily flux of the sun's energy, technology and the goods they produce become both magical and yet mere things—fetishized objects divorced from the social contexts and ultimately solar-powered energy flows of which they are an inextricable part. Hal complements this angle by examining how practices are both agentive and social. He illuminates how such magical/mystical things and processes are always already domesticated, used, and understood. Through an anthropological lens, we see people, not technology, return to center stage; the social circuitry and shared meanings of people in community shape the flow of energy in terms of substance, scale, and rate of consumption. Though our volume is not intended to bear directly on policy, the policy implications of such robust anthropological theorizing are enormous and compelling.

HAL WILHITE: These observations on "energy slavery" open an interesting debate. The permeation of life everywhere with new energy-using technologies does lead to a replacement of labor with mechanical power, making the execution of some of the simplest chores (cleaning, washing, cooking) technology dependent. However, this substitution is not all negative. The use of energy technologies has reduced drudgery and saved time. For example, washing clothes by hand involves the soaking, rinsing, and repeated beating of clothing. Women in South India use up to two hours a day for this. Washing machines lead to a drastic reduction in clothes-washing time and drudgery. As has the electric iron, which saves time and energy associated with gathering fuel and heating up the iron. Also, the manual grinding of herbs and spices for food preparation is monotonous and time-consuming. The advent of the electric mixer has reduced daily time used in food preparation enormously. However, in affluent societies everywhere, the permeation of energy appliances has led to a form of alienation. Technology mediates between ecosystems and use to the extent that people lose touch with the interrelationship between consumption practices and consequences. Also, if technologies break down, either at the infrastructural level or in the home, most people are helpless to do anything about it. In other words, know-how associated with the most banal of practices is externalized. In this sense, people are slaves to the machines and the technology experts.

ALF HORNBORG: Yes, the way new energy-using technologies facilitate life for the poor in the periphery remains an essential component in the ideology of "development." True, mechanization is no longer, as in the nineteenth century, as obviously a case of displacing work efforts (redistributing human time)—for instance, to the cotton slaves or to British coal miners—but technology and energy consumption continue to mystify unequal global flows of embodied labor, land, and energy. It suffices to scrutinize how we are fooled by our own use of cars: some have estimated that the *real* speed of our cars (if we divide distances traveled by all the time spent caring for them, looking for parking lots, and so on) is about the same as biking. Even John Stuart Mill suggested that there was never a time-saving machine that had saved a minute of human time. . . . Technologies tend to become symbols of modernity and development, but do they actually liberate human time and creativity, or do they fetter us in new constraining habits and dependencies?

HAL WILHITE: The truth is that technologies are both liberating and constraining. I doubt that any of us would propose that the way out of the energy dilemmas put forward in this volume is a total de-technologization of everyday life. The point is that we need to think about how to reinvent and reconfigure technological regimes in ways that reinstate human agency, demystify the relationship between everyday practices and its consequences, and drastically reduce the energy and other resources needed to operate them. This will only be accomplished through a radically different vision of both what it means to be developed and what it means to be modern.

MIKE DEGANI: It's interesting that an anthropology of energy puts the idea of demystification back on the theoretical agenda. Clearly the dynamics that allow us not to know or to ignore the consequences of our actions are of enormous power, and the image of a world system built on energy slaves is an arresting, revelatory one. How everyday cultures will eventually relate to that macro-level understanding is a rich ethnographic question. Too much focus on technocratic fixes and their habit-generating scripts and we are somewhere near a Skinner behaviorism; too much focus on rituals and subjectivities that surround energy and we risk mistaking phenomena that are interpretively rich with ones that are socially significant (cf. Graeber 2006). As we shudder out of the era of cheap, asymmetric energy, the relevance of "demystified" concepts and technologies will surely only increase.

PART 2
Culture and Energy: Technology, Meaning, Cosmology

CHAPTER THREE

Considering Energy: $E = mc^2 = (\text{magic} \cdot \text{culture})^2$

Stephanie Rupp[1]

> Is it magic or science to tell a student that $E = mc^2$? . . . While each definition tells us how to use energy, none of them tell us what it is. Energy is expressed only in terms of other abstractions, and yet we *know* it exists, that energy, the magically convertible phenomenon behind all phenomena, is REAL.
>
> —Dethlefsen 1978:21, emphasis in original

Urban America is drenched in energy. Energy in New York City is omnipresent yet invisible; ambiguous; contested. While energy has precise scientific definitions—for example, $E = mc^2$—such precision evaporates in contemporary public discourses of energy. Beginning with the hypothesis that New Yorkers rarely stop to consider what energy is, the ways in which it enables their lives, and how it is generated, transmitted, distributed, and quantified, this chapter seeks to understand the conceptual infrastructure that undergirds New Yorkers' assumptions and expectations about energy. Just as the technical infrastructure of energy—cables, substations, power stations, generating plants—underlies our daily activities in invisible but fundamental ways, so too does our conceptual infrastructure organize and orient our expectations of energy. In this chapter, I argue that lacking accessible technical knowledge for thinking about energy and its uses, New Yorkers turn to multiple and hybrid images—magical, spiritual, corporeal, social, political, as well as technical—to explain the forces that enable their everyday lives.

Striding with purpose down a busy Manhattan sidewalk, one dodges all manner of people and obstacles. Preoccupied pedestrians steer around each other, anticipating and avoiding awkward intersections without seeming to notice one another. On a crystal-clear autumn afternoon, two young energy activists set up their dragnet along the west side of Broadway, trying to interrupt the flow of pedestrians to proselytize their message of conversion to alternative-energy philosophy, politics, and applications. Finally intersecting a passerby with a sliver of time to spare, the young representatives of Environment New York deliver a clear and concise agenda to decrease New York City's energy usage: cut consumption of oil by 33 percent by 2020; generate at least 25 percent of electricity from renewable sources by 2020. Their bright-eyed message exudes alternative-energy enthusiasm in a city that takes its supply of energy for granted. But when the target of their sermon asked them what, in the first place, energy *is*, the messengers of alternative energy were stumped. Blinking in the glittering sunlight, momentarily speechless, one of the young activists suddenly gushed, "Energy is whatever powers whatever we do. Energy is . . . so, so *much*. Energy is . . . . That's a good question. Energy is . . . is . . . *everything!*" (twenty-one-year-old resident of Queens and energy activist, New York, 2009).

New Yorkers experience energy as a force that is ubiquitous yet invisible, uncontrollable yet indispensable. City residents speak of other forms of energy, such as the forces that animate our bodies (witness the proliferation of energy bars and energy drinks in corner stores), illuminate and lubricate our spiritual lives (see the number of psychic experts who offer to balance your chakra), and propel our city (hear comments about the energy of the city itself). In these senses, the notion of energy refers to the vitality of human endeavor, a qualitative, socially embedded force that can be neither measured nor reliably controlled.

At the same time, city residents are dimly aware that energy—a material substance that provides physical force—fuels their hectic pace of life and ensures that everything from subways to taxis, internet connections to cellular phones, computers to refrigerators function smoothly—thus enabling their lives to function smoothly. In addition to noting the social and spiritual dimensions of energy, New Yorkers may speak of energy as a force that is quantifiable, material, and usable, even if they do not understand the science of energy systems, do not comprehend the different forms of material energy, and do not consider how energy is harnessed or quantified. In such technical capacities, New Yorkers understand that energy exists as a force and can be discussed as kilowatt-hours or calories without necessarily understanding what a kilowatt-hour or calorie is, or how such forces are harnessed in complex physical systems.

New Yorkers' discussions of energy reveal that city residents perceive energy through both formalist and substantivist models. Formalist models refer to quantifiable energy systems that provide the technical infrastructure on which high-energy, high-technology, information-saturated city residents depend. Substantivist models of energy are reflections on energy as a qualitative force that is socially embedded and mediated by people's relationships with each other and with the conditions of their daily lives. Borrowing the contrasting, complementary models of "formalism" and "substantivism" from economic anthropology and applying these theoretical frameworks to contemporary understandings of energy in New York City reveal a multiplicity of concurrent models that *together* shape how urban residents perceive and manage energy, the forces necessary to get things done. These models of energy—quantitative, rational, formalist on the one hand; qualitative, relational, substantivist on the other—are distinct but are complementary rather than contradictory. Both New Yorkers' scientific and cultural models of energy involve an understanding of energy as the "ability to do work" and also reference underlying notions of "currency" and "connectedness." There is fundamental importance in considering Americans' models for energy, whether the models are formalist (rational) or substantivist (relational) or *both*; if Americans, as people who *believe* in the functionality of energy in both technical and social contexts, are to become more conscientious users of energy resources, it is essential that we consider what they think energy *is*.

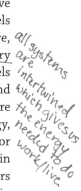

SLIDING REGISTERS OF ENERGY

Midmorning on Forty-Second Street near Grand Central Station: in a chain retailer of electronics, fluorescent tubes illuminated dozens and dozens of devices that facilitate connections between people and generate social energy—cell phones, cameras, computers—even as the machines also consume physical energy in the form of electricity. A salesclerk asked if I have found everything I need, and asked with just-opened-shop enthusiasm what I plan to do with the digital voice recorder I have selected. Not satisfied with my vague response, "It's for research," the clerk pressed further in a cheery voice:

> ***Salesclerk:*** "Oh, what's your research about?"
> ***Anthropologist:*** "I'm studying energy. I'm curious about what people think energy is."

The salesclerk looked startled for an instant, eyes fixed on the recorder, then quickly turned back to the cash register to complete the sale. This time it was the researcher who pressed further, bouncing the conversation back:

> ***Anthropologist:*** "So, what do you think energy is?"
> ***Salesclerk:*** "Ummm. . . . Energy is physical . . . force. It's what physics is."
> ***Anthropologist:*** "Do you know where the energy that runs your shop comes from?"
> ***Salesclerk:*** "No—I don't have a clue. I've never thought about it before. . . . (Passing the credit card invoice) . . . Energy comes from God—it's a force that God gave us. Energy is hot and cold. You can see energy. You can feel

energy. You can hear energy. It's the wind in the trees. It's the songs of the birds. Energy is in us (tapping his forearm with his index finger).

Energy is a vibe—it's what makes me feel good or bad."

Anthropologist: "So, how do you know when you come into your shop in the morning and flip on the switches that the lights will come on, the music will come on, all of the electronics will come on?"

Salesclerk: "Well, I pay for it, so it better come on!"

Anthropologist: "So do you know where the energy—this stuff (pointing to fluorescent lights overhead—comes from?"

Salesclerk (hesitating): "No—I have no idea."

In this brief exchange, several themes emerge. When confronted with the idea that energy is something under investigation, the salesclerk's initial reaction was a combination of confusion and avoidance. As with many New Yorkers whom I have asked to respond to the basic question of what energy is, the clerk seemed caught off guard by the underlying realization that energy is something to be considered at all. Composure regained, the clerk's first attempt to articulate what energy is references a physical force. However, the formalist scientific explanation of energy ended quite abruptly, as the clerk seemed to hit a conceptual dead end with the idea of energy as a "physical force." In other conversations, too, although New Yorkers certainly are aware that energy exists in a scientific, physical capacity, their explanations in this formalist framework tend to be truncated. People's attempts to answer questions about what energy is in the context of technology (such as lights or electronic devices) or where energy comes from often end with an admission of bafflement: "I have no idea" or "I have never thought about it before." By contrast, New Yorkers' discussions of energy are animated, robust, and detailed when they speak about the qualitative, cultural explanations of what energy is and how it works in their lives. In the case of the salesclerk, even in the midst of such a brief conversation his descriptions of energy quickly turned to qualitative, cultural models that provide numerous "handles" for thinking about energy as it is socially construed. The origins of energy are spiritual: "Energy comes from God." Energy is ubiquitous: we can hear, see, and feel energy around us. Energy is ambiguous: energy is both hot and cold. Energy is in nature: it is present in wind, in birds' songs. People embody energy: energy is both part of our physical bodies and is a "vibe" within us, a dynamic quality that makes us "feel good or bad." Energy is also a commodity, a good that people have a right to access because they pay for it. In his spontaneous expressions of what "energy" means to him, the salesclerk was reaching for conceptual handles that allow him to grasp energy, to think, perhaps for the first time, about how energy is part of his life.

Encountering explanations that roll together quasi-scientific definitions of energy ("energy is physical force") with social models for thinking about energy ("energy is a vibe"), experts who work with energy in a scientific capacity might be tempted to dismiss social explanations of energy as misinformed and misguided. Indeed, surveys designed to gauge Americans' attitudes and beliefs toward energy tend to measure whether Americans understand technical applications of energy in everyday life and conclude by emphasizing Americans' poor comprehension of energy science and the need for energy education on a vast scale (for example, NEETF/Roper Report Card 2002). However, although their technical understandings of energy in a formal sense may be muddled, Americans hold a wealth of cultural models for thinking about energy.

BRONX ENERGIES

New York City residents apprehend energy using hybrid, cultural frameworks that reference invisible powers in multiple contexts. Among young adults in the south Bronx,[2] cultural conceptions of energy integrate scientific and technical, cultural and magical models of energy; this multiplicity of models provides a variety of conceptual handles for grasping the invisible, elusive materiality of energy in everyday life. Ethnographic discussions of energy pivot around notions of agency: whether the energy is technical, social, magical, spiritual, or corporeal in its essence, energy is what enables people to get things done and to achieve their goals. Cultural conceptions of energy also consistently reference dynamics of currency, connecting individuals to each other, to their inner spirit and ambition, to (and through) technology, to nature, and to higher orders of power.

In articulating what energy *is* in its most basic, elemental sense, young people in the south Bronx reached for a palette of images and concepts to illustrate their notions of power. The ethnographic canvases that were provided—in the form of guiding questions during interviews—were intentionally broad and blank: What is energy? Where does energy come from? What kinds of energy do you use in your everyday life? People depicted power in a variety of contexts in hybrid forms and working in multiple ways. At the most general level, respondents expressed energy as "power" and "force," a source of movement, light, or the creation of objects. Capturing this sense of overarching omnipotence, one respondent said, "Energy is the powerhouse that creates everything everywhere. Although there are many types of energy, these factors depend on how energy is used such as kinetic energy, the sun, or mitochondria." This response illustrates several themes that reverberate through New Yorkers' conceptions. The sense of ambiguous omnipotence—that energy creates everything everywhere—reflects New Yorkers' resounding confidence in the general power of energy to create everything, and their palpable lack of certainty about how energy actually accomplishes this feat. The response identifies "many types of energy," reflecting the tendency toward multiplicity of models of energy. The response also suggests that energy's "type" is dependent on its uses; energy emerges as a force that can shape-shift, for example from kinetic to solar to mitochondrial energy.

As demonstrated in conceptual maps of ethnographic responses (see Figures 3.1 and 3.2), a small minority of people articulated energy in a sin-

FIGURE 3.1. What Is Energy?

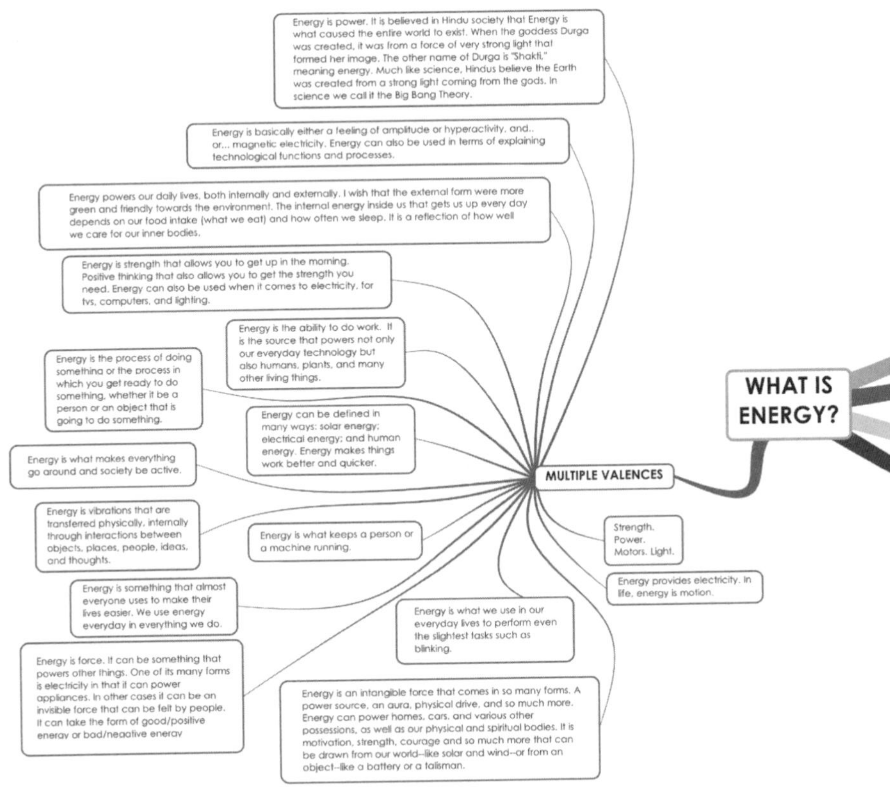

gular technical capacity. These responses reflect the role that energy plays in practical functions: "Energy is electricity or voltage. Whenever you turn on your light, that is energy" and "Energy is the power which can light up bulbs and move vehicles, the power of every technology and the main source of mechanism for machines." Several responses illustrated energy as a purely physical force, apart from necessary social applications: "Energy is a chemical that can only be transferred. It can never be removed"; "Energy is the amount of power that is used to move something"; and "Energy is a substance that gives off or absorbs heat when being used." Such explanations of energy hint at scientific properties of energy. But the overall paucity of explanations of

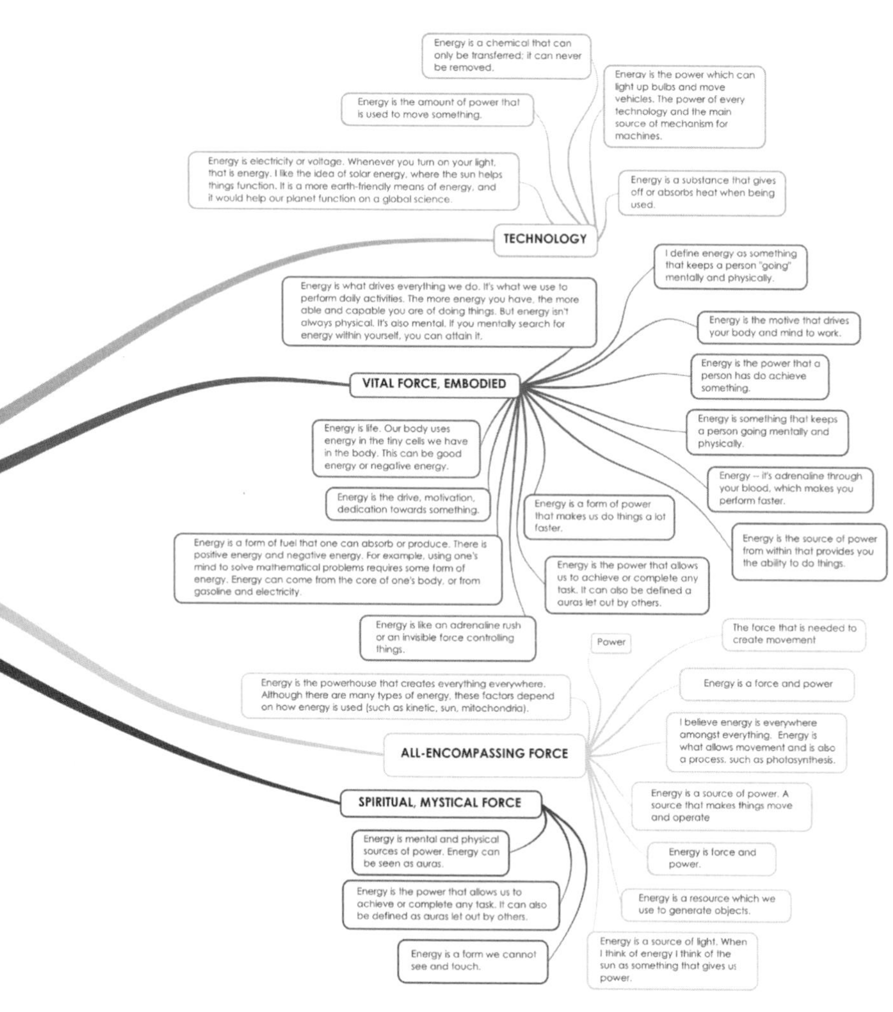

energy that reference science and the shallowness of such technical models of energy are striking, particularly when contrasted with the many diverse, textured models of energy that draw on bodily, spiritual, mystical, and omnipotent valences of energy, and on combinations of these valences.

A significant subset of responses depicted energy as a vital force, a power within the human body that animates the individual. But even in the body, energy has multiple meanings: energy is the physical force necessary for life, a force that simultaneously holds psychic value. “Energy is life. Our body uses energy in the tiny cells we have in the body. This can be good energy or negative energy.” In other responses, energy emerges as a force that enables a person to be active both physically and mentally, providing the inner power to achieve one’s goals.

> Energy is what drives everything we do. It’s what we use to perform daily activities. The more energy you have, the more able and capable you are

FIGURE 3.2. What Kinds of Energy Do You Use in Your Everyday Life?

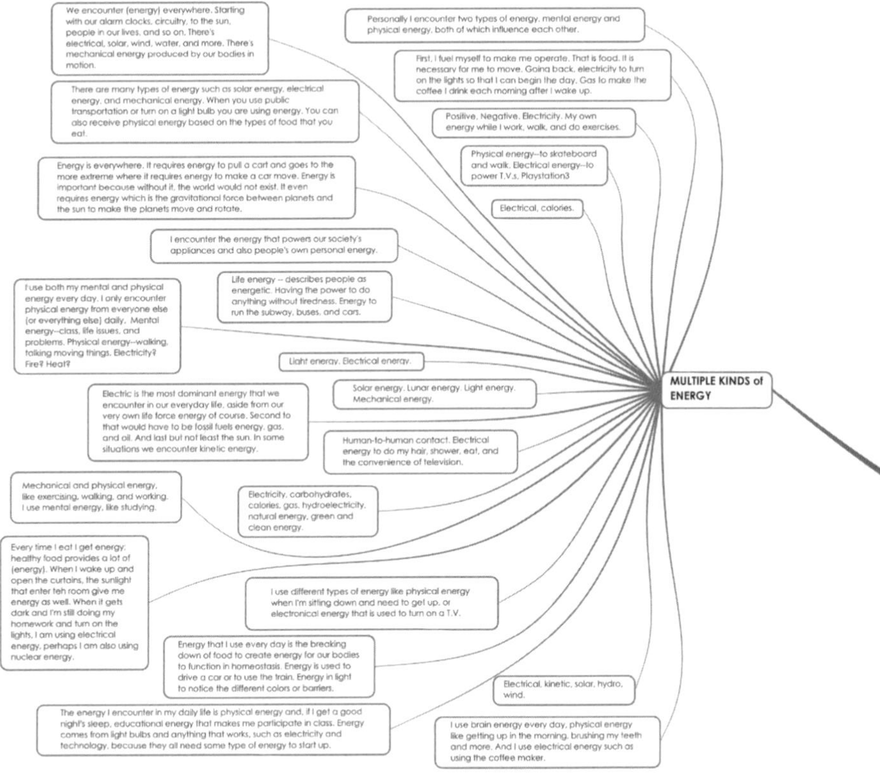

> of doing things. But energy isn't always physical. It's also mental. If you mentally search for energy within yourself, you can attain it.

> Energy is the power that allows us to achieve or complete any task. It can also be defined as auras let out by others.

In this cluster of responses, energy is portrayed as human potential in physical, mental, and spiritual capacities.

The majority of responses integrated multiple frameworks of power in expressing "energy," linking them together in a cascade of connected concepts. Taken together, these responses indicate that energy is a force with multiple manifestations that serves varied functions simultaneously.

> Energy is an intangible force that comes in so many forms. A power source, an aura, a physical drive. Energy can power our homes, cars, and

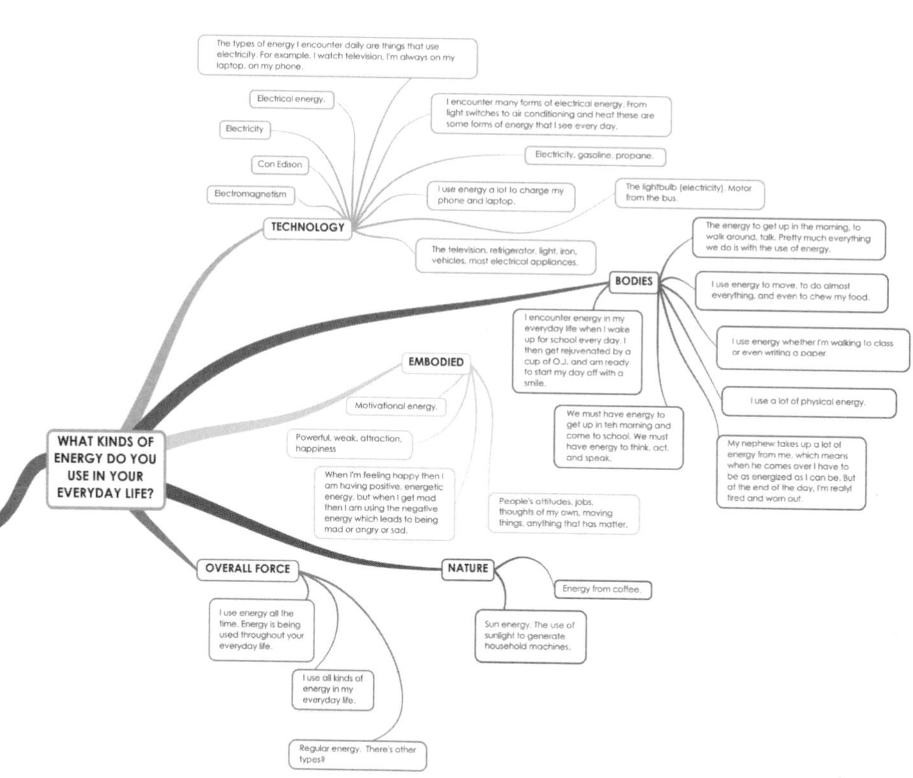

> possessions, as well as our physical and spiritual bodies. Energy is motivation, it's strength, and courage. It can be drawn from our world, like solar and wind, or from an object like a battery or talisman.

In this comment, energy is a force that transcends physical, technological, social, and spiritual boundaries. From wind to auras, sources of energy found in nature and in a person's inner core, and from a battery to a talisman, mundane and occult receptacles for energy, energy itself is evocative in its pervasive, cross-cutting power. Another response similarly integrates distinct registers of energy, folding together religion and science:

> Energy is power. In Hindu society we believe that Energy is what caused the entire world to exist. When the goddess Durga was created, it was from a force of very strong light that formed her image. The other name of Durga is "Shakti," meaning energy. Like science, Hindus believe that Earth was created from a strong light coming from the gods. In science we call it the Big Bang Theory.

This respondent draws an explicit comparison between energy in religious and scientific contexts, suggesting a parallel between the origins of energy in these two spheres. This mixing of mystical, spiritual, and scientific notions of energy is evident in the responses of Americans from a variety of cultural backgrounds. As an Irish-American woman put it,

> I think energy is paired with nature and the cycle of life, which transfers souls, spirits and thoughts generated from people's minds and actions, through the creation of matter, which cannot be created nor destroyed.

Rather than compartmentalizing different notions of energy and prioritizing some (formalist) notions of energy as "true" and other (substantivist) notions of energy as "false" or misguided, respondents in the Bronx clearly integrated their diverse notions of energy into cultural, qualitative models that account for the physical realities as well as the social embeddedness of energy in contemporary society.

While this multivalence in cultural expressions of energy may seem at odds with technical definitions of energy that are precise and narrow, such diverse, qualitative models share two fundamental premises with scientific notions of energy: energy as the ability to do work, and energy as formed through currents and connections. A concise definition of energy from the physical sciences is the "ability to do work." New Yorkers perceive energy as an essential component in their agency, their ability to do work, to achieve their goals, to strive for success, and to be motivated. The notion of "work" is itself embedded in the social context of individuals and requires individuals to engage their bodies, minds, and spirits. As a result, energy also animates these diverse components of the human experience of work. According to young Bronxites,

> Energy is the ability to do work. It is the source that powers not only our everyday technology, but also humans, plants, and other living things.

> Energy is what keeps a person or a machine running.

> Energy makes everything go around, and lets society be active.

> Energy comes from power and enthusiasm of all sorts.

> I think that energy comes from within a person's heart. Even though you think you're not capable of doing something, there's this feeling in your heart that keeps you determined and dedicated.

In these comments about the nature of energy, work is a central theme that emerges as young people consider how energy animates their lives. While in formal, scientific senses of energy-as-work there is no analytical space for a spirit of motivation or enthusiasm, it is apparent that in the social context of energy, the ability to work *is* motivation or enthusiasm.

Currency is the second scientific characteristic of energy that seems to flow easily into the social concepts of energy held by New Yorkers. Just as electrons form chains to create currents or flows of energy, Bronx residents intuitively link energy to social, creative, ideational, and technical connections.

> Energy is vibrations that are transferred physically, internally, through interactions between objects, places, people, ideas, and thoughts.

> Energy comes from cooperation.

> People in our lives make energy.

> Energy is human-to-human contact.

In this sense, energy enables individuals to connect to each other, just as energy is simultaneously generated by connections between and among individuals. When New Yorkers speak of the energy of the city itself, they refer to the sense of connectedness of human effort and creativity, the social effervescence generated by people working with, around, and through other people. Just as energy is crucial to the functioning of technological devices, social interactions are necessary in the creation, consumption, and utilization of communication and information technologies of urban America today. Energy is both technical and personal, a force that mediates social relationships between people face-to-face and device-to-device. Fundamentally, circuits and flows constitute the dynamism of agency; in urban society, success is achieved through accessing and maintaining circuits and flows of energy, in the form of material power such as electricity and technologies, *and* in the form of social relationships, networks, and a collective, connective spirit of human endeavor.

COMPREHENDING INVISIBLE FORCES: SCIENCE, RELIGION, AND MAGIC

Across time and space, people have developed frameworks for thinking about forces in our lives that we can neither see nor understand. Science, religion, and magic have emerged as three legs of the stool of causality, explanations for the relationships between cause and effect. New Yorkers' concepts of energy appear to straddle all three legs of the stool, integrating elementary technical understandings of energy (science), a faith in the functioning of energy systems in America (religious belief), and a sense of enchantment with energy as a power that enables individual objectives in invisible, effortless ways (magic).

Over two centuries of debate among philosophers, theologians, and anthropologists, the theoretical relationships between magic and science have shifted in content and emphasis. Rather than attempting the impossible task of recapitulating this lengthy and entangled debate here—a debate that has been thoroughly analyzed by Stanley Tambiah (1990) in his now-classic overview of magic, science, religion and rationality—I modestly seek to highlight the changing trends in the theoretical relationships between magic and science, and to suggest what these shifting trends might have to do with our understandings of energy. The changing theoretical relationship between magic and science can be condensed by emphasizing the preposition between them: magic *versus* science; magic *into* science; magic *alongside* science; magic *with* science; magic *in* science.

Prominent nineteenth-century scholars considered magic and science to be frameworks for understanding causality that were distinct from, and even opposed to, each other. Furthermore, magic and science were considered to be emblematic of societies that were distinct from, and even opposed to, each other. Early anthropologists such as Edward Tylor and James Frazer (1911) analyzed magic and science as oppositional frameworks of causality. For Tylor (1871), belief in magic was a hallmark of primitive society, whereas scientific reasoning characterized civilized society. Magic *versus* science, primitivity *versus* civilization: the elements in both pairs were cast as opposing and irreconcilable. At the turn of the century, Emile Durkheim (1912) brought the debate about the relationship between magic and science to a new position, suggesting that magical frameworks of understanding are natural and reasonable precursors to scientific thought. In Durkheim's analysis, magical thought evolves *into* scientific thought.

In the first half of the twentieth century, anthropologists still considered distinct "ways of knowing" through conceptual frameworks of magic and science; however, scholars began to appreciate that people from a particular society could turn to both magical and scientific frameworks of causality to explain events. For example, Bronislaw Malinowski (1935) and E. E. Evans-Pritchard (1937), writing about very different cultural contexts of Melanesia and northeastern Africa, came to similar conclusions about the existence of magical and scientific modes of thought *alongside* each other. Malinowski and Evans-Pritchard demonstrated that both Trobriand Islanders and Azande rely on a combination of scientific and magical thinking in their explanations of cause and effect. Where events could be explained through direct observation and could be controlled through people's direct actions, Trobriand Islanders and Azande explained the happenings in terms that anthropologists identify as scientific reasoning. Where events or their circumstances could only be surmised, Trobriand Islanders and Azande turned to terms that anthropologists identified as relational or magical, linking happenings to interrelationships and to the emotional energies that underlie social interrelationships.

Lucien Lévy-Bruhl and Ludwig Wittgenstein advanced theories about the relationships between frames of causality by making theoretical space for magic and science as coexisting ways of thinking in all people, including people in contemporary society. Lévy-Bruhl argued that "in every human mind . . . there subsists an ineradicable fund of primitive mentality . . . . It is not likely that it will ever disappear. . . . For with it would also disappear, perhaps, poetry, art, metaphysics, and scientific invention—almost everything, in short that makes for the beauty and grandeur of human life" (Lévy-Bruhl 1949, cited in Tambiah 1990:91). In this expression of the mental potential of people everywhere, Lévy-Bruhl alluded to the common source of creative energy that generates cultural expressions, metaphysical beliefs, and scientific insight. Tambiah underscores two important notions that emerge from Lévy-Bruhl's work. First, that magical and scientific thought can be considered as ideational systems that coexist in the same society and, second, that people are able to switch between scientific and magical ideas in quick succession (Tambiah 1990:92). Wittgenstein (1969) brought this line of reasoning further, proposing a model for thinking about rationality that specifically brings magic and science into shared conceptual space. Furthermore, Wittgenstein argued that "we" ("civilized") people are actually like those "savages" in the ways that we construct symbolic and ritual meaning and activate these values in our lives. Lévy-Bruhl and Wittgenstein advanced their theories of causality to underscore that magic coexists *with* science in all societies regardless of their place in evolutionary developmental schemes. Indeed, the coevalness of magic and science effectively collapses distinctions between "primitivity" and "civilization," allowing us to investigate the place of magical thinking in contemporary society and science.

Numerous anthropologists, especially working in studies of science, technology, and society, have underscored the influence of cultural models that undergird scientific research and knowledge (among others see Franklin 1995, Gusterson 1996). Laura Nader, in particular, has looked at the interweaving of cultural models and expectations in the work of scientists and technologists working in the field of energy science and policy. She argues that the constellation of magic, science, and religion is "likely universal, and therefore observable among scientists and technologists in the most developed of contemporary science and technology" (Nader 1996:261). In her research on cultural constructions and "irrational" beliefs held by energy scientists about science itself, Nader provides contemporary material for rethinking Edmund Leach's investigation of the intermingling of magic *in* science (Leach 1957). If, as Nader argues, scientists hold beliefs about their own scientific work that are based on cultural models and expectations, it should come as no surprise that laypeople without advanced training in science would apprehend invisible forces such as energy through frameworks that reference multiple models of invisible power, including magic, mysticism, and religion—areas of life that similarly engage invisible powers that shape our destinies.

CONCLUSION: $E = \int [agency + currency] \cdot [magic + faith] \cdot [science + society]$

New Yorkers articulate ideas of energy that integrate models of agency and currency, frameworks such as magic and faith for thinking about invisible power, and references to science and society. Energy is a cultural issue at the same time that it is a technical substance. New Yorkers tend to perceive energy through a multiplicity of frameworks and models for understanding. Energy is a force that is simultaneously intimate and external; energy is deeply personal—contained within the body and soul—and also connects each individual to concrete, external realities that are social as well as technical.

Considering the interfaces among science, religion, and magic is helpful in framing the cultural attitudes and interpretations of energy held by contemporary New Yorkers. At the same time, assessing how twenty-first-century New Yorkers think about energy offers interesting contemporary material for rethinking categories such as science, religion, and magic, interrelated concepts that have been discussed and debated in philosophy, theology, and anthropology over the course of two centuries. This research suggests that different theories of science and magic—and different understandings of how science and other forces work in society—coexist and overlap without necessarily creating cognitive dissonance for ordinary people. Rather than identifying one "true" understanding of energy, it is productive to investigate how multiple ideas of energy flow together to create contemporary perceptions, understandings, and expectations of energy. Making analytical space for multiple conceptual understandings of what energy is and how it works in everyday life will be indispensable if we are to begin to understand what energy means for everyday people, and if we hope, eventually, to shape how everyday people utilize energy and energy resources.

Wittgenstein argues that the truth of empirical observations belongs to our frame of reference. All testing, confirmation, and disconfirmation of a hypothesis takes place within a system—a system that cannot validate claims to "truth," but instead provides the context in which hypotheses are generated and advanced (Wittgenstein 1969, in Tambiah 1990:63). Thus, if Americans' frames of reference for energy are qualitative, relational, and substantivist, then people make their decisions about energy—how they value and use it—in this qualitative framework that involves more than making "rational" economic or "rational" scientific choices (cf. Gell 1988). The notion of sustainability itself integrates multiple aspects of life, including science and the environment, the body and spirituality, social relationships and community (cf. Kelly 2010). As we have seen, New Yorkers' conceptions of energy involve the integration of similar sliding registers of meaning. Allowing space for this multiplicity of understandings of energy, and laying this multiplicity alongside the interwoven variables that constitute a sustainable system, might offer a preliminary step toward energy conservation in an American context where energy means everything.

Because energy is meaningful to people in the social contexts of their lives, it is worth considering energy as a socially embedded notion, idea, model, and substance. While energy is a material force in a technical, scientific sense, it is only made useful through its harnessing in the social world of machines, people, and work. As a result, *not* to consider energy in its qualitative, social manifestations and understandings is to limit our analytical ability to see how energy actually operates—or how it is presumed to operate—in the daily lives of people. And it is in the context of people's daily lives that we seek to manage energy more reasonably and rationally, for the good of us all.

NOTES

1. I thank the City University of New York and the Professional Staff Congress for their generous support of my ongoing research on cultures of energy in New York City.

2. Representing a preliminary sample of ethnographic informants, the individuals who offered ideas about energy for this paper were randomly sampled on the campus of Lehman College, City University of New York, where I am an assistant professor.

REFERENCES

Dethlefsen, E. Stewart
1978 On redefining energy: A humanistic proposal. *Anthropology and Humanism Quarterly* 3(4): 21–5.

Durkheim, Emile
1965 [1912] *The elementary forms of religious life*. New York: The Free Press.

Evans-Pritchard, Edward Evan
1976 [1937] *Witchcraft, oracles and magic among the Azande*. Oxford: Oxford University Press.

Franklin, Sarah
1995 Science as culture, cultures of science. *Annual Review of Anthropology* 24: 163–184.

Frazer, James G.
1911 *The golden bough: A study in magic and religion*. London: Macmillan.

Gell, Alfred
1988 Technology and magic. *Anthropology Today* 4(2): 6–9.

Gusterson, Hugh
1996 *Nuclear rites: A weapons laboratory at the end of the Cold War*. Berkeley: University of California Press.

Kelly, Ingrid
2010 *Energy in America: A tour of our fossil fuel culture and beyond*. Burlington: University of Vermont Press.

Leach, Edmund
1957 The epistemological background to Malinowski's empiricism. In *Man and Culture*, edited by Raymond Firth. London: Routledge and Kegan Paul.

Lévy-Bruhl, Lucien
1949 *Les carnets de Lucien Lévy-Bruhl*. Paris: Presses Universitaires de France.

Malinowski, Bronislaw
1948 [1925] Magic, science and religion. In *Magic, science and religion and other essays*. Garden City, NJ: Doubleday.
1966 [1935] *Coral gardens and their magic: A study of the methods of tilling the soil and of agricultural rites in the Trobriand Islands*. London: Routledge.

Nader, Laura
1996 Anthropological inquiry into boundaries, power, and knowledge. In *Naked science: Anthropological inquiry into boundaries, power, and knowledge*. New York and London: Routledge.

NEETF/Roper Report Card
2002 *Americans' low "energy IQ": A risk to our energy future. Why America needs a refresher course on energy*. Washington, DC: National Environmental Educational
Training Foundation and Roper ASW. http://www.neefusa.org/pdf/roper/Roper2002.pdf (accessed 15 December 2011).

Tambiah, Stanley J.
1990 *Magic, science, religion, and the scope of rationality*. Cambridge: Cambridge University Press.

Tylor, Edward B.
1871 [1970] *Religion in primitive culture*. Gloucester, MA: Peter Smith.

Wittgenstein, Ludwig
1969 *On certainty*. G. Anscombe and G. von Wright. Oxford: Blackwell.

CHAPTER FOUR

Multinatural Resources: Ontologies of Energy and the Politics of Inevitability in Alaska

Chelsea Chapman[1]

During recent negotiations over whether or not to develop energy resources in the Yukon Flats region of central Alaska, those involved appeared not only to disagree about the value of potential energy but also to deploy different ways of thinking about what energy is. The following account of these negotiations asks how multiple ontologies of energy animate such debates over natural resources. The term *ontology* used here helps to signal the presence of alternative and indigenous epistemic spaces (Watson and Huntington 2008) in such conflicts as it indicates ways of knowing and acting toward energy sources that, like water, land, and wildlife, are all too often considered neutral and static commodities. Oil, gas, and renewable fuels are curious hybrids that are at once social constructions of value and unruly objects with unique, place-specific, biophysical properties (Bakker and Bridge 2006; Whatmore 2006; Kaup 2008; Bridge 2009). They are multinational: material substances extracted and circulated among countries, metrics, economies, and other sorts of petro-capital alliance. But they are also multinatural in the ways that they exist in many natures, diverse cosmologies of resources, society, and environment. In Alaska, such curious hybrids—especially oil and gas—have a long and storied presence as a recurring gold rush, as sources of phenomenal wealth and indigenous empowerment and/or dispossession, and as harbingers of ecological collapse. These mythic narratives are underpinned by historical conceptions of what energy is and fields of knowledge of how it works that hover uneasily around the interactions of corporations, state and federal regulatory groups, tribal governments, scientists, and others involved in making decisions about regional development.

A recent upswing in studies of energy in anthropology, political ecology, and economic geography coincides with a broad interdisciplinary reconceptualization of constructions of the human (posthumanism) and its inverse, of nature. Beyond just addressing the social construction of nature and natural resources, the imperative of this "social nature" is to ask "who is currently empowered to define what counts as 'nature'—discursively and materially—

and what implications do accepted or hegemonic definitions have?" (Castree 2001:xiii). A long-standing tradition in anthropology (particularly of northern North American hunting societies) urges us to take seriously the cosmologies of indigenous people not just as metaphors but also as statements of truth about what the world is made up of and how it works. Accordingly, this chapter attempts to highlight plurality among the conceptions, definitions, and evaluations with which people in Alaska understand energy. Studies of relational knowledge systems in which animals and landscapes are more-than-human (Hallowell 1960; Brightman 1993; Bird-David 1999) and of the political implications of such ontologies (Povinelli 1995; Nadasdy 2007) help to connect indigenous intellectual tradition to the constructivist socionature work of unsettling supposedly "natural" resources. Thus, moose, willow trees, oil, and gas can be considered to constitute a topography of social relationships. The interplay of different cosmologies of oil and gas, together with the qualities of light, heat, and movement they produce, might also be conceived as components in a relational universe (Gibson-Graham 1996).

A relational ontology of energy is, here, one in which energy exists not in the form of neutral, fungible "natural resources" but in circulating social obligations among people and other nonhuman beings. This way of knowing energy surfaced in the efforts of coalitions of Alaska Native tribal communities and regional conservation groups to prevent oil drilling in the Yukon Flats National Wildlife Refuge as they attempted to unsettle a "politics of inevitability" (Heynen and Robbins 2005:6). The ecological politics of inevitability present in Alaska depends in part on defining energy resources as at once scarce (existing in a state of geographic and economic precariousness that demands ever-greater production efforts) and stranded (trapped in the landscape but ready to be liberated by newly emerging technological or economic arrangements). Together, these qualities help make the rhetorical case for oil, gas, and renewable development as a shared private and civic obligation. This chapter begins with an account of such premises in the 2002–2010 land trade negotiations between staff of the United States Fish and Wildlife Service Yukon Flats Refuge, residents of Yukon Flats villages, environmental justice advocates, and Doyon Limited, a regional Alaska Native corporation. Next, it turns to the transposition of these ontological premises in local biomass energy projects, using pelletized willow trees, to address the political expedience of "scarcity" and "stranded resources" rhetoric across energy sources. Finally, the chapter concludes by considering the ways in which villagers, backcountry hunters and guides, and conservationists deployed conceptions of energy in ways that presented an alternative to the ecopolitics of inevitability.

THE YUKON FLATS LAND TRADE

Although Alaska Native land claims were settled in 1971 and crude oil began moving through the resulting Trans-Alaska Pipeline from Prudhoe Bay to Valdez in 1977, political conflict over these resources has not subsided since then. Alaska is a landscape where energy conflict is usually played out in courtrooms, the state legislature, shadowy negotiations between officials and oil company executives in Anchorage hotels, and in editorial columns by state oil experts. There are no easily defined factions: as a recent political economic account of oil in Alaska notes, "oil interests have become so successful that in the state political arena, the interests of multinational corporations, local governments, state agencies, legislators and executives, local economic actors and the federal government have become intricately interwoven" (McBeath et al 2008:13). One such complex economic actor is Doyon Limited, a corporation whose shareholders are also citizens of Athabascan Indian tribes from throughout the interior. Doyon is the largest private landowner in Alaska and its subsidiary companies, Doyon Drilling and Doyon Utilities, hold contracts for many projects including drilling rigs on the North Slope, managing a power plant on the military base in Fairbanks, and running tourist facilities in nearby Denali National Park. Doyon, twelve other regional corporations, and some two hundred small village corporations exist because of the 1971 Alaska Native Claims Settlement Act (ANCSA). ANCSA resolved Alaska Native land claims by extinguishing aboriginal rights to land and resources and forming a suite of corporate entities designed to profit from property title conferred by the state.[2] Native corporations are complex cultural and economic entities and exist uneasily alongside other oil companies, tribal governments, and state agencies. Unlike other parts of the world, oil conflict in Alaska is not overt and plays out in these precarious convergences of state and private interests. Oil production in countries surrounding the North Pole is relatively lacking in petro-violence. But quiet violence, structural inequality, symbolic struggle, and "accumulation through degradation" (Johnson 2010) are part of the energy landscape on the North Slope of Alaska (Wernham 2007), in the long-contested prospect of drilling in the Arctic National Wildlife Refuge (Banerjee 2009), and Canada's tar sands (Zalik 2011). In what follows, I offer an account of such conflict in interior Alaska in which ontological pluralism—a multiplicity of ways of knowing and of being oriented towards energy—begins to surface.

The oil and gas underlying the Yukon Flats has long been an object of interest for prospectors, surveyors, and producers of Alaskan oil. Physical traces of exploration in the 1980s—miles of lines left by underground explosives used to create seismic pictures of hydrocarbon deposits—are still present on the land itself. Consultants and corporate officers at Doyon Limited have, for some thirty years, wrangled geological information, market fluctuation, and state and national political environments, waiting for an opportune moment to develop corporation lands in the Yukon, Nenana, and Minto Flats. The Yukon Flats itself constitutes a boreal watershed in the northeast corner of Alaska. The land is forested with black and white spruce, birch, and diamond willow, and is inhabited by lynx, marten, beaver, and wolves. Its wetlands are braided by the Yukon and its tributary rivers and are destinations for massive global migrations of swans, ducks, geese, and other birds. A complex patchwork of land tenure characterizes most of Alaska since Native claims were settled and this region is no exception: landowners include Doyon Limited and several tribal corporations, Native allotment holders, the Alaska Bureau of Land Management, and the federal Fish and Wildlife Service. At eleven million acres, the Yukon Flats Wildlife Refuge is the third largest wildlife refuge in the United States and overlays the ancestral territory of Koyukon and Gwich'in Athabascan nations. It also contains the wildlife and landforms on which residents of local tribal communities depend for spiritual and physical sustenance, and for which it is highly valued by backcountry travelers, conservationists, guides, and ecologists.

In 2002, Doyon proposed a swap with U.S. Fish and Wildlife in which the corporation would relinquish some of its land within the refuge and in return would acquire about two hundred thousand acres of federal refuge lands underlain by potentially profitable deposits of oil and gas. A United States Geological Survey report quantified these deposits in 2004 and speculated they could produce up to 173 million barrels of oil and 5.5 trillion cubic feet of natural gas. During the following six years, the refuge and the small, mostly indigenous communities within it became sites of contest and negotiation between Doyon's staff, state regulatory agencies, regional environmental groups, and tribal governments, and between competing rhetorics of energy. Accounts of the early days of land trade negotiations described the corporation's lobbyists' close relationship to Alaska Senator Ted Stevens in the production of a 2005 Omnibus Appropriations Bill that allotted funds to complete the exchange and expedite an environmental impact assessment process. Over the spring of 2008, the Fish and Wildlife Service collected thousands of public comments from communities in and near the Flats, some of which appears below. In the end, they collected more than one hundred thousand overwhelmingly opposed responses during the environmental impact assessment period. Rural and urban Alaska Natives, trappers and guides, environmentalists, and scientists spoke in a series of contentious Fish and Wildlife hearings that sometimes lasted into the night. These were held in tribal halls, libraries, and auditoriums in Anchorage, Arctic Village, Beaver, Birch Creek, Central, Chalkyitsik, Circle, Fairbanks, Fort Yukon, Stevens Village, and Venetie. Others, unable to travel to the hearings, wrote letters and opinion pieces. Many worried that the cumulative damages to human and environmental health caused by North Slope oil development would be repeated in the Flats, specifically air, water, and ground pollution; contamination of vegetation and wildlife; seismic disturbance; increased rates of cancer, diabetes, and respiratory illness; and harm to culture and spiritual practices. Others emphasized the exceptional importance of healthy animals (ducks, moose, salmon) and other Native foods on which their cultural identity, spiritual practice, and physical well-being depend. Some professed intimate knowledge of a watershed for which they had been successful "resource managers" for, according to a tribal leader from Fort Yukon, twenty-five thousand years. They contrasted this experience with observations of the poor health of animals and people on the North Slope. They had seen or heard of cancerous animals and asthmatic children living near oil production pollution, of residents in Nuiqsut and elsewhere traveling great distances to find animals on a hunt, and of people eating fewer traditional foods and gradually losing their health and vitality. Finally, many people pointed out the irony that fuel used in the extraction of hydrocarbons in Alaska contributes to lake and ice loss, forest fires, permafrost erosion, and changing weather and animal activity. Against the notion that energy is a local product with relatively narrow impacts (such as pollution in and near oil fields and pipelines) in contrast to climate change, a vast and global problem with no way to assign blame, they insisted on the direct causal relationship between drilling projects and the very local outcomes of anthropogenic climate change.

People involved in the negotiations also opposed legal and economic arrangements in the land trade proposal. Built into the deal was a 1.25 percent wellhead tax on oil pumped by Doyon's rigs, but the Fish and Wildlife Service allocated those revenues for federal authorities to purchase Native land throughout Alaska. Tribal opponents of the land swap saw these buyouts as the latest in a decades-long process of dispossessing sovereign tribes of ancestral lands. Residents and environmentalists were suspicious of collusion between Doyon and the federal government, especially given the sudden speed with which the trade was moving. They thought that, with the Bush Administration leaving office, pro-drilling politicians would have little time left to move on the issue. It was said that Doyon was secretly contracting some communities to support the trade using promises made for a greater share in future profits. Doyon officers, they complained, were offering small satellite pipelines that would send cheap local gas to communities near the proposed wells, but distrustful residents remembered the same offer being made—and never fulfilled—in the Trans-Alaska Pipeline construction days. People were angry that the environmental impact document of some seven hundred pages, supposedly drafted to inform communities of the land swap details, only reached them days before their testimony was collected, making it difficult to study and comment on the densely worded document. Again, in the words of a leader from Fort Yukon testifying at a public hearing in February 2008, his tribal council felt that eight hours of public testimony did not adequately reflect twenty-five thousand years of professional land management. The document was not translated into local languages, rendering it incomprehensible to those elders and ecological experts for whom English is not a first language. And as those opposed to the trade and intimately familiar with the ecology of the Flats noted, many of the studies used to forecast environmental impact to wildlife were from oil and gas developments in Canada, Wyoming, and the coastal arctic North Slope of Alaska, having only tenuous biological relevance to a boreal and riverine landscape.

STRANDED RESOURCES:
OIL, GAS, BIOMASS, AND THE POLITICS OF INEVITABILITY

In making the case for oil and gas development in the Flats, Doyon presentations and promotional materials emphasized economic security (oil revenue for its shareholders, many of them residents of the villages) and social well-being (jobs for people living near the proposed development sites). But another prominent line of reasoning echoed related conflicts over developing fossil and renewable fuels: the specter of stranded resources. These are sources of energy perceived to be languishing untapped in Alaskan landscapes, substances or phenomena like trees, wind, currents, natural gas deposits, or hot springs, that could be converted to energy but are geologically, geographically, or politically inaccessible. While these languish, the narrative goes, citizens starve for energy: Alaskans use three and a half times as much fossil fuel per capita as the next most energy-intensive state (Wyoming). Yet according to popular rhetoric, the federal government holds Alaska's oil captive by environmental protection laws and the state's adverse climate and geography traps its resource treasures. Precariousness is assumed to be a natural part of Alaska's energy landscape. Village communities are especially prone: market fluctuations, seasonal demand, and transportation expense (diesel and gas are often delivered by barge to remote villages) are used to justify exceptionally high costs of energy in rural communities (Martin, Killorin and Colt 2008; Hamilton et al. 2011).

In rural communities where heating oil costs, at the time of writing, around $6.50 per gallon and gasoline costs up to $10 in Arctic Village,[3] residents are acutely aware of a scarcity of hydrocarbon energy that has yet to impact most of the rest of the country as severely. Energy scarcity is linked to eating fewer traditional foods (as hunting and other subsistence activities that use boats and snow machines become prohibitively expensive) and to shrinking village populations, as people are forced to move into Fairbanks or Anchorage. According to commentary by Alaskan legislators and Yukon Flats residents supporting development, oil and gas from the region would offset this rural energy poverty and stem urban flight. But hydrocarbon scarcity is not just a natural outcome of difficult geography, insufficient technology, or self-regulating markets but rather a social phenomenon (Huber 2011). The lived experience of such scarcity is energy poverty, "a geographical assemblage of networked relations of various kinds, including flows of energy, infrastructures of production and distribution, the properties of the built environment, and the social and economic networks that sustain communities" (Harrison and Popke 2011:950). It bears emphasizing, however, that many people in Native communities would disagree with being labeled poor when they feel themselves to be rich, if not in cash or carbon fuels but in healthy spirit, food, and land, interwoven qualities that together produce vital individual and social energy.

In rural Alaska, then, hydrocarbon energy poverty is certainly real—not just a forecasted menace—but it is not inevitable or a "natural" outcome of market competition and geological scarcity (Bridge and Wood 2010). Popular narratives of "peak oil" and dwindling supply are used to propel changes in the "how and where" of hydrocarbon energy extraction, portraying the damages wrought by tar sands and the British Petroleum Gulf of Mexico spill a regrettable but inevitable outcome of that growing scarcity. If "scarcity is the meta-narrative of oil" (Bridge and Wood 2010:567), Alaskan government and industry make frequent use of this image of menacing shortage and employ rural fuel crises and high energy costs in urban Alaska as justification for ramping up development in the Arctic Refuge and Outer Continental Shelf.

Resource scarcity is not neutral and neither is energy physics, which grew out of a moral exchange economy in which humanity, guided by a Christian imperative to avoid wasting the divine gifts of nature and to thus avoid the sin of dissipation, reciprocated God's gifts of natural energy by dutifully redirecting it (Smith 1998). The aftereffects of this epistemology resonate in the imperative of twentieth-century fossil-energy developments and in the present-day renewable energy rhetoric of capturing ambient and otherwise wasted energies of the sun and wind. If energy is trapped, unused, in the forests, as it is in the tar sands or the Arctic National Wildlife Refuge, then resource development is obligatory—and inevitable (Sawyer 2009). At the Chena Hot Springs Renewable Energy Fair outside Fairbanks in 2011, a series of politicians made rousing speeches about renewable resources. The mayor of North Pole, assorted state legislators, and finally Lieutenant Governor Mead Treadwell described dark days in Alaska, when economic recession and excessive federal regulation hindered the release of both "stranded renewables" and hydrocarbons. Speaking at the 2011 Business of Clean Energy in Alaska conference in Anchorage, Treadwell again urged that, to keep the country strong and to fulfill "our Alaskan destiny," government and industry should together free the renewables trapped by insufficient technology and the offshore oil reserves imprisoned by federal environmental regulation. In what follows, I describe some of the ways in which a cosmology of trapped energy (here, woody biomass pelletized for use in village boiler systems) relates to hegemonic ways of being toward resources in Alaska.

Scarcity and entrapment were prominent discourses in the land trade negotiations but are also present in renewable energy projects in the Yukon Flats. This section discusses how such rhetoric creates new "fixes" for energy capital. I preface this discussion with the caveat that many renewable energy projects are thriving in rural Alaska, whether initiated by tribal governments, local entrepreneurs, or state agencies. But renewable energy projects in Alaska—as elsewhere—are often driven by private interests, and their recent upsurge is also a way to deflect attention from the fact that oil, heavily subsidized by the state, is still the cheapest everyday option, if the vast and as-yet-unquantified costs of its consumption on global ecologies are not considered. Renewable energy development in Alaska is said to aim at offsetting the catastrophic impacts of climate change in the Arctic, but it is clearly no panacea for regional or global warming. In Alaska, a cottage industry has sprung up around renewable energy, in which state and federal grants are shuffled between a cadre of oil energy experts, technical entrepreneurs, researchers, and nongovernmental organizations. They work from a premise that renewable energy could be abundant, indeed virtually limitless: running out of sunlight or currents is harder to picture than the vivid image of oil derricks running dry like a straw sucking at the bottom of a glass. Yet renewable energy development also leans on a perception of scarcity in which fossil fuel is inevitably in short supply and renewable fuel is stranded in the landscape. The cottage industry of experts and entrepreneurs depends on an economy of thinking about ways to liberate these sources of fuel. Once renewable energy sources are actually accessed, this flourishing economic niche—currently filled with consultants, researchers, and engineers—becomes redundant.

Renewable energy experts within state and nongovernmental organizations suggest that biomass fuels might be rural interior Alaska's best renewable energy option. The Flats are far away from ocean currents and coastal wind farms, too dark during much of the year for photovoltaic power, and without accessible geothermal resources. These ecological factors are said to leave communities interested in alternatives to fossil fuel with the prospect of bioenergy. In practice, bioenergy entails pelletizing and burning the willow and alder shrubs that grow relatively quickly in the short seasons and shallow soil of the interior. Studies of bioenergy elsewhere in the world show it to be a successful alternative. But according to Alaskan biomass studies, Yukon Flats communities would need about ten thousand tons of wood a year, harvested from some five hundred acres surrounding villages. Although some communities have successful biomass furnaces in operation, heating a washeteria in Tanana and a school in Tok, for example, cutting and transporting all that wood presents a problem for others. And while willow shrubs have little economic value as anything other than pellets for boilers, they have multiple other purposes for many people, especially feeding and housing healthy and abundant animals. The concerns raised about woody biomass projects confirm that there is no easy solution to the question of fueling rural Alaska. Alternatives to hydrocarbon fuels are laudable, but renewable energy development is presently funded and undertaken by a quasi-privatized governmental apparatus that relies on the rhetoric of scarcity and stranded resources in all its fuel industries. Further, among the side effects of renewable development appear pernicious manifestations of the ongoing convergence of state and private interests, especially in a hydrocarbon enclave where the Alaskan public and its environmental commons are not necessarily the priority.

AGAINST INEVITABILITY: PEOPLE OF THE YUKON FLATS SPEAK

This chapter focuses on how residents of central Alaska create, contextualize, and respond to proposals to develop regional hydrocarbon and renewable energy sources in order to understand concepts of energy within Alaska Native and non-Native intellectual traditions. Energy capital depends on a robust conception of oil, gas, and renewables as fungible material objects latent in a neutral and external physical environment. At a public meeting in February 2008, an Episcopal minister from Beaver said that the greatest fault of the Yukon Flats land exchange document lay in "its pursuit of Aristotelian categories . . . from counting the number of swans to types of soils to whatever, it misses the overall conflict between the different world views, and it doesn't seem to be aware that there's any conflict at all." The contrasting world view he mentions, arguably held by many of those opposed to the land trade, includes a conception of energy as a manifestation of moral relationships among animals, people, and land. Of course, Yukon Flats people are acutely aware of global oil and gas markets, regional energy economies, and other embodiments of hegemonic concepts of what energy is. But this alternative conceptualization was rendered visible—and, eventually, politically salient—by the repeated deployment of relationality among people, land, and animals. Dacho Alexander, chief of the Fort Yukon tribal government, testified at the same meeting that, as a result of the land trade negotiations,

> [t]hat tree over there, it's got a price tag. That blade of grass, that has a price tag now. That muskrat that's swimming down the creek, that's got a price tag. Every single thing within our area, our traditional area here that Doyon owns has a price tag now.

The Yukon Flats hearings echoed with gustatory claims about the cultural, personal, and political significance of eating healthy moose, fish, and migratory birds, contrasted with the popular quip that "you can't eat oil." Athabascan interlocutors spoke of eating moose and fat ducks as a practice that identifies them as a people and upon which their sovereignty as unconquered Indian nations depends. But others went beyond cultural identity to speak of a banquet of earthy substances—clean air, water, and even carbons underground—and how one should orient toward them for the circulation of vital strength and energy. In Fort Yukon's public comment session in 2008, one man put it this way:

> What we eat, in our stomach, that later gives us strength, in our body. And what we do to the land, it is much the same, and whatever we work into the land it effects the land. . . . Our land, water, every thing that we breathe, that's how we have strength, that's how we live.

In making claims against Yukon Flats energy as a neutral substance whose extraction and profitability are natural and unavoidable, Alaska Native communities and others allude to a world view in which energy exists in quite another register. Energy here is relational, circulatory, and manifest in healthy interactions between humans and other beings. The tension that I have emphasized between dominant and "Other" conceptions perhaps overstates contrast where really it should highlight fluidity and convergence among ontological traditions. But positing an Alaskan indigenous concept of energy suggests the multiple ontological footings on which accounts of oil make sense and are acted upon as truths, whether trapped and languishing beneath the surface of real estate or, as a Gwich'in activist told me, circulating as the earth's blood.

In the ecological politics of inevitability, construing a particular mode of action toward nature and natural resources as the only possible way forward effectively removes alternatives from collective debate. In the case of development in the Yukon Flats National Wildlife Refuge, the argument that oil and gas drilling is inevitable worked to move the debate toward questions of where and when oil resources would be developed, and away from questions of whether they should be developed in the first place. Employing rhetoric of God-given natural resources, local legislators and civic leaders participating Doyon's campaign indicated that expanded oil and gas development in the Yukon and Tanana Flats is not only in the best interests of society but an obligatory move toward Alaska's destiny. Doyon officials engaged Fish and Wildlife field scientists and staff in a snowstorm of paperwork, assessments, and environmental studies, helping to shift the agency away from a mandate of resource conservation toward one of profit. But alternative ontologies were deployed against that inevitability as Native and rural people were able to indicate the presence of an alternative to systematic capitalism and dominant conceptions of energy. An elder at the Fort Yukon hearings invoked an older coalition to halt the proposed Rampart Dam on the Yukon River in the 1950s: *Gwichyaa Gwich'in Ginkhye*, or Yukon Flats People Speak. And speak they did. By some accounts, it was only because village activists felt that development *wasn't* inevitable that they were able to rally the outpouring of opposition that halted the trade. Those voices against the inevitable encouraged Fish and Wildlife to return to its original mandate of resource conservation and to end negotiations with Doyon in 2010.

Against the popular decoupling of energy extraction and climate change, the coalitions working against oil development in the Yukon Flats also argued that local fossil production contributes directly to local climate change. They insisted on the relationality among oil and natural gas drilling, outlandishly high fuel prices, boreal drought and shrinking lakes, glacial melt and flooding, and ever-larger wildfires. This productive and precarious landscape is not a passive outcome of inevitable market demand but an active and fertile site where phenomena like natural gas and biomass fuel are both political historical projects and entities with unique material influences on the social relations of their production. When contextualized in the cultural politicsof capitalism, the micropolitical processes by which conceptions of energy are produced and deployed help to illuminate the ways in which normative conceptions of energy have accrued such omnipotence and universality, even while it is known and experienced in varied ways.

NOTES

1. Support of research for this chapter from The Wenner-Gren Foundation and the National Science Foundation Office of Polar Programs Arctic Social Science Program is gratefully acknowledged.

2. The legal status of ANCSA remains contested some thirty years later (e.g. *Alaska v. Native Village of Venetie Tribal Government* [96–1577] 522 U.S. 520 [1996]) and the social, ethical, and political outcomes of these congressional acts are widely explored by indigenous communities, policy makers, academics, and the public (Berger 1985; Dombrowski 2001, 2002; Brown 2007; Peter 2009; see also the September 2010 *Washington Post* series of articles "Two Worlds").

3. Current Commodity Conditions Alaska Fuel Price Report January 2012, Alaska Department of Commerce, Community and Economic Development.

REFERENCES

Bakker, Karen and Gavin Bridge
2006 Material worlds? Resource geographies and the "matter of nature." *Progress in Human Geography* 30 (1): 5–27.

Banerjee, Subhankar
2009 Terra incognita: Communities and resource wars. In *The Alaska Native reader: History, culture, politics,* edited by Maria Shaa Tláa Williams, 184–191. Durham, NC: Duke University Press.

Berger, Thomas R.
1985 *Village journey: The report of the Alaska Native review commission*. New York: Hill and Wang.

Bird-David, Nurit
1999 'Animism' revisited: Personhood, environment and relational epistemology. *Current Anthropology* 40: 67–79.

Bridge, Gavin
2009 Material Worlds: Natural Resources, Resource Geography and the Material Economy. *Geography Compass* 3: 1217-1244

Bridge, Gavin and Andrew Wood
2010 Less is more: Spectres of scarcity and the politics of resource access in the upstream oil sector. *Geoforum* 41: 565–576.

Brightman, Robert A.
1993 *Grateful prey: Rock Cree human-animal relationships*. Berkeley: University of California Press.

Brown, Caroline L.
2004 Political and legal status of Alaskan natives. In *A companion to the anthropology of American Indians*, edited by Tom Biolsi, 248–267. Malden, MA: Blackwell Publishing Ltd.

Castree, Noel
2001 Socializing nature: Theory, practice and politics. In *Social nature: Theory, practice and politics*, edited by Noel Castree and Bruce Braun, 1–21. Malden, MA: Blackwell Publishers.

Dombrowski, Kirk
2001 *Against culture: Development, politics and religion in Indian Alaska*. Lincoln: University of Nebraska Press
2002 The praxis of indigenism and Alaska native timber politics. *American Anthropologist* 104 (4): 1062–1073.

Gibson-Graham, J. K.
1996 *The end of capitalism (as we knew it)*. Minneapolis: University of Minnesota Press.

Hamilton, Lawrence C., Daniel M. White, Richard B. Lammer, and Greta Myerchin
2011 Population, climate, and electricity use in the Arctic integrated analysis of Alaska community data. *Population and Environment*. Published online May 2011. Springerlink DOI: 10.1007/s11111-011-0145-1.

Hallowell, Irving
1960 Ojibwa ontology, behavior, and worldview. In *Culture in history: Essays in honor of Paul Radin*, edited by Stanley Diamond, 19–52. New York: Columbia University Press.

Harrison, Conor and Jeff Popke
2011 "Because you got to have heat": The networked assemblage of energy poverty in eastern North Carolina." *Annals of the Association of American Geographers* 101 (4): 949–961.

Heynen, Nik and Paul Robbins
2005 The neoliberalization of nature: Governance, privatization, enclosure and valuation. *Capitalism Nature Socialism* 16 (1): 5–8.

Huber, Matthew T.
2011 Enforcing scarcity: Oil, violence, and the making of the market. *Annals of the Association of American Geographers* 101 (4): 816–826.

Johnson, Leigh
2010 The fearful symmetry of Arctic climate change: Accumulation by degradation. *Environment and Planning D: Society and Space* 28(5): 828–847.

Kaup, Brent Z.
2008 Negotiating through nature: The resistant materiality and materiality of resistance in Bolivia's natural gas sector. *Geoforum* 39: 1734–1742.

Martin, Stephanie, Mary Killorin, and Steve Colt
2008 Fuel costs, migration, and community viability. Institute of Social and Economic Research, University of Alaska Anchorage.

McBeath, Jerry, Mathew Berman, Jonathon Rosenberg, and Mary F. Ehrlander
2008 *The political economy of oil in Alaska: Multinationals vs. the state*. Boulder, CO: Lynne Rienner Publishers, Inc.

Nadasdy, Paul
2007 The gift in the animal: The ontology of hunting and human-animal sociality. *American Ethnologist* 34 (1): 25–43.

Peter, Evon

2009 Undermining our tribal governments: The stripping of land, resources and rights from Alaska Natives." In *The Alaska Native reader: History, culture, politics,* edited by Maria *Shaa Tláa* Williams, 178–183. Durham, NC: Duke University Press.

Povinelli, Elizabeth

1995 Do rocks listen? The cultural politics of apprehending Australian aboriginal labor. *American Anthropologist* 97 (3): 505–518.

Sawyer, Suzana

2009 Human energy. *Dialectical Anthropology*, Published online 28 August. Springerlink DOI 10.1007/s10624-009-9122-9.

Smith, Crosbie

1998 *The science of energy: A cultural history of energy physics in Victorian Britain*. Chicago: University of Chicago Press.

Watson, Annette and Orville Huntington

2008 They're here—I can feel them: The epistemic spaces of indigenous and western knowledges. *Social and Cultural Geography* 9 (3): 257–281.

Wernham, Aaron

2007 Inupiat health and proposed Alaskan oil development: Results of the first integrated health impact assessment/environmental impact statement for proposed oil development on Alaska's North slope. *EcoHealth* 4: 500–513.

Whatmore, Sarah

2006 Materialist returns: Practicing cultural geography in and for a more-than-human world. *Cultural Geographies* 13 (4): 600.

Zalik, Anna

2011 Protest as violence in oilfields: The contested representation of profiteering in two extractive sites. In *Accumulating insecurity,* edited by Shelley Feldman, Charles Geisler and Gayatri A. Menon, 261–284. Athens: University of Georgia Press.

CHAPTER FIVE

Siting, Scale, and Social Capital: Wind Energy Development in Wyoming

Sarah Strauss and Devon Reeser

> Driving along I-80 west from Cheyenne, Wyoming, you arrive in Laramie via the "Gangplank," a gradual route across the Front Range of the Rocky Mountains that made the transcontinental railroad possible. It is a barren and windswept landscape, described by some as almost lunar. You see some antelope and cattle, a few oil and gas rigs on the way, but the dominant new feature is a forest of towering white wind turbines. As you drive, large tractor-trailers pass by at high speeds, carrying the disassembled pieces of the giant machines—a blade here, a tower piece there; dwarfed by their transports, you begin to appreciate the massive size of these machines—and they are only middling in the scale of wind turbine possibilities.

INTRODUCTION

Wyoming's energyscape (cf. Appadurai 1990) is among the most productive in the nation, yielding the majority of U.S. coal and uranium, a fair amount of natural gas, and up to 50 percent of the best and most accessible wind in the western part of the country. Wyoming is among the top ten state producers of wind-powered electricity; its wind generation capacity is limited primarily by lack of transmission lines to get it to market (Wiser and Bolinger 2010). In addition to supplying a significant proportion of the U.S. energy stocks, Wyoming is home to the first regional wind energy cooperatives, new forms of collaborative engagement to support local management of wind resources in the face of wider development interests. Because of their success, these wind co-ops have been written up in the *New York Times* (Barringer 2008),[1] featured on National Public Radio, and even researched by China; they are models for other states seeking to balance local interests against regional or multinational forces, and for others in Wyoming seeking solidarity in alternatives to petroleum-based energy resources in the face of the new Niobrara oil shale boom (Farquhar 2009; Thompson 2009). In this chapter, we consider how the creation of wind co-ops in Wyoming, and the opposition generated in response to their efforts, reflects and resonates with larger questions of siting and scale in wind energy development more generally, and examine how different types of social capital might be employed in service of such decisions.

Wyoming's citizens support many views about how our vast energy resources should be developed, and which kinds of energy development support the quality of life they seek (Heart and Mind Strategies 2009). In a recent study, 80 percent of the population supported fossil fuel extraction in Wyoming, and the same percentage supported wind resource development; about 60 percent agreed with the notion of continued uranium mining (Nelson et al. 2010). While wind energy development might seem to be the next step forward for an energy-producing state's economy, some oppose full-scale development of the state's wind resources. Under the auspices of preserving culture, mountain vistas, wildlife, and recreational integrity, these groups seek increased zoning regulations. Such laws would infringe on property rights and increase government control, changing Wyoming culture and its traditional land-use patterns.

In the past decade, Wyoming has been forced to examine its values, seeking balance between preservation and prosperity. The complex flows of wind energy that circulate through Wyoming and beyond highlight the many challenges to moving beyond cheap petroleum. Questions of scale and siting are essential to understanding the perceived risks and benefits of wind energy production, and therefore their impact on the reception of local communities to wind energy development. In order to make sense of the choices facing Wyoming citizens, as well as Americans across the country, we need to first discuss the communities whose lives are intertwined through the energyscape that has brought so much development to this vast Western state.

TALKING WIND IN WYOMING

In the spring of 2010, Devon Reeser drove around the state of Wyoming, interviewing ranchers and other landowners in the south-central and southeastern counties,[2] seeking to understand how members of the same communities could differ so radically in their attitudes toward wind power production in Wyoming, especially in relation to global climate change. This corner of Wyoming is at the center of the wind energy boom and is the site of significant contestation regarding appropriate siting or even existence of new wind farms and power lines to carry the resource out of state. Global climate change threatens to alter earth systems drastically in the coming decades. In many instances, it can already be linked to altering precipitation patterns and intensifying drought within regional ecosystems (IPCC 2007). People dependent on natural resources and agriculture for their livelihood, like ranchers, are thus vulnerable to climatic changes.

In the past decade, persistent drought has caused the Wyoming cattle population to drop 21 percent, gravely reducing economic returns for many ranchers (Joyce 2009). As one rancher shared with us, during a particularly bad drought in 2002 he "was forced to sell half [his] cow herd," and he's still in a "slow rebuild process" as a result. For an already economically tumultuous business, even slight climate variability can have drastic repercussions. And the regional climate-change scenarios suggest that even drier conditions are likely to become the norm. Simultaneously, in the past few years, western states have pledged to increase sustainable energy use, largely through subsidizing wind energy development. Wyoming has some of the best onshore wind in the country, and much of this wind resource is in southeast Wyoming on private ranches. Wind energy developers have thus rushed to lease property rights from ranchers over the past decade. This wind boom has provided an opportunity for ranchers with threatened livelihoods to economically diversify.

Studies indicate that communities that pool their social resources can build effective networks to decrease their vulnerability to such environmental changes (Adger 2003; Adger et al. 2007). This idea of "social capital," or a community's ability to respond collectively to change or crisis, can be broken into three distinct dimensions: bonding (family ties), bridging (connecting friends and acquaintances), and linking into external resources (Sabatini 2007; Dudwick et al. 2006; Pelling 2003). Our research with Wyoming landowners asks: how do communities adapt their social capital to cope with environmental changes, and how can we document strengths and weaknesses in social capital dimensions within particular communities, especially within those most vulnerable to climate change? Furthermore, how does culture contribute to social capital formation? We have sought to gain insight into these questions by assessing why and how Wyoming landowners, whether working ranchers or recreational owners, are using the different dimensions of social capital—bonding, bridging, and linking—to make choices about land use for or against wind energy that may ultimately increase their resiliency to environmental change. The locations of specific proposed wind developments, and the scale of those developments in relation to the landscape and the existing uses of that place, matter a great deal for how people perceive the value of such resources.

We found two different groups of Wyoming citizens engaged in the wind development debate, the members of REAL—the Renewable Energy Association of Landowners—and the members of the NLRA, or Northern Laramie Range Alliance. Many ranchers have welcomed wind energy development as a tool for economic diversification. However, in contrast to historical wind energy deals with individuals in California and other states, and in stark contrast to typical Wyoming rancher independent business practices, most landowners are not acting independently to secure wind deals. Instead, landowners are forming associations in order to increase their marketing potential, increase their economic returns from leasing to developers, minimize legal costs, and elicit broader public support. Furthermore, these local landowner associations have now banded together to create one conglomerate, REAL, that can exert greater influence in political spheres and guide transmission policy. Wind developers need transmission infrastructure to make wind farms viable, and individual landowners and landowner associations have much more power to influence Wyoming state transmission policy as a larger group. REAL now represents thirteen landowner associations, three hundred individual members, and eight hundred thousand acres across six Wyoming counties. Landowner associations and REAL have been successful. As its chairman shared with Devon, "While we're not a formal group, people don't ignore us anymore." Six associations have already secured agreements with wind energy developers.

This successful deal-making has awakened some demons, though; not everyone in Wyoming welcomes turbines and the transmission lines they require. In March of 2009, a few landowners along the Northern Laramie Range received postcards from Rocky Mountain Power indicating its interest in building a high-voltage transmission line through their backyards. A number of concerned citizens from southeast/south-central Wyoming went to two public scoping meetings.[3] The group of concerned citizens ranged from prominent six-thousand-acre ranch owners to "a lady with a hip replacement going to Douglas and down two flights of stairs because her husband had died and she wanted to protect their little cabin." Neighbors got together and quickly formed the NLRA, which now includes nearly eight hundred members who oppose the "industrialization of the mountains." Rocky Mountain Power withdrew its bid to place lines directly through the Laramie Range, but the Alliance is still concerned about energy companies, especially Wasatch, that have signed deals with the neighboring landowner association to place a wind farm in the Northern Laramie Range, an hour or two north of Laramie. The Alliance is anxious over connector transmission as well, especially in light of eminent domain laws that allow "taking" (with payment) of property for transmission corridor development that is deemed necessary; in Wyoming, unlike in some other states, there is currently a moratorium on wind energy developers themselves using the right of eminent domain to take over private property, though no such protection exists against oil or gas development. The NLRA formed a political action committee and has been using grassroots, collective action to advocate for new zoning laws for wind energy placement. "Converse County welcomes wind energy development in the right places. Help keep it there, and not in the mountains" (NLRA 2010). The NLRA advocates zoning and more careful placement of transmission lines and wind farms. Both "sides"—REAL and the NLRA—formed from conglomerates of neighbors, most of whom had never worked together on a specific project before. But many of the individuals were known to each other, both within and across the two movements; the total population of Wyoming is 563,000, and within the communities we are describing here, likely fewer than fifty thousand were affected directly by the proposed development. As three different people expressed to Devon, "Wyoming is just one small town with really long streets." The members of these different groups "rode the school bus together" and "still say 'hi' in the grocery store." How might these people from the same communities use different strategies to harness social capital for collective benefit, especially when the goal of each movement was perceived in oppositional terms? The success of each group can be traced to particular components of each group's social capital matrix (see Figure 5.1), particularly to the driver that invoked

FIGURE 5.1. Social capital matrix

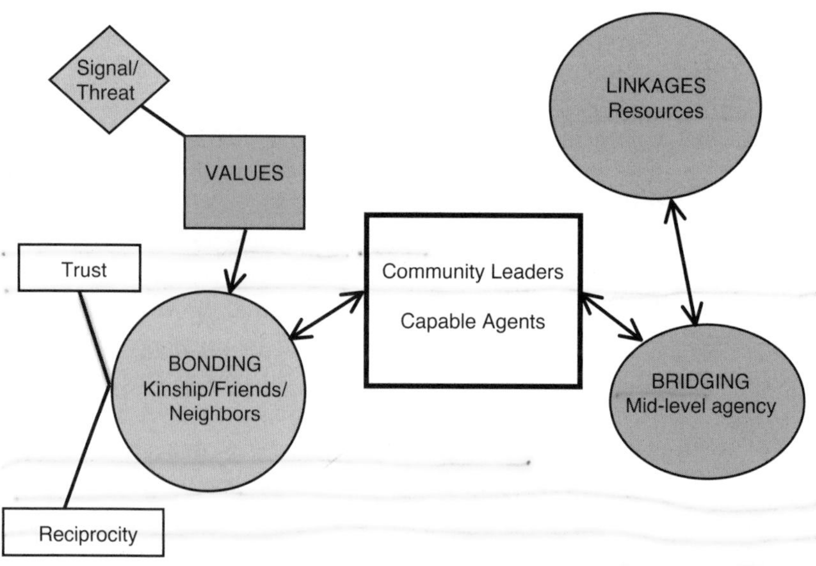

group formation and to the leadership involved in each group. A detailed analysis of each group's matrix aids in greater understanding of the complex interplay between social capital dimensions, cultural values, and individual and group actors endeavoring to use collective action to bolster adaptation to change. First we need to analyze the context of wind energy development in Wyoming to understand the complex interplay of social capital dimensions and cultural values involved in these groups' formations.

UNDERSTANDING THE PLACE OF WIND ENERGY PRODUCTION IN WYOMING

In 2009, EPA administrator Lisa P. Jackson gave a speech at the Happy Jack Wind Farm, between Cheyenne and Laramie, Wyoming, saying:

> Wyoming is pioneering the implementation of clean, renewable wind energy and exploiting their abundant wind resources. They're also a top exporter of our nation's coal and have significant natural gas and oil resources as well. . . . Wyoming is the heart of our energy future—that's why we're here today. (Jackson 2009)

The "heart of our energy future," Wyoming has historically been known as an internal energy colony for the rest of the United States, akin to Alaska or West Virginia. Vast trainloads of coal leave the state twenty-four hours per day, seven days per week, headed for processing and value-added sales in other places. And by vast, we mean to convey an almost unimaginable and relentless flow of coal trains leaving the state: imagine one hundred rail cars, each filled with one hundred tons of coal, at least one hundred times per day, moving this resource from Wyoming to at least thirty different states as well as to other countries. It is a staggering amount of carbon, representing just under 5 percent of total annual global CO_2 emissions.[4] Oil and gas are also exported in their natural states, though at a lesser rate. These are, of course, "raw" energy resources that require processing to be valuable for end-user consumption.

But in the past decade, Wyoming's energyscape has been transformed by the advent of a new resource. Wind resources differ from fossil fuel–based energy sources in that wind is captured and transformed immediately, moving from air to electric transmission lines almost instantaneously. Wind's processing occurs as it is being harnessed, much like solar and hydropower; the resulting electricity, if it is not used directly on site, can be made immediately available to end users through the power grid. Wind is thus a "value-added" commodity instead of a raw material whose value increases only after it has been refined, often in other states.[5] From a meaning-centered perspective, wind itself is an invisible force that converts more magically to usable energy in the form of electricity than do the tangible fossil fuel resources more familiar in the state. Prior to wind energy development, the wind in Wyoming was viewed more as a cross to bear, the thing that grabbed your car door out of your hands as you opened it or blew up dust storms so that you could barely open your eyes when you went out to feed your livestock. Jokes about Wyoming wind abound; its transformation into a more valuable energy resource turns the humor on its head. Still, the utility of the wind was not lost on early settlers; the use of wind energy to pump water for stock purposes in remote areas has been a staple technology for ranchers for one hundred and fifty years. But as several Wyomingites have told us, using the wind as a personal resource is one thing; generating it at scales designed for export to other states, with attendant power lines, is quite another.

Wind energy generation technology has another major benefit—it offers a great deal of flexibility at scales ranging from a single household to massive industrial complexes. From one-to-two-kilowatt household-scale turbines, to the massive five-megawatt wind turbines in the North Sea off the German and Danish coasts, wind has the potential to rival fossil fuels in ways that have largely eluded renewable sources until now. This wide range of scales for wind production is very important; one of the largest concerns for transitioning from fossil fuels to renewable resources is the difference in energy density. The fact is that a relatively small quantity of fossil fuel can be used to generate a large quantity of energy. But in comparison with fossil fuels, a vast geographic space is required to produce an equivalent quantity of energy using renewable sources. Because it is possible to scale renewable sources like wind and solar up and down with relative ease, however, these sources are much more flexible in creating local and regional systems that are appropriate to the specific terrain and community needs.

Engaging the needs, desires, and interpretations of both values and risks associated with different energy sources as they are produced and consumed in local communities is therefore critical to understanding how we can most effectively respond to the twin needs of climate-change mitigation and regional/national energy independence with the fewest negative consequences for the environment and for human health. For this reason, anthropological research on energy transitions and the discourses surrounding them is very important. Wyoming is an interesting place to explore such issues, both because it has a long history as an energy-producing state and because it is home to iconic wildlife and landscapes; these contexts have generated a meaning-centered rift between environmentally focused groups who might normally be allies in the pursuit of non-carbon-based energy sources and created other sets of strange bedfellows. In the sections that follow, we examine the differences between the two main groups concerned with wind development in southeastern Wyoming, REAL and the NLRA.

REAL AND THE LANDOWNER ASSOCIATIONS

> There's nobody more independent than ranchers. Don't want anything to do with anybody. To get them to buy in, I don't know how we did it. Historically, producers have never joined co-ops; they work independently.[6]

But join they did. More than three hundred ranchers now work together through associated cooperatives to improve their own livelihoods and better their communities. What made these fiercely independent ranchers abandon their tradition of autonomy and associate? And how did they achieve this integration? The REAL phenomenon manifested from convergent drivers threatening ranchers' values to such a degree that these ruggedly independent individuals decided to act in unison with others. Even though most people had never really worked with their neighbors "anywhere near the extent of this project," community leaders emerged to start bolstering bonding capital—to get neighbors talking to each other. However, while some of these leaders formed cooperatives years ago, or had been thinking about forming cooperatives, none succeeded without a push from a midlevel agency. The quote above came from a capable midlevel agent, the coordinator of a vertically linked affinity organization, who created this concept for Wyoming wind development.

He came up with the idea and then handpicked the leaders for each one of the cooperatives. "When you put those steering committees together, you have to pull strings, but they have to be strong leaders. Otherwise the whole association will fail. . . . I talk to them beforehand. I say, 'When I ask for volunteers, you have to raise your hand.' And they do." Every association leader indicated that this coordinator was his or her "go-to guy," his or her primary resource. Not only did he develop a model for forming a legal corporate entity (LLC), which nearly all thirteen associations followed, he linked landowners with educational and technical resources, such as anemometers to measure wind quality; by generating specific data about their local area, they are better able to market themselves. Hence, while associations might have formed on their own, they received a big push from one person, one forward-thinking community leader who has provided a link to additional shared resources such as technical devices and legal services. And this support is not limited to wind-development cooperatives but has recently been extended to help other landowners in the region negotiate terms for a cooperative response to proposed oil pipelines for the new Niobrara shale development using the same model. The reasons for seeking collaborative action in Goshen and Laramie Counties were the same as for the initial wind co-ops in Converse and Albany Counties (Kessler and Kessler 2011): to ensure that landowners were able to negotiate fairly with industry representatives instead of being forced into a position of pitting neighbor against neighbor, as has been the experience with California and Texas wind development.

While leadership enabled cooperation, three hundred people are participating because impending changes in the regulatory and investment landscape triggered signals that action had to be taken. The signal that catalyzed action was wind energy development, which has the power to improve or threaten livelihood for all of these people in different ways. For most of the association leaders, threats to their ranching lifestyles have been pressing for years; they constantly seek opportunities for economic diversification in order to stay afloat. "Not a rancher in Converse County that doesn't have an outside income. Contractor. Wife working. Feed salesman. Truckers. Isn't one person that doesn't have oil or gas lease income, retirement income, something that supplements the ranch."[7] Ranchers can run cows right up to a turbine. And "turbines don't need water."[8] Hence, "wind and agriculture are kind of like a hand in a glove."[9] Each turbine could bring $20,000 in annual revenue to a rancher, which is a fortune in a business where one is happy to break even at the end of the year. Others are reacting because they know development will happen regardless, and if they associate, they can benefit. Some members will not get turbines, but they have agreed to let their land be used for transmission connector lines or other infrastructure needs. With a "community-based payment" system, all landowners in the cooperative split the proceeds from a wind energy deal—40/60—with 40 percent of proceeds allocated based on landholding acreage and 60 percent of proceeds going to ranches that allow wind farms to be built on their property. This system allows individuals as well as towns and counties to benefit from midscale development, to a greater degree than could be achieved on each rancher's land independently. Because these REAL members are primarily ranchers, they view the wind farms as a way to ensure multiple uses of their own lands; the presence of turbines does not impinge on number or quality of cattle, as stock can continue to graze by the turbines with no ill effects. While there are still concerns about sage grouse, raptors, and other environmental impacts, Wyoming has been quite proactive in creating a wind development road map hat has been adopted by the state to help minimize these siting risks (Biodiversity Conservation Alliance 2008).

NLRA: SAME VALUES, DIFFERENT OUTCOME

> If we were on a ranch, and we depended on it, it would make a difference. We're looking at it from a purely historical, emotional connection to the land. That's all we have. We all work indoors, but we were raised on ranches outside. Our cabin is our escape. It's getting back to our roots. My great-grandparents homesteaded it in the twenties, and my grandmother put it in a trust for every legal and blood heir into perpetuity. My grandmother set it up to prevent road construction and restrict cabin building, grazing, hunting, everything. There's two track roads, no electricity, no running water, no nothing. I will always live within three hours of the property.[10]

Northern Laramie Range Alliance members are part of the same "small town with a long street," but they have used social capital in distinctly different ways to achieve their goals. Most importantly, the driving forces of collective action among NLRA members are emotional and aesthetic as opposed to economic; wind turbines will not significantly affect their livelihoods. This fact has led to accusations of NIMBY-ism (not in my back yard) on their part, which NLRA members strongly deny. To NLRA members, wind energy development threatens sentiments of deep-rooted heritage that is shared by the REAL members; the difference is that for the majority of the NLRA, the intangible heritage and the viewshed is all that remains (or all that ever existed, in the case of newcomers to Wyoming), while for the REAL ranchers, their heritage is tied to the ability to continue running cattle, which in turn is tied to a need to find alternative sources of capital. The key difference between these groups is in the distribution of bridging ties, which has been determined by the nature of the various organizational goals, the leadership involved, and access to midlevel agency and vertical linkages (see again Figure 5.1); their aesthetic and emotional concerns, though in many ways shared by the REAL ranchers, take on heightened significance for the NLRA community because they have nothing else to tie them to the land.

> Last year—I can't give you a date—I got a little postcard from Rocky Mountain Power saying that they were interested in building a corridor over the Northern Laramie Range. So there were a number of people that voiced concerns at that meeting. . . . And [Wasatch Energy] wanted to lease some of our property for a wind farm. They were very aggressive in their presentation. They suggested that they had already signed up some of my neighbors. I thought, "Is this my biblical neighbor or

> my actual neighbor?" So I called my contiguous neighbors up, and they hadn't signed up any of them. In addition they gave me a contract, and they suggested that since they signed up some of my neighbors there was no need for negotiation. And now I thought, "Now I'm really concerned about what they are doing."[11]

NLRA is truly a grassroots organization. While REAL formed through the efforts of capable agents and community leaders and was linked together by a midlevel coordinator anchored at a federal outreach agency, NLRA formed spontaneously from a perceived threat to a community's heritage. "The group is loosely formed—you're a member if you say you're a member."[12] A steering committee of a dozen or so members does exist and provides leadership for the nearly eight hundred neighbors who are involved in the movement, but these "members" formed the association simply by raising their hands at a meeting. No voting took place.[13]

The NLRA's goal is to form a large collective voice to oppose "industrialization of the mountains." They say that they are not against wind energy development; they are only interested in proper siting of turbines and transmission lines. This position, however, leads to accusations that it is only their backyard mountains that should be protected. The main issue is that to get energy from Converse County to grid substations, new transmission lines would have to be built somewhere in the vicinity of the Northern Laramie Range. The eight hundred members of NLRA love the Northern Laramie Range and do not want it altered in any way. Members of the steering committee who had never worked together before are now close friends. Unlike REAL leaders, they have very diverse backgrounds—from a World Bank treasurer to a prominent fossil fuel developer to a dental hygienist. What they do have in common, however, is a sincere love of their Wyoming heritage (whether a deep-rooted one carried through generations or one newly purchased), symbolized in their minds by the Laramie Range. "You just don't want that [industrial stuff] in the mountains. That's why people live in Wyoming. That's who we are. And that's why there has been such a public outcry—everyone thinks they own a piece of the mountains."[14] Many NLRA members grew up on ranches but do not live on them anymore. They love the ranching lifestyle but were forced or chose for various reasons to go into business or professional fields for their livelihoods, so many who originally grew up in rural communities now live in one of Wyoming's largest "cities"—Casper, population fifty-five thousand in 2010—or have moved out of state. The only thing remaining of their rural roots is a cabin or homestead with a view.

THE CODE OF THE WEST: TRUST, RECIPROCITY, AND SOCIAL CAPITAL

> It seems like a lot of folks are awful busy telling ranchers how they should be taking care of things. Ranchers have a Code of the West that was based on some pretty steadfast rules, many years ago. It's still an ingrained part of rural society.[15]

These lines reflect Devon's conversations with cowboy poet Terry Henderson (1992) about the wind cooperative development. The Code of the West directs ranchers to take care of themselves within a strict but simple normative code of trust and reciprocity. The rancher lives by a principle of honesty, handed down from her parents and instilled through a Christian evangelical fear and love of God. The primary value influencing Wyoming ranchers to join cooperatives through REAL and develop wind energy is a desire to keep the heritage of their ranches intact. People's different ways of building and leveraging social capital shape an important set of dynamic processes, and understanding these processes helps us to understand the decisions that ranchers made to join cooperatives or to abstain from participation. The siting and scale of wind developments also comprise crucial factors in ranchers' decisions to join cooperatives or to remain independent. In other contexts, it has been noted that social capital is strongest in cultures characterized by high levels of trust and reciprocity that also have capable agents who can connect others with vertical resources (Adger 2003; Pelling 2003; Dudwick et al. 2006; Khan et al. 2007; Sabatini 2007). One reason that landowners give for their decision to join associations is to bolster economic returns from their businesses, despite the deep historical trend among ranchers to "go it alone" in their business ventures. Even those who bought their ranches later in life still hold the same strong value of keeping the heritage of ranching alive for their children or grandchildren. Wind energy development also creates jobs that enable more young people to stay in Wyoming and manage family ranches. While NLRA members also live by the Code of the West, and their cultural values are virtually identical to REAL values, bonding capital has blossomed for different reasons based on the fact that the driver is emotional and aesthetic as opposed to economic.

The ranching community in southeastern Wyoming has a strong face-to-face community imbued with values we have identified in interviews and through other sources to include heritage/lifestyle, the right to private property, pragmatism, injustice, and entrepreneurship. These values are further supplemented by a strong moral code of trust and reciprocity, one of the key

want to keep heritage alive but development will bring in younger ppl to continue ranching & heritage of the land mythologies of the modern West. Along with a combination of threats from outside development to loss of their ranching heritage, these values helped generate the social bonding capital needed to support the development of collaborative action through REAL. The community's value base, inscribed through the Code of the West, created the preconditions for bonding, and they are holding the bonds together still. Values have also created preconditions for strong community leaders, who, coupled with strong midlevel agencies, are extremely capable agents to bridge the "masses" to wind markets and other resources.

As the comparison of REAL and NLRA shows, even though groups might have similar values, including a sense of heritage and emotional attachment to a particular landscape, those values can manifest in divergent ways. A community's relationship to place is a fundamental factor shaping the expression of, and action on, communally held values. People who live in the same state and share similar values may react differently to a change not only because of differences in their livelihoods but also because of differences in their relationship with the landscape. While many landowners, ranchers or not, may not object to a single stock tank windmill, or even a single modern wind turbine, the scale of vast wind farms is more than they can bear. But for others, the silent[16] white sentinels permit the preservation of lifeways and land use that will allow heritage practices to continue. In both cases, the game is won or lost through the collaboration of neighbors and like-minded others, using all of the social capital at their disposal to achieve their preferred futures.

NOTES

1. We thank the Wyoming members and participants in REAL and the NLRA, as well as others around the country who have given graciously of their time to help us understand the meaning of wind development for them and their communities.

2. This has so far been a very limited pilot study, but we expect to continue work in this area in years to come, both in the United States and India.

3. Public scoping meetings are a standard part of the federal NEPA environmental impact statement process, in which the agency, organization, or entity proposing the development describes the actions and possible alternatives and invites public and other agency comment regarding potentially affected resources, environmental issues that must be considered, and other factors that must be analyzed.

4. This ballpark figure was calculated using EIA figures for Wyoming coal production relative to global carbon.

5. The variability of wind also makes it somewhat more difficult to use; because it is not currently cost effective to store wind energy, it cannot be controlled for release at high-demand times (as is the case with hydropower), and some energy is inevitably lost as a result of these fluctuations.

6. Coordinator, affinity organization.

7. Discussant 6.

8. Discussant 2.

9. Discussant 3.

10. Discussant 10.

11. Discussant 8, NLRA steering committee member.

12. Discussant 8.

13. However, although the NLRA members would attest to the grassroots nature of their organization, new data has recently come to light to indicate that the Heartland Institute has been infiltrating "grassroots" anti-wind organizations around the United States in an effort to discredit climate-change-related rationales for renewable energy development. See http://m.guardian.co.uk/environment/2012/may/08/conservative-thinktanks-obama-energy-plans?cat=environment&type=article.

14. Discussant 8.

15. Discussant 2.

16. Large turbines can be surprisingly quiet, but this again is a function of siting. In Wyoming, where the wind blows constantly and the space is wide open, one can stand under a 2.1 megawatt turbine and not hear anything but the wind blowing (though that is pretty loud itself). But in a narrow valley with limited space, perhaps in New York or Georgia, the sound might reverberate strongly, bouncing off the landscape.

REFERENCES

Adger, W. Neil

2003 Social capital, collective action, and adaptation to climate change. *Economic Geography* 79 (4): 387–404.

Adger, W. N., S. Agrawala, M. M. Q. Mirza, C. Conde, K. O'Brien, J. Pulhin, R. Pulwarty, B. Smith, and K. Takahashi

2007 Assessment of adaptation practices, options, constraints and capacity. In *Climate Change 2007: Impacts, adaptation and vulnerability*." Contribution of Working Group II to the Fourth Assessment Report of the Intergovernmental Panel on Climate Change, edited by M. L. Parry, O. F. Canziani, J. P. Palutikof, P. J. van der Linden, and C. E. Hanson, 717–743. Cambridge: Cambridge University Press.

Appadurai, Arjun

1990 Disjuncture and difference in the global cultural economy. *Public Culture* 2 (2): 1–24.

Barringer, Felicity

2008 A land rush in Wyoming spurred by wind power. *New York Times*, November 28. http://www.nytimes.com/2008/11/28/us/28wind.html (accessed February 12, 2010).

Biodiversity Conservation Alliance

2008 *Wind power in Wyoming: Doing it smart from the start*. Laramie, WY: Biodiversity Conservation Alliance.

Dudwick, Nora, Kathleen Kuehnast, Veronica Nyhan Jones, and Michael Woolcock
2006 *Analyzing social capital in context: A guide to using qualitative methods and data*. Washington, DC: The International Bank for Reconstruction and Development/ The World Bank.

Farquhar, Brodie
2009 Landowner associations rope wind. *Wyoming Business Report*, January 12. http://www.wyomingbusiness report.com/print_article.asp?aID=98087 (accessed March 28, 2010).

Henderson, Terry
1992 Livin' the Code of the West. In *It's been one of those days! Cowboy poetry describing real Wyoming ranch life*. Cougar Valley Enterprises.

Heart and Mind Strategies
2009 Building the Wyoming we want: Priorities and values study. http://www.building wyoming.org/wp-content/uploads/2009/12/2009_06_BW3_Priorities_Values_ Study.pdf (accessed March 20, 2010).

Intergovernmental Panel on Climate Change (IPCC)
2007 Summary for policymakers. In *Climate change 2007: Impacts, adaptation and vulnerability*. Contribution of working group II to the fourth assessment report of the Intergovernmental Panel on Climate Change, edited by M. L. Parry, O. F. Canziani, J. P. Palutikof, P. J. van der Linden and C. E. Hanson, 7–22. Cambridge: Cambridge University Press.

Jackson, Lisa
2009 Remarks at Happy Jack Wind Farm, April 20, as prepared for delivery. http://yosemite.epa.gov/opa/admpress.nsf/12a744ff56dbff8585257590004750b6/57dea49c 682d5d6a852575f50057b707!OpenDocument (accessed April 6, 2012).

Joyce, Matt
2009 Drought thins cattle herds. *Casper Star-Tribune*, August 11. http://www.trib.com/news/state-and-regional/article_ba9e253b-f9c1–5d24-a032-b02acdb080ba.html (accessed February 12, 2010).

Kessler, John and Nancy Kessler
2011 Interview, October 2. Wyoming's Recent Energy Boom: An Oral History Program. UW American Heritage Center's Wyoming Energy Booms Oral History Project, Accession #11749.

Khan, Shahrukh Rafi, Zeb Rifaqat, and Sajid Kazmi
2007 *Harnessing and guiding social capital for rural development*. New York: Palgrave Macmillan.
2008 Social capital and economic development. In *The handbook of social capital*, edited by Dario Castiglione, Jan W. van Deth, and Gugliemo Wolleb, 438–466. Oxford: Oxford University Press.

Nelson, Nanette, Brian Harnisch, and Bistra Anatchkova
2010 Public opinion in Wyoming about in situ uranium recovery. Statewide Survey, 2010. WYSAC Technical Report No. SRC-1004, Revised October 2010. Laramie, WY: WYSAC (Wyoming Survey and Analysis Center), University of Wyoming.

Northern Laramie Range Alliance (NLRA). 2010. http://www.nlralliance.org/.

Pelling, Mark
2003 Social capital and institutional adaptation to climate change. Rapid Climate Change Project Working Paper 2. http://www.rcc.rures.net/resources.html (accessed April 6, 2010).

Sabatini, Fabio
2007 The empirics of social capital and economic development: A critical perspective. FEEM Working Paper 15.06, January 2006, Eni Enrico Mattei Foundation, Milan. http://www.socialcapitalgateway.org/NV-eng-measurement.htm (Accessed February 12, 2010).

Thompson, Jonathan
2009 Wind resistance: Will the petrocracy—and greens—keep Wyoming from realizing its windy potential? *High Country News*, December 21. http://www.hcn.org/issues/41.22/wind-resistance (accessed April 25, 2010).

Wiser, Ryan and Mark Bolinger
2009 *Wind technologies market report*. Golden, CO: National Renewable Energy Laboratory.

CHAPTER SIX

Cartel Consciousness and Horizontal Integration in Energy Industry

Arthur Mason

> Energy is not a commodity in the sense that societies can switch away from it. Modern energy operates akin to a collective tax gathered by energy companies who themselves are feudal lords.
>
> —Octavia Shadowz, cocktail waitress and fashion designer

INTRODUCTION

The culture of power surrounding large-scale energy systems over the past century can best be described as forms of collusion whose decision-making authority relies on structural positions of bureaucratic- and capitalist-led industry organization. In this chapter, I depart from this model by drawing attention to the increasing role played in energy policy decision-making by one group of experts, intermediaries (consultants) whose authority is based not on their structural position but instead on their theoretical knowledge and independent stance within the energy sector.

In the past, energy systems were highly regulated by a national political community in which expertise was embedded as part of the originary political organizational form. Wrestling civilian control of nuclear power from the military, for example, resulted in the establishment of a core set of experts embedded within U.S. congressional politics. Atomic scientists and expectations of nuclear power as too cheap to meter were present in the popular imagination. However, the transparency of expertise was not autonomous from government nor did experts view themselves as independent of any sector of the industry. This is the case even after the 1970s, when expansion in the scope of conflict and interested publics led to bureaucratic fragmentation and reorganization of nuclear power.

In fact, one need only draw attention to popular catchphrases of collusion and government capture throughout the twentieth century to realize that prior to restructuring of energy markets in the 1980s, the culture of power and political decision-making was based upon structural position in industrial organization. The notion of *iron triangles* or *subgovernments*, for example, draws attention to the closed-circle partnerships of industry leaders, congressional members, and technocratic elites involved in promoting nuclear power from the postwar years to the 1970s (Temples 1980). *Managerial consensus* reflects the backroom arrangements of public utility officials and industry leaders that results in expansion of electricity transmission from the Depression era to the restructuring of the 1980s (Hirsh 2001). *Natural monopoly* and *negotiated settlements* refer to growth of the natural gas industry to, in the case of the former, a government selection process, and in the latter, pre-agreements that forestall litigation among pipeline builders, natural gas producers, and distributors (Doucet and Littlechild 2006; Tussing and Tippee 1995). The government-sponsored *project*, as in the Manhattan Project that exemplifies an alliance of military and managerial expertise, was not limited to the advent of the nuclear era but inclusive of other federally sanctioned megaprojects (Rochlin 1994). *Interest group* may be included here, especially the forms of claims-making across civil and governmental spheres to remediate environmental insult (Tugwell 1980; Wapner 1995). All such phrases call attention to a crucial feature of twentieth-century styles of collusion: the forces that influence and indeed authorize political and economic arrangements are based on decision-making authority in which possessors of theoretical knowledge are the dominated faction of the dominating group.

Curiously, the most pervasive arrangement of collusion in which the dissembedding of expertise becomes transparent is the *cartel*. A cartel refers to a group of sellers whose intent is to fix prices and production outputs in concert to maximize wealth, usually by strategy of trial and error. The cartel arrangement is associated with oligopolistic industries in which the presence of few sellers facilitates coordination. Oligopoly means few sellers in the marketplace, often with strategic interaction among rival firms. While each firm may independently decide its strategy, its actions anticipate the reaction of rival firms.

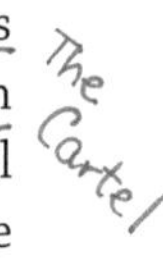

Among students of cartel theory, anticipation and reaction represents a "consciousness of interdependence" (Dibadj 2010:595). That is, even without intent to agree on specific conditions, oligopolies are marked by coordinated conduct across industries where prices are suspiciously similar or change in rapidly parallel ways (gasoline, airline tickets, cell phone rates, credit-card fees, movie tickets). This coordinated conduct has given rise to the phrase *conscious parallelism*, to describe a tacitly collusive conduct in which firms engage in parallel behavior in order to gain collusive profits but where a cartel is not set up explicitly. The absence of explicit agreement is consequential in antitrust law, where the cartel fulfills a "contract," "combination," or "conspiracy" requirement (section 1 of the Sherman Act). In the legal profession, conscious parallelism is restricted to "probable reactions of competitors" in setting their prices (Turner 1962). "Although it is hard to find a precise definition," conscious parallelism refers to "tacit collusion in which each firm in an oligopoly *realizes* that it is within the interests of the entire group of firms to maintain a high price or to avoid vigorous price competition, and the firms act in accordance with this *realization*" (Hylton 2003:73, emphases added).

In this chapter, I highlight the role of independent experts in energy policy decision-making by focusing on the forms they employ for *realizing* interdependence among energy companies. I draw attention to representational strategies used by consultants (workshops, commodified forms of knowledge, expert advice) for translating information into knowledge that becomes the collective property of energy industry elites. I argue that the advisory services of firms such as Wood Mackenzie, Cambridge Energy, and others structure the location and content of high-level conversations within the newly privatized and globalized energy markets. Through mastery of skill gained through experience, competency in education and employment, the discrimination they perform as it pertains to judgment between knowledge claims based on one's involvement in certain social networks, consultants reduce the complexity of facts into the kinds of simplicity that can form the basis of decision-making. In so doing, experts disentangle themselves from political and economic rights in techno-economic decision-making by addressing, at least on the surface of things, reasons for adopting their advice in virtue of the things that they do and know rather than as members of institutions.

To characterize their role, I begin by outlining the ascendency of energy consultants and then identify media representations, such as brochures and advertisements produced by consultant firms, through which clients become witness to a detailed interplay of images about global modernity. These images establish a relation between consultants' intended audience (energy executives), those defined as outside this audience (energy consumers), and the future. Because of its resonance with risk, the future is open to contestation. Contestation is the norm in the energy decision-making arena where a cartel alliance and cartel-like consciousness are reconstructed continuously. The temporary stabilization of an alliance relies upon a sustained perception of the credibility of a given future. To be sustained, it is incumbent to replicate an image of the future that is both believable and authoritative.

Three brief examples suffice. Through the use of graphics, consulting firms portray energy forecasts completely bereft of varying energy demand, thereby simplifying the future as an increased trend in demand, providing an absence of detection to changes in sales volumes. Designs of the market without demand fluctuation lessen the requirement for individual entrepreneurial actions, and thereby strengthen a cartel-like consciousness. Temporalizing energy demand twenty or thirty years into the future coordinates a diversity of rivals by placing their horizon of expectation on the one and same plane (Koselleck 2004). Furthermore, visualizing the oil and gas industry in the form of a high frequency of sales where future demand for product (both high and low values) does not fluctuate secures an image of regular profits so that firms become more likely to have interests in a cartel-like structure.[1] As cartel theorists argue, the high-volume sales of products with low value creates long-term gains from acting in concert. By contrast, breaks in cartels occur often in cases where firms can benefit greatly in the short term from a small number of high-value contracts (Fog 1956). Finally, the greater the variation in firms, the more probable that each will individually pursue aggressive and independent pricing strategies. Thus, even the nomenclature used in analyses is crucial for providing an appearance of limits, as when reports collectivize the oil and gas industry into two groups on the basis of national oil companies (NOCs) or independent oil companies (IOCs). A reduced variance in firms is a factor in the success of cartel arrangements because it allows the perception of members' behaviors as capable of being monitored by other members.

That the future plays such an important role in the energy industry can be related more broadly to the logic of risk society in which, unlike premodern times where risk was typically personal in nature, today's society is continuously exposed to risk stemming from complex sociotechnological systems. Further, modern society is so preoccupied with responding to the consequences of technology that a reflexive process similar to that utilized by consultants has become the process of modernization itself. In this way, future-oriented reflexivity is a scientized tool to further separate ourselves from scarcity society. Executives become witness to rising trend lines to anticipate future energy productivity, without really understanding what anchors these trend lines. Industry practioners embrace the idea of energy futures, divorced as they may be from events and ideas of the past, and weave new imaginaries of how to get there. The energy industry requires this new vision, this new identity, to prevent decisions from succumbing to the anxieties produced by a truly uncertain future. Thus, consultants totalize and subsume all these uncertainties into a ken, a scope, a collective identity to stave off self-dissolution into paradox.

EMERGENCE OF FIRMS

Energy consulting firms exercise powerful, albeit complex, forms of influence on energy markets. Their ability to process information, promote conformity of analysis, and build communication networks provides a mediating role between calculation and judgment. Such firms emerged in North America and Western Europe in the mid-1980s, during a period of market restructuring. Their initial duties included collecting, analyzing, and distributing information of relevance to buyers and sellers, including information about weather, future prices, fuel switching, demand patterns, and more. By the beginning of the twenty-first century a more elaborate system of advisory service has emerged, in which consultants rank future energy projects through combining technical prediction with new modes of communicative exchange, making available what might be described as a "community of interpretation" on a commodified basis (Mason 2007:374). That is, through soliciting the opinions of a broad sector of industry, analysts begin to act as organizers of community knowledge for executives and government leaders about the future of energy systems and the viability of particular projects within these systems. Such knowledge begins to form the basis of decision-making strategies and generates profits through client fees for access.

By enabling systematic and commodified access to community interpretations, analysts today provide the grounds for more formalized assessments of energy development projects. They have organizational significance in the way government and industry leaders stabilize future perspectives. Specialist industry analyst organizations such as Cambridge Energy Research Associates have taken center stage in global market forecasting (Banerjee 2002). The growth of Cambridge Energy is no doubt a response to deep uncertainties surrounding the future of supply-and-demand interactions, but at the same time the founding of this organization also provides an opportunity created by and for experts to enhance their own expansion and prestige (Brooks 2002:146).

Calculating future risk in energy markets is an open-ended, future-oriented project, the goal of which is to anticipate all loci of uncertainty while increasing the chance of economic success. The open-endedness of risk assessment is especially the case since the 1980s, when market restructuring resulted in the adoption of institutions by the financial industry so that prices could be based on competition rather than regulation. But the industry's competitive structure has raised problems for an older market segment of energy producers who seek to develop new sources of supply. By renouncing control over energy price, government dismantled a structured risk environment surrounding the high-stakes, high-costs uncertainty of investing in large energy systems. As such, market risk has become critically privatized and it is increasingly difficult to synchronize the long-term horizon of energy production with the long-term stability of primary markets because of uncertainties around climate and energy policies and increased competition, among other issues.

Tackling these uncertainties is generating interest among industry and government leaders in a market for information that can create perspectives that are fundamental for social coordination surrounding issues of risk. For example, through social technologies—scenario planning, executive roundtable meetings, and Internet-based analyses—consulting firms translate the uncertainties of a variety of stakeholders into their own network. By absorbing the fragmented understandings of their clients, consulting firms can provide them with an objectivized view of how the industry operates, including the risks (Mason 2006, 2008). In short, intermediaries are increasingly important actors in forming post-restructuring (neoliberal) energy regimes. The sets of connections they produce—from communication networks linked to complex financial instruments—disrupt an industrial-based form of production with its emphasis on relations between a national regulatory regime and sovereign bordered economy (LiPuma and Lee 2004).

POWER SUMMIT

Histories of the international oil industry provide examples of collusion on price fixing among companies or by governments (Sampson 1975; Yergin 1991). One example of an oil cartel structure is the As-Is Agreement of 1928, in which leaders of Exxon, Royal Dutch/Shell, and British Petroleum met at Achnacarry Castle, Scotland, to devise a collective strategy to defend their companies' profitability from problems of overproduction and low oil prices. According to news reports at the time, executives of three of the so-called Seven Sisters, an oligopoly of global oil producers that lasted through the 1960s, gathered at an aristocratic site to hunt deer, shoot grouse, hike across the moors, and entertain each other over cocktails. More recent narratives refer to the occasion as a holiday entourage including secretaries and advisors housed in a secured cottage several miles away (Sampson 1975; Yergin 1992). Nevertheless, their activities included a planned limit to commercial rivalry through control that was outlined in a seventeen-page document called the Achnacarry As-Is Agreement. The central features of the As-Is Agreement allocated each company a quota in various markets and identified efficiencies by driving down production costs, agreeing to share facilities, as well as exercising caution in building new refineries and developing new supply.

Thus, the As-Is Agreement specified the details for integration of a system of international oil production and marketing. Profits were secured horizontally across the industry by controlling efficiencies of collaboration, marketing, and associations. In these details, the built environment fell under a new form of social organization. This new form of corporate collusion implied descriptions, negotiations, and affirmations of meaning (concepts and consciousness of space) and was rendered legible through a distinct form of communication (seventeen-page document). This historical example of the establishment and elaboration of a cartel brings to light the significance of horizontal integration in providing a mode of regulation and control, and producing the effect of consciousness.

One version of the As-Is story is related within the pages of *The Prize* (1991:263–269), a best-selling book about the oil industry by Daniel Yergin, cofounder of Cambridge Energy Research Associates. With a dozen offices worldwide, Cambridge Energy is a leading consulting firm in providing anticipatory knowledge to retainer clients. The renown of Cambridge Energy can be traced to changes brought about in risk assessment and operational models by the restructuring of the energy industry. This development constitutes a space of uncertainty and an economic niche for Cambridge Energy whose products for redressing the uncertainties of clients range from graphics depicting the future to glossy advertisements for round tables.[2]

In the late 1990s and early 2000s, Cambridge Energy established the Global Power Summit, inviting industry leaders to take part in "a private, neutral, club-style setting" to explore issues facing the international energy community (GPF n.d.). The venue offered formal and informal exchange in a focused but relaxed atmosphere. All details—ground transportation, meals, Internet access, language translations, entertainment—were arranged by Cambridge Energy. Attendees included executives of major oil companies. One venue of the biannual Power Summit was the Westin Turnberry Resort in Ayrshire, Scotland, a luxury accomodation several hours' drive from Achnacarry Castle mentioned above. Like Achnacarry, the Westin Turnberry offers hunting, fishing, hiking, and seclusion in an aristocratic Scottish countryside (GPD 2002). The high cost of attendance—$15,000 for three days—ensures that participants are elite members of their organization.

In a promotional brochure titled "Global Power Development: Competitive Realignment in Changing Times," a photograph of the Turnberry Resort shows a stark white manorial building facing the sea. Its stately appearance set against a manicured countryside reflects the historical aura of the 1928 Achnacarry Castle meeting as described in *The Prize*, that of a decaying aristocratic life between the two world wars as popularized in such novels as Evelyn Waugh's *Brideshead Revisited*.

I suggest that the Power Summit venue is a technique for providing "horizontal comradeship" (Anderson 1992:15) among disparate leaders who seek to internalize an image of energy industry not from their own hierarchically differentiated organizations but instead by participating in a context that can best be described as an exclusive club. As such, the Power Summit serves as an integral constant of the twenty-first century in that industry leaders remain enlightened and aggravated by the collective memory of historical associations, which repeatedly are strengthened by new systems of communication (Luhmann 1998). Cambridge Energy's Global Power Summit provides a discursively shaped historical force. Its form takes shape not only by the way historical facts of the Achnacarry As-Is Agreement have become managed by Daniel Yergin but also by the manner in which Cambridge Energy replicates the event—the Achnacarry meeting—as an executive round table oriented toward managing concepts of risk. In this way, today's special character of interrelated values of oil, natural gas, and coal is reaching a complexity analogous only to an earlier period of mercantilist speculation over coinage and the interrelated values of gold, silver, and copper (Foucault 1970). The restructuring of energy markets has intensified the need not so much for empirical research on ways to move forward but rather, borrowing a phrase from Jürgen Habermas, for attempts to establish a "rational infrastructure of action oriented toward understanding" (1985:106).

But historical associations of the oil industry as "a very exclusive club" have also been attributed to Pérez Alfonzo in reference to his conception of the sovereign cartel Organization of Petroleum Exporting Countries, or OPEC (Sampson 1975:4). Often described as "founder" and "architect" of OPEC (Coronil 1997:353), Alfonzo served as Venezuela's Minister of Mines and Hydrocarbons in the 1950s. He is noted to have spent considerable time studying the papers of the Texas Railroad Commission (TRRC), which he credits for the original sovereign oil cartel design. During the 1920s, in an effort to bring stability to a market laden with oversupply, the TRRC perfected a cartel-like structure of "prorationing" (to each according to his prior ability), raising it to an art of "regulatory form," capable of being transplanted into different sociocultural milieus (Prindle 1981:34).

IMAGE OF GLOBAL MODERNITY

In the years following 2001, Cambridge Energy acknowledges a series of "key events"—Enron bankruptcy, terrorist attacks, volatility in gas and power markets—for which they began providing clients with interpretations about onsets of danger (CES 2002). In 2002, "The New Face of Risk" was the topic of Cambridge Energy Week, an executive conference that the *New York Times* referred to as a gathering where "leaders of the world's largest energy companies go to think big thoughts" (Banerjee 2002).

At the conference, a game of chess symbolized "The New Face of Risk." In particular, a figure of the king appeared on program guides, wall posters, Internet key cards, security reminders, and conference hall backdrops. Chess is a game requiring reflection with neither dice nor a stake—that is, outcomes are not governed by rules associated with games of chance. Early modern descriptions note that several figures, including the queen and bishop, are types of advisors or administrators to the king. As to strategy, the straight moves of the king are associated with his legal power in collecting rents while the oblique moves with extortion (Yalom 2004).

Closer inspection of Cambridge Energy's ubiquitous chess symbol connects the king to an image of the earth as seen from the perspective of space. This apparition is displayed most notably on an Internet key card repeatedly used by conference participants at computer kiosks. On one side of the key card a glass earth rested in the hands of an Asian child, suggesting that strategic planning belongs to a future generation of leaders, presumably those living in Asia where strong economic growth reflects increased energy capture, particularly in China. Here inscribed into the features of a living being was an interplay of signs that relate future worth, productive capacity, and investment opportunity. But the scenario conveyed one of several possibilities. Another suggested that perhaps *this* child was "the new face of risk." What was on display here is a symbol of population growth in Asia, increasing demand for higher living standards, greater requirements of energy capture, and anthropogenic climate change.

The image of a transparent earth is itself significant. It marks the hidden presence of fossil fuel (coal, oil, natural gas) and serves as a collective reminder that its buried potential is a visible signature of wealth in the world. The image is necessarily a metaphor for accounting practice in which corporate valuation is measured against a reserve's buried potential and its economic recovery. Finally, its significance also finds meaning by its inversion as an ironic reminder of a lack of transparency surrounding individual proprietary knowledge among companies.

A spherical glass earth in order to peer into what is hidden calls attention to the fortune teller and the crystal ball, both of which are prominent contemporary icons. Such images during this period are employed in the advertisements of global communications companies such as Comcast, representing the company's ability to identify new systems of communication, global insight, and anticipatory knowledge. The image serves as a reminder that the ability to look ahead to anticipate whatever lies in the future is a desire that dates back as far as Mesopotamia. In the past, the practice of divination—appeals to a deity who is believed to reply through significant tokens—served as a mode of inquiry into future events or matters obscure. Today, the imperative to know the future is no less important than it was in premodern times. Unlike the diviners of yore, obtaining knowledge of secret or future things by mechanical means and manipulative technique no longer depends on the aid of spirits or deities. Differentiating knowledge systems, the rise of nonhuman forces of regulation, cybernetic systems, and probability calculations stand in for and procure the aura of the superhuman powers they replace.

In the context of these images, distance from the world relates confidence and perspective of the future. The child on the Internet key card, for example, exists outside the earth. Outer spatial distance is a style of thinking that is neither unique nor the invention of Cambridge Energy. In 2004, the *New York Times* captured an image of Philip Watts, executive of Royal Dutch/Shell, arriving on stage in a spaceship and an astronaut suit and declaring to six hundred executives, "I have seen the future and it was great" (Labaton, Gerth, and Timmons 2004). These remarks functioned as a rejoinder to accusations that Shell was pumping oil out of the ground faster than it could find new supplies. Watts sought to regain control over corporate valuation not by means of traditional accounting practice but by positioning the uncertainties of the future in the past, behind him. His language and dress proclaimed he had already seen what reserves lie in the future by visiting the future, through his space ship.

Similar images used by Wood Mackenzie aim to draw the energy future into the present through a global economic model (GEM). A crystal sphere image symbolizes this software program that appropriately is named after its acronym GEM. In a related promotional brochure, an anonymous, omnipotent hand supports an image of the world.

CONCLUSION

The most prominent discourses of energy capture from the previous century—society's dependence on laws of thermodynamics (Soddy 1920; Odum 1971), civilization's level of achievement based on rates of energy capture (Smil 2000), democracy's undoing by increased energy use (White 1949)—are strangely absent in the sentiments by which today's energy executives strive to realize the value of their commodity. The transitory existence of these earlier discourses is a reminder that in the established order of things, destined oblivion is immanent (Foucault 1971:8). It is also a reminder that the unabashed economic motivation behind the rise of intermediary knowledge reflects a postwar expansion of expert systems as part of a broader movement to a knowledge economy. The growth of this type of economy itself provides justification of an apparent contradiction, on the one hand, of increased democratization of expertise and, on the other, its privatization (Mason 2005).

Consulting firms, buoyed by venture capital, operate like transnational entities in which their power relies on the strength of their networks. Consequently, emphasis in energy development increasingly is placed on global financial markets, instead of structural positions within national political systems. For the elative isolation and elitism of these deciders who think big thoughts, squirreled away in jaw-droppingly expensive conferences, located in elite resorts, the performativity of knowledge creation suggests knowledge artifacts seem to materialize out of thin air. The use of images of strategy and transparency to ensure control over information is complete, suggesting knowledge is occluded and manipulable by the companies themselves. Thus, cartel consciousness is the reproduction of oligopoly through horizontal integration, a type of "clubbiness" that is strategically beneficial to participants and impenetrable to nonparticipants, who remain vulnerable and at risk in the new world of energy insecurity.

NOTES

1. For example, the average operating expenditures (OPEX) for costs of production ($/bbl) among top producers across major territories (where OPEX is predominantly lifting and transport) is presented in analyses as $6 (2010 prices), but nearly $11 excluding OPEC producers. Therefore, prices set by OPEC have a stabilizing effect for non-OPEC producers (Deutsche Bank 2010:80).

2. Yergin's discourse on market transition is widely recognized (see Wilson 1987:141; Yergin and Hillenbrand 1982).

REFERENCES

Anderson, Benedict
1992 *Imagined communities*. New York: Verso Press.

Banerjee, Neela
2002 Energy industry gauges the Enron damage. *New York Times*, February 18.

Brooks, Andrew
2012 Radiating knowledge—The public anthropology of nuclear energy. *American Anthropologist* 114(1): 137–140.

CES
2002 *New realities, new risks: North American power and gas through 2020*. Client report. Cambridge Energy Research Associates.

Coronil, Fernando
1997 *The magical state: Nature, money, and modernity in Venezuela*. Chicago: Chicago University Press.

Deutsche Bank
2010 *Oil and gas guide*. London.

Dibadj, Reza
2010 Conscious parallelism revisited. University of San Francisco Law Research Paper No. 2011-17.

Doucet, Joseph and Stephen Littlechild
2006 Negotiated settlements and the national energy board in Canada. Research paper, School of Business, University of Alberta.

Fog, Bjarke
1956 How are cartel prices determined? *The Journal of Industrial Economics* 5(1): 16–23.

Foucault, Michel
1970 *The order of things*. New York: Vintage Books.
1971 Orders of discourse. *Social Science Information* 10(2): 7–30.

Global Power Development (GPD)
2002 Global power development: Competitive realignment in changing times. Promotional material. Cambridge Energy Research Associates.

Global Power Forum (GPF)
n.d. Global power forum. Cambridge Energy Research Associates. http://www.cera.com/aspx/cda/client/knowledgeArea/serviceDescription.aspx?KID=52#17759 (accessed October 19, 2008).

Habermas, Jürgen
1985 *The theory of communicative action*. New York: Beacon Press.

Hirsh, Richard
2001 *Power loss: The origins of deregulation and restructuring in the American electric utility system*. Cambridge, MA: Massachusetts Institute of Technology Press.

Hylton, Keith
2003 *Antitrust law: Economic theory and common law evolution*. Cambridge: Cambridge University Press.

Koselleck, Reinhart
2004 *Futures past*. New York: Columbia University Press.

Labaton, Stephen, Jeff Gerth, and Heather Timmons
2004 At Shell, new accounting and Rosier oil outlook. *New York Times*, March 12.

LiPuma, Edward and Benjamin Lee
2004 *Financial derivatives and the globalization of risk*. Durham, NC: Duke University Press.

Luhmann, Niklas
1998 *Observations on modernity*. Stanford: Stanford University Press.

Mason, Arthur
2005 The condition of market formation on Alaska's natural gas frontier. *European Journal of Anthropology* 46: 54–67.
2006 Images of the energy future. *Environmental Research Letters* 1(1): 20–25.
2007 The rise of consultant forecasting in liberalized natural gas markets. *Public Culture* 19(2): 367–379.
2008 Neglected structures of governance in U.S.-Canadian cross-border relations. *American Review of Canadian Studies* 38: 212–222.

Nye, David E.
1990 *Electrifying America: Social meanings of a new technology, 1880–1940*. Cambridge, MA: Massachusetts Institute of Technology Press.

Odum, Howard T.
1971 *Environment, power and society*. New York: Wiley.

Prindle, David F.
1981 *Petroleum politics and the Texas railroad commission*. Austin: University of Texas Press.

Rochlin, Gene
1994 Broken plowshare. In *Changing Large Technical Systems,* edited by Jane Summerton, 231–264. Boulder, CO: Westview.

Sampson, Anthony
1975 *The seven sisters*. New York: Bantam Press.

Smil, Vaclav
2000 Energy in the twentieth century: Resources, conversions, costs, uses, and consequences. *Annual Review of Energy and the Environment* 25: 21–51.

Soddy, Fredrick
1920 *Science and life*. Whitefish, MT: Kessinger Publishing.

Temples, James R.
1980 The politics of nuclear power. *Political Science Quarterly* 95(2): 239–260.

Tugwell, Franklin
1980 Energy and political economy. *Comparative Politics* 13: 103–118.

Turner, Donald F.
1962 The definition of agreement under the Sherman Act: Conscious parallelism and refusals to deal. *Harvard Law Review* 75: 655–671.

Tussing, Arlon R., and Bob Tippee
1995 *The natural gas industry: Evolution, structure, and economics*. Tulsa, OK: PennWell Books.

Wapner, Paul
1995 Politics beyond the state. *World Politics* 47(3): 311–340.

White, Leslie
1949 *The science of culture: A study of man and civilization*. New York: Farrar, Straus and Giroux.

Wilson, Ernest J.
1987 World politics and international energy markets. *International Organization* 41(1): 125–149.

Yalom, Marilyn
2004 *Birth of the chess queen: A history*. New York: HarperCollins Publishers.

Yergin, Daniel, and Martin Hillenbrand
1982 *Global insecurity*. Boston: Houghton Mifflin.

CONVERSATION 2

Technology, Meaning, Cosmology

Chelsea Chapman, Arthur Mason,
Devon Reeser, Stephanie Rupp, Sarah Strauss

The chapters that constitute the second section of *Cultures of Energy* share several orienting themes concerning technologies, meanings, and cosmologies of energy. Authors Stephanie Rupp, Chelsea Chapman, Sarah Strauss, Devon Reeser, and Arthur Mason address energy issues as they intersect with notions of power in the senses of both agency and control. Across the diverse contexts presented in these chapters, people's perceptions of energy are entangled with views of their rights as individuals to electricity, to land and landscapes, to a way of life, and to expert knowledge. The value(s) and cost(s) of energy are not evenly distributed within communities or across a larger society or nation; the formal, economic value of energy resources themselves shift, and so too do cultural, substantive values of energy change over place and time. Energy seems to be perceived as an invisible substance that nevertheless exerts a powerful force. While energy seems to be the invisible, taken-for-granted lubricant of contemporary Euro-American society, energy is suddenly visible and valuable in its scarcity and impending—or actual—absence. In the following discussion, authors in this section offered wide-ranging responses to the following questions (and to each others' discussions): What kind of rationality do people engage when they talk about energy? Is energy something that is absolute in its measure, quality, or quantity? Or is energy something relative—and if relative, then relative to what? How is energy valued? If there are multiple different groups of "energy users" or "energy actors" in your field of research, can you describe the different kinds of paradigms of energy that are in play across the spectrum of users?

DEVON REESER: In my research it became apparent that people have an easier time forming a "rational" connection to physical, tangible energy resources. The ideas of oil and gas are more "rational" than wind, for example, because they are more visible. Burning oil and gas produces fire; we connect with that kind of energy. As proven energy sources, material energy resources seem more "rational" than alternative energy. Wyoming residents are wary of wind energy because they doubt that wind can be a tangible, reliable, "rational" source of power, and its development would break the tradition of energy use that is ingrained in the state's culture. A grassroots political action group protesting wind energy development uses language such as "industrialization of the mountains," indicating a cultural sentiment that using wind—something valued as clean and pure—for energy is irrational and even toxic. The inability to control the wind itself makes it a less "rational" energy source than the traditional physical sources of energy that are extracted and sold by the barrel for human consumption.

Perhaps the "rationality" of energy has deeper connections to a sense of security in cultural traditions of energy. For example, Paraguay shares the largest hydroelectric dam in the world with Brazil; nearly all electricity is generated at this plant and another dam shared with Argentina. In rural Paraguay, where I have also been conducting research, while most locales have access to electrical power the people have no control over its reliability. Frequent outages from infrastructural and political issues drive people to rely on alternative energy sources, mainly gas-powered stoves or forest resources. Most households in the town where I live and conduct my research rely on three different sources of fuel for cooking: gas, wood/charcoal, and electric. People cannot rely on a constant provision of electricity, and gas supply to the country is often cut in the winter months due to increased usage of gas for heating in richer neighboring countries. People rely on wood and charcoal because they can harvest those sources of energy locally. Hence, stoves that use wood and charcoal have a higher value because of their reliability. People need energy, and traditional, proven sources of energy that they can control and quantify, and resources that have an immediate connection to direct energy output (resources that they can burn) are more "rational" to people in rural Paraguay.

SARAH STRAUSS: Part of the problem, at least in terms of research on energy in my experiences in the U.S. and Switzerland, is that people *don't* talk about energy just as they don't talk about sewers or other parts of our domestic infrastructure. It is a given that when you turn on the light switch, the darkness evaporates. When there is an absence of energy flowing through the system, as in a blackout, then that is of course noted with annoyance. But otherwise, I have found that many of my conversations about energy—even with people who are otherwise quite environmentally aware—end up quite quickly with a sense of both puzzlement and intrigue, as people respond, "I never thought about it [energy] before." The absence of the vocabulary of energy is striking. Even when speaking directly about energy production—wind versus coal, for example—the conversation is usually about the impact of the physical structures or substance (towers, blades, coal dust/particulate air pollution) and not about the output, the energy itself that is produced by the infrastructure. Even in my Swiss field site of Leukerbad, where a direct link is visible between the flipping of a light switch and the flow of a stream to generate hydropower, people did not seem to make note of or be concerned about their relationship to energy resources.

CHELSEA CHAPMAN: I agree with you, and also want to add that energy might be more or less visible in its absence depending on one's socioeconomic position. The energy poverty I encounter here in Alaska reminds me of teaching at public schools in Berkeley and Oakland and hearing about the hardships of unpaid electric bills that caused blackouts and shortages at the household level, alongside the regional rolling brownouts happening statewide at the same time. Not being able to make breakfast before school because "power's out" for some kids, but not all, got me thinking about how absence and visibility are uneven, even in high-consumption societies like the United States.

I'll share a few ways in which energy is valued in its presence or absence here in Alaska. Energy prices are among the highest in the country, and people consume the most hydrocarbon fuels per capita nationwide (three times more than residents of Wyoming, the next biggest consumer). So energy is made visible by its (socially and economically produced) scarcity. Also, many people in and near Fairbanks—where I am based—live in off-grid homes or quasi-off-grid cabins like mine. I haul water in five-gallon jugs once per week and receive a (very expensive) tank load of kerosene heating oil once every winter. I don't experience energy poverty, but I guess, to state the obvious, things we consume are more visible because we have to carry them.

But the real energy poor tend to be rural indigenous residents living in villages of a few dozen to a few hundred mostly Alaska Native people, which are accessible by air or water but are off the road network. Here, cost is part but not all of energy's visibility. Rural fuel prices are really, really high: eight dollars per gallon of heating oil, and up to ten dollars per gallon of gasoline in some communities in the wintertime. Fuel prices go up when it's more expensive to transport; these seasons coincide with the coldest winter months, when the most fuel is consumed and it's the hardest to move around. The weather renders energy movement more visible and more tenuous. Rural residents often bring up the additional costs of energy (burned elsewhere but sourced in Alaska) on their local environments; as the media and academic lens has focused so closely on Arctic climate change, a lot of people are acutely aware not only of local climate impacts (forest fires, eroding riverbanks, melting permafrost) but also of global consumption arrangements in which energy comes from their land and resources but fuels lifestyles and communities in other parts of the nation. And they don't hesitate to add the unquantified, "real" cost of energy as it directly impoverishes their lands and subsistence food sources.

I wonder too if, alongside the burden of energy poverty, the coinciding (moral and economic) obligations to conserve aren't also unequally distributed. In some communities in rural Alaska, NGOs and state agencies have installed meters inside houses so that residents can watch their consumption go up or down, with the idea that this quantification will make energy more visible and promote conservation. This effort is certainly worthy, but is it unfair to impose this kind of energy self-surveillance on such a small fraction of energy consumers, while people in Fairbanks or Anchorage aren't compelled to do the same? Here in town, a contingent of pro-development folks proudly leave their trucks idling for hours while they work or shop, as an act of defiance against "the eco-police" and the ongoing "theft" of Alaska's oil resources by the EPA and the Obama administration. Their somewhat theatrical use of energy to display an unnecessarily running machine in political protest makes it visible, like the meters, but perhaps in a way that's most resonant in a place where energy and its sources are constantly contested.

ARTHUR MASON: I want to jump in with a quick reminder that the intersection of visibility and rationality that we keep returning to was also of specific interest to Bronislaw Malinowski. Recall, for example, his demonstration of how ritual lends a meaningful structure to the uncertainties of open-ocean fishing (where both fish and land are nonvisible) yet is absent altogether in spaces of high visibility, such as inner-lagoon fishing (where both fish and land are visible). Malinowski used this example to illustrate that magic functions to protect the rational productive enterprise in contexts of insecurity. I have found the same use of magical thinking and magical practices to offset insecurity in my work with energy consultants. Certain kinds of visibility serve as forms of technique for advancing thought on what is rational productive activity in capitalist energy production. By *rational*, I mean that productive activity must always strive for a greater intensity of separation between what is discretionary from what is routine activity. And here I refer perhaps simply to an efficiency of thought that extends to all facets of capitalist life, whether separating managers (whose "thoughts" are qualitative) from laborers (whose rewards are quantitative and based on rates of production), to separating facts ("known knowns") from risk ("known unknowns") and uncertainty itself ("unknown unknowns").

STEPHANIE RUPP: In my research on energy in New York City, there seem to be very distinct groups of energy "actors," and these roughly disaggregated clumps of people seem to have very distinct paradigms for what energy is and how it works in their lives. Here are some examples. The city is full of people who understand and use energy in technical ways: electricians, linesmen for utility companies, people who come to install electronic devices. The city is full of people who think about energy in metaphysical ways: people who go to have their chakras balanced, people who practice yoga. People very often think of the city itself as having energy, as if the people and their creative juices vibrate together to produce some kind of collective buzz that makes the city alive, vibrant, and endlessly charging ahead. Most people don't really think about energy in the mundane, technical sense at all. New Yorkers are saturated in energy and seem to consider it right and normal that there should be lights and electronic gadgets all around them. But unless there has been a recent blackout or some other event that has suddenly made energy (and therefore individual agency) disappear, New Yorkers seem to disregard energy altogether. When there *is* a blackout, people seem to think of the institutions or corporations that control energy as somewhat sinister. ConEd, the primary utility company in New York City, seems to evoke a deep public suspicion of institutions that have too much power—the power to provide or curtail power itself, in the form of electricity, to individual, ordinary people. And then there are people who trade "energy futures" and other energy investment instruments on Wall Street: how much more abstract can you get?

But these various registers of power almost seem like nonintersecting circles on a Venn diagram. People don't necessarily think that energy in the form of electricity is the same *thing* as energy in the form of a chakra, for example, but metaphorically both forms of energy accomplish similar ends: enabling individuals to realize their potential, to achieve their goals, either immediate or long-term. Although the conceptual circles of "energy actors" and their beliefs about energy seem to be quite disparate, one of the things that brings them together is the underlying idea that energy is power as a force, as an ability, and involving (promoting or preventing) dominion or control over something.

CHELSEA CHAPMAN: There are a few kinds of "energy actors" that surface in energy conflict here in interior Alaska. Energy workers at the local refineries, power plants, and support industries for Prudhoe Bay and the Usibelli coal mine, whose livelihoods depend on production of power from fossil resources, tend to convey an idea of latent underground treasures alongside their technical energy expertise. Renewable energy developers tend to share this paradigm of latent energy, albeit in their vision energy is "stranded" in different substances. Policy makers depend on the production of (pipe)dreams about energy, including natural gas pipelines, expanded development, fracking proposals, socially/environmentally responsible development, energy access, energy inequality, and developing stranded renewable energy re-sources. Neo-Pentecostal Christians depend on a spiritual energy rising from Alaska to compel a nationwide revival movement resulting in theocratic takeover of government, civil society, and economics. This paradigm is often shared by the policy makers and often shows up in talk about "God-given" physical energy resources divinely placed in Alaska to support an emergent Christian dominion. People—usually but not always indigenous—who live near energy-producing landscapes depend simultaneously on fossil energy (and the accompanying environmental contradictions) to live, hunt, and travel. At the same time, they also depend on the energy they draw from eating wild meat and plants, which they see as being endangered by fossil production. Finally, Alaska Native elders and others speak of a personal and social energy (*tyea* in Lower Tanana Athabascan) that circulates via correct and balanced relationships among humans, animals, and land.

Thinking of Stephanie's nonintersecting Venn diagram, perhaps the overlap for all these paradigms (beyond the nomenclature of energy) might be their relationship to place and identity. The various individuals involved in these groups have particular ways of imagining Alaska—or specific cultural landscapes within it, or even geological formations below ground—that are connected to ways of imagining themselves and that have political implications. Not only are many of Alaska's energy regulators and legislators involved in neo-Pentecostal churches that proselytize energy resource development and consumption at any cost, but the broader rhetoric of stranded energy (and the obligation to liberate and harness it) continues to justify the corruption, social inequality, and environmental damage in the Alaskan fossil energy scene, as it has since the pipeline days in the 1970s. I'm interested in thinking more about metaphors of energy as a force or as a form of control—something to harness—versus energy as something that can't be controlled or harnessed but for which people themselves are conduits and transmitters. Without overstating the relation of energy to capitalism, it seems relevant that Western societies tend to naturalize their control over ever-grander sources of energy (coal, oil, atomic) as an aspect of dominion over nature. But other paradigms exist that emphasize relationality and, while people who are concerned with *tyea* wouldn't say that it is the same thing as coal energy or crude oil energy any more than a New Yorker might claim electricity and chakras to be the same, they do point out that both forms physically come from the land, just via different ways of relating to it.

PART 3
Electrification and Transformation

CHAPTER SEVEN

Electrifying Transitions: Power and Culture in Rural Cajamarca, Peru

Thomas Love and Anna Garwood[1]

Declining fossil fuels and increasing volatility in energy provisioning promise accelerating change in the configuration of energyscapes for people around the world. In this chapter, we examine a different type of change—one not of looming shortages but of peasants' cultural understandings, socioeconomic condition, and the energo-politics in which they become enmeshed as they acquire electricity for the *first time*. In the process, rural identities are reconstituted in ways that highlight locals' agency even as they remain off grid and energy-poor.

Electricity is arguably the central symbol of globalizing modernity and its associated urban consumption styles. Lacking it, rural dwellers are placed in situations of both material and symbolic poverty. A resulting maze of binary tropes—urban/rural, rich/poor, modern/traditional, progress/backwardness, light/dark—powerfully yet largely unconsciously frames their and our understanding of rural life. Rural electrification challenges and undermines these strained dichotomies, in the process raising a tangle of new and largely unexamined issues and tradeoffs regarding cultural autonomy and change, within which rural actors are central players.

Since the early 1990s, high-altitude agropastoral communities in Cajamarca, Peru, have been installing micro-hydro and, more recently, wind and solar renewable energy systems to generate electricity as part of larger NGO-driven programs of rural community development. Since installation of a two-kilowatt wind electricity turbine system, villagers of Alto Perú have added a growing mix of solar photovoltaic (PV) and pico-hydro installations to provide all households with electricity. The development of the project in different phases with different technologies exposed long-standing rivalries in the community even as it empowered new forms of community organization.

Highland Cajamarca is not just the site of locally controlled rural electrification, however. Looming large over Alto Perú and villages like it are Latin America's largest gold mines. These open pit cyanide heap leach mines threaten water supplies and rip through *páramo* grasslands and thin soils even as they provide well-paying jobs and are boons to construction, transportation, and other business sectors in the region. State-facilitated large-scale mining has rapidly expanded during the same time frame as these renewable electrification projects (Bury 2004). Mining casts a long shadow over and poses dilemmas for peasant participants in these projects, in Cajamarca and elsewhere in Peru, not only because of such tradeoffs between economic benefits and environmental and other impacts, but also because infrastructure developed by these large mines, such as paved roads and cell phone towers, can provide ancillary, if unintended benefits to villagers. The energy issues in Alto Perú mirror competing discourses on "development" at a national and global scale: large-scale "development" represented by multinationals, extraction-based economies, and electricity grid extension competes with rural development inspired by a "small is beautiful" philosophy, as represented by community-controlled decentralized renewable energy systems. The energo-politics of locally controlled versus grid-supplied electricity are complex and portend very different development paths, as they imply different loci of ownership, scale, and purposes of electricity.

BACKGROUND

Peru has enormous energy resources—world-class solar incidence, substantial hydro potential, oil, and natural gas—and over the past decade there has been a dramatic quickening in assessing and developing this potential. About 60 percent of the country's electricity is hydroelectric, with a growing 40 percent generated by natural gas–powered thermal plants (Ministerio de Energía y Minas 2010a).

Despite this overall energy wealth, an estimated 6.5 million people—24 percent of the population—still had no access to electricity as recently as 2007 (Ministerio de Energía y Minas 2007). Though the total unserved has been impressively halved to 12 percent since then, Peru still has the lowest overall electrification rate in Latin America (Ministerio de Energía y Minas 2010b). Cajamarca is one of the poorest regions in Peru (64.5 percent of the population lives below the poverty line) and has the lowest electrification rate in the country at just over 40 percent (Ferrer-Martí et al. 2010), all pointing to the region's predominantly rural character. Peru's extreme geography, dispersed rural population, and neoliberal political economic priorities tilted toward large-scale mining all pose huge challenges to extending the conventional energy grid to places like Alto Perú, leaving a large percentage of rural Peruvians still lacking electricity.

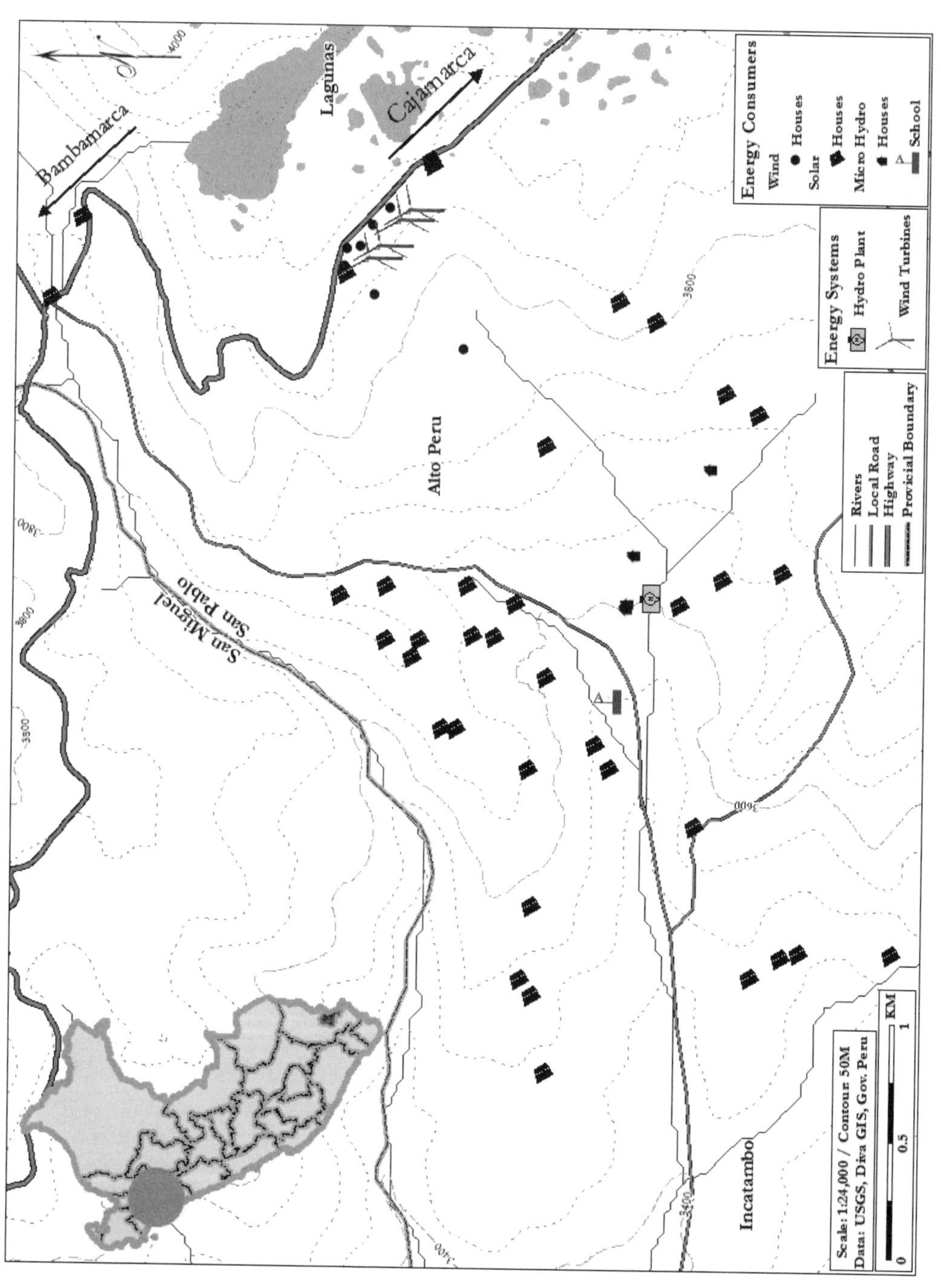

FIGURE 7.1. Map of Alto Perú (courtesy John Steed, Brandeis University)

By the 1990s, renewable energy technologies came to be seen as one way to address rural poverty by bringing the benefits of electricity to off-grid villages. Implementing them has constituted a core program of two decades of work in rural development in Cajamarca by Soluciones Prácticas (SP), the Peruvian affiliate of UK-based Practical Action (formerly Intermediate Technology Development Group), assisted in the last several years by U.S.-based Green Empowerment. As a politically left-center, technocratic, non-governmental entity committed to environmentally sound rural community development in Cajamarca (and elsewhere in Peru) for over twenty years, SP is highly regarded in the region.[2]

We use the community of Alto Perú as a case study of how implementation of a two-kilowatt wind project has affected village social processes and cultural understandings in these larger contexts.[3] Wind turbines had been installed in 2008 in Alumbre, a high-altitude hamlet above the nearby Yanacancha basin, one of the first projects in the world to install multiple small-scale wind turbines and the first community in Peru to have small-scale wind-powered electricity (Ferrer-Martí et al. 2010). The Minister of Energy and Mines personally helped inaugurate the project on July 28, 2008—national independence day—and national television as well as CNN did news stories on the groundbreaking project.[4] The project in Alto Perú was the third village wind installation but the first to integrate micro-grids of connected households and then add solar and micro-hydro in the same community.

ALTO PERÚ

Despite its tropical location, cold and wind dominate the landscape year-round at this high altitude. Annual temperatures vary from 10 to 15 degrees Celsius, with seasonably strong winds. Alto Perú is the highest altitude *caserío* (hamlet) of the *centro poblado* of Ingatambo, District of Tumbadén, Province of San Pablo in the Region of Cajamarca. Alto Perú lies on the very west side of the continental divide between 3,500 and 4,130 meters elevation, and several large lakes and many smaller ponds dot the rolling grasslands, constituting the headwaters of a major affluent of the Río Jequetepeque. The lagoons are located on land on which the nearby Yanacocha gold mine owns mineral rights. There have been long-term disputes over the ownership of the lagoons and associated water rights, escalating to national courts.

Appropriately named, Alto Perú is the recent name for the sparsely settled, high-altitude grazing lands of the ex-hacienda Ingatambo, expropriated during the 1960s agrarian reform.[5] Starting even before the reform era, such vast yet poorly capitalized ex-haciendas underwent successive fractur-

FIGURE 7.2. Alto Perú (Photo by Anna Garwood, November 20, 2009)

ing. Desiring their own elementary school, in 1995 Alto Peruanos initiated a formal process to split off from Ingatambo.

Alto Perú is about fifty kilometers (90 minutes) north of Cajamarca on the road to the Bambamarca, provincial capital of Hualgayoc. The population of 345 is organized into eighty- five households, thirty-one of which are located in the upper part rather concentrated along the road, and fifty-four of which are more dispersed in the lower part. Of these, seventy are homeowners and permanent residents, the rest being second houses or houses occupied by temporary workers. The *caserío* has an elementary school in the lower part, but the nearest medical post is in an adjacent settlement.

Nearly everyone depends on cultivating potatoes, barley, broad beans, and some *oca* and *olluco* (Andean tubers), without fertilizer or pesticides, using traditional methods and hand tools, as well as raising cows, sheep, pigs, guinea pigs, chickens, rabbits, and a horse or two for hauling *porongos* (milk jugs). The lower part is less windy, slightly warmer, and not quite as steep, with higher productivity and supporting more people.

The post–World War II emergence of dairying in the region opened up many opportunities for villagers, especially after the agrarian reform (cf. Deere 1990). The shift to a cattle economy ensued: "Obliged to participate in the money economy, with insufficient resources (and probably little desire) to retreat into autonomous subsistence agriculture, yet with each day more limited sources of income, life for these peasants has become increasingly difficult. Under these circumstances cattle have become crucial" (Gitlitz and Rojas 1983:171).

Each household owns several hectares of *ryegrass* pasture (perhaps a third of the village's total area of *jalca*, high-altitude grassland, elsewhere *páramo*), fruit of a succession of Ministry of Agriculture pasture-improvement schemes. On average, a family sells twenty liters of milk every day to either Gloria or Nestlé in Cajamarca, for about S/. 0.80 per liter, for a total monthly income of about S/.465 (about US$155). Given the harsher climate, upper households maintain fewer cows with lower milk production; one woman reporting selling only one-half *porongo* per day from a few cows. Occasional sale of cattle provides some additional income. Many households hire local young men as *mitayos* (resident laborers, usually on two-year contracts, which may be renewed many times) to manage cattle.

The community identifies an irrigation canal as the dividing line between the upper and lower parts of Alto Perú. Each part has its own *ronda campesina*, or community patrol, which walks the land each night on guard for theft, cattle rustling, or other crimes.[6] The primary rationale people offered for dividing what had been a unified *ronda* was that the whole area was simply too big for the men to cover in one night. There is some truth to the claim, as it is a vast area to cover. But underneath this were tensions between upper and lower parts stemming from a grievance over the school's solar panel. When a large company had constructed the line of high-voltage towers to take grid electricity over Alto Perú to the new Goldfields mine to the north, a payment was made to Alto Perú authorities, who decided to spend this on a solar photovoltaic panel on the elementary school to power a computer and DVD/TV for instructional purposes. There was still resentment on behalf of individual landowners for never having received adequate compensation from the power company for the installation of the huge high-voltage towers.

Echoing the earlier split of Alto Perú from Ingatambo, several factors seemed to be driving this shearing of Alto Perú itself into upper and lower parts. The upper part is increasingly commercial, tied very closely to the highway, which is experiencing a sharp increase of mine-related traffic.[7] The family that now dominates the roadside string of households clearly stated that they relocated here for commercial purposes. Led by Don Aurelio and his siblings, the fewer households in the upper part are more entrepreneurial than any others in Alto Perú, united by kinship and faith. Aurelio and his brother own the only two shops that provide basic necessities in Alto Perú, adjacent to each other along the road in the upper half. Aurelio, an amazingly entrepreneurial figure, drew on his experience on the coast and his Seventh Day Adventist faith to advocate for change and development in the area and became the de facto leader of the wind project.[8]

Despite the undercurrents of disunity, there is an array of existing community-wide organizations, including the Subcommittee of Development of San Cirilo–Alto Perú (related to Yanacocha gold mine's interest in exploiting the San Cirilo hill and lakes above the community), the *ronda feminina* (which did not split, perhaps because of its ties to the school-based *vaso de leche* program), Property Owners United (related to the road work that may have expanded onto community land), the Irrigation Committee (though with upper and lower sections), and the Association of Family Parents of the Educational Center.

Apart from a very few generators, nobody in Alto Perú had electricity before this project; in fact, only 13 percent of households in the entire Province of San Pablo have electricity (Soluciones Prácticas 2007). Accordingly, the rhythm of life is shaped by such things as day length, passing of the milk truck, children coming and going from the elementary school or a few who walk farther down valley each day to the secondary school in Ingatambo, and especially the daily rounds tending cattle. Cargo trucks from the milk plants in the Cajamarca valley labor in the wee cold hours of the morning up and down these remote slopes collecting *porongos* of milk from each cluster of households. The milk trucks also provide the only mode of public transportation below the graveled highway, so passengers and goods are often piled on top of the metal milk jugs in typical Andean open-air truck fashion.

On first visiting the area, then, one might assume from its extremely rural aspect that people in Alto Perú would consider even the small amounts of new, renewably generated electricity with wonder. Villagers are hardly strangers to electricity, however. Most Alto Peruanos go to Cajamarca to collect biweekly payments for their milk, visit family and friends, and make purchases at the market. Upwards of 33 percent of the population, mostly young men, has been emigrating seasonally for work in the mines or elsewhere, or some for schooling in Cajamarca (cf. Deere 1990). With all this and decades-long engagement with the cash economy via dairying, then, Alto Peruanos' isolation is deceptive, for they have substantial experience with both electricity and fossil-fueled transportation. Alto Peruanos' apparent isolation is additionally belied by the fact that they now have to deal with gold mines both to the north and south as well as the emerging energo-politics surrounding state investment in grid electricity for the mines—matters we take up after the next section.

To bring prosperity to Alto Perú, Aurelio and his siblings envision a variety of development projects—stocking the lakes with trout and building a tourist hotel. He takes credit for promoting many of the completed projects, such as drinking water, as well as taking the lead in working with SP to promote the wind electricity project. Aurelio and his siblings dominate the several formal organizations in the community, such as the Ronda Campesina (upper part), the irrigation committee, and the Subcommittee of Development of San Cirilo. The selection of the upper part of the *caserío* by SP for installation of wind turbines additionally made great sense on social grounds; kin and church ties as well as physical proximity among the households there facilitated installation of the new micro-grid (multihousehold) model. The road access also promised opportunities to use electricity for income generation, such as roadside stores or restaurants, and residents of the upper part of Alto Perú seemed more connected to urban life, regularly going back and forth to Cajamarca. In short, they had more social capital and, as we will see next, deployed it in the wind project.

THE WIND TURBINE PROJECT

This was the setting into which small-scale wind-generated electricity was introduced in July 2009. Community selection was preceded by a district-wide, NGO-driven process to develop a Rural Electrification Plan, the intention of which was to transfer SP's experience of doing energy projects one at a time to the government structure to allow scaling up through state funding and implementation.[9] This plan, funded by GE and implemented by a team of SP engineers and sociologists and representatives from the provincial government, aimed to identify what type of renewable energy (hydro, wind, solar) would be most appropriate given each community's different endowment of energy resources. The process consisted of technical evaluations of the natural resources, surveys of a random sample of households, and focus groups with local leaders. In the plan, Alto Perú was identified for its obvious wind resource. It was also a strategic village to start with, due to its location on the main road from Cajamarca to Bambamarca; it serves as a showcase for the possibilities of rural electrification.

In the run-up to launching the wind turbine project early in 2009, NGO-related activity was viewed with a mix of admiration and skepticism. After months of seeing the curious anemometer spinning in the wind, some people in the upper part suggested just selling it off, but Aurelio advised against this. People throughout Alto Perú wondered exactly *who* would benefit from the program. The lead elementary teacher, though residing locally and related to the siblings up by the road, averred that they were creating unnecessary disunity in raising this issue about the school's PV panel. But implicit was the teacher's concern that electrification was going to benefit only the families up on the road—the very ones who were breaking down Alto Perú's unity through their *ronda* secession.

NGO technical staff had put up two anemometers in mid-2008 and early 2009 to create a database for modeling where best to eventually locate the wind turbines[10]—one down at the elementary school and one up by the chapel. Resulting data confirmed what everyone already knew—that wind frequency and intensity were higher in the upper half. This provided a convenient technocratic reason to install the wind project (with a finite budget) in only one part of the community. Staff explanation that the upper part was chosen "scientifically," based on the wind resource, not on favoritism, seemed in subsequent meetings to allay suspicions. SP was also currently implementing two other projects (water and micro-hydro) in nearby villages, so there was some confidence that the promises would be kept.

It was inevitable, then, that on technical, environmental, economic, social, and location grounds, the NGOs would work with Aurelio and his siblings as the project developed. While rhetoric among people from both upper and lower parts, as expressed publicly (mostly by men) during at least two community meetings, was all about benefiting the *entire* community in whatever development projects might come about, everyone clearly wanted to know when they too would get electricity. Despite the underlying grievances, talk of unity was not entirely hollow. Shared history and economic and family ties between the upper and lower halves of Alto Perú are strong. Despite the bid for more political autonomy by the upper part, the political center of gravity remains downslope—in Ingatambo and, more importantly, in the district and provincial capitals farther down the valley. Yet everyone from below also passes through the upper half to get to the highway to catch a *combi*, interurban bus or truck to Cajamarca. While the electricity project thus risked exacerbating the divide between upper and lower residents, NGO workers, well aware of this danger, had emphasized that the wind project in the upper part was to be the *first* phase of a larger project. The fact that the previously completed wind project in Alumbre, which everyone knew about, had been implemented in two phases lent credibility to the proposition.

Based on the community *diagnóstico*, energy demand in Alto Perú was estimated at 2729.3 Wh/day (= 2.73 kW/day), taking account of the total reported daily use of these most commonly used appliances: radios—2250 Wh/day (young men often walk around with portable radios hanging from a neck strap), radio/tape player—240 Wh/day, TV/DVD units—200 Wh/day, and cell phones—39.3 Wh/day (Soluciones Prácticas 2009).

All these data formed the basis for deciding to install four five-hundred-watt wind turbines in July 2009, carried out on the last day of a wind workshop in Cajamarca with an estimated fifty participants from around Latin America. Two teams of course participants and villagers assembled electrical boxes and, despite a few mishaps and missteps, raised the turbine towers. Though visitors and technicians ran the show, young men (mostly from the upper part of the community) participated in the work and slowly gained confidence in the technical work. With cables connected and batteries installed, the blades started whirling and the turbines started to generate light. After some technical adjustments and turbine replacements (the strong winds required even more robust systems to be built by the small Peruvian manufacturing company in Lima), the installation was finally completed in November 2009. The final design was two micro-grids, each one with four five-hundred-watt wind turbines, controller, inverter, and battery bank. The electricity is delivered to thirteen houses and the church.

Thus technological circuitry intersects social circuitry. A contrasting example is just as revealing. Our SP colleagues went to another part of the community (below the school) where there is another small grouping of houses to investigate the possibility of installing another micro-grid. The owner of the hill (i.e., high windy point) said that they could install the turbine there, but only on condition that he would get to say who the electricity

TABLE 7.1. DECENTRALIZED ENERGY SYSTEMS IN ALTO PERÚ

Technology	**Funder**	**Households with electricity**	**People with electricity**	**Number of students**
Wind–solar micro-grid	Green Empowerment / Toyota Environmental Activities Program	13	65 + 2 small businesses	–
Solar panels	Green Empowerment / Toyota Environmental Activities Program	18	90	–
Solar panels	Ingeniería Sin Fronteras / Valencia, Spain	23	115	–
Pico-hydro	Ingeniería Sin Fronteras / Valencia, Spain	4	20 + school	60
Solar micro-grid	Ingeniería Sin Fronteras /Valencia, Spain	4	20 + medical post with vaccine refrigeration	–
TOTAL		**62**	**310**	**60**

would be delivered to—and who not. He didn't want to give electricity to the house of his niece but would give it to the other relatives living in the area. We don't know the family story here, but it does bring up interesting questions of property and shared public services, contrasting with the unified family situation on the road above.

In 2010, solar panels were installed for the medical clinic (north on the road) and for all of the rest of the homes not connected to the wind-powered grid, and a pico-hydro system was installed, completing village electrification, reinforcing community solidarity, and making Alto Perú an emerging demonstration site of wind, solar, and hydro power together.[11]

LOCAL CULTURAL ASPECTS: STRATEGIC USES OF RURAL IDENTITY

Depending on both supply and demand, villagers could count on one to four hours of electricity per day with the wind power—enough to power electric lights instead of burning candles, both at home and for *ronda* meetings. They were now able to run radios and TV/DVD players to be and feel connected with larger regional, national, and international events and issues. They charge cell phones, mostly to call family members everyone has in other rural towns, Cajamarca, Lima, and elsewhere. Many people from neighboring villages with no electricity are coming to Alto Perú to charge their cell phones (one *sol* per charge). The teacher of the school reports noticeably higher performance of the students since the solar panels were installed and is optimistic about the newer pico-hydro project for powering a computer. People remarked on improved air quality and lower cash outlays from having replaced candles and kerosene lanterns with CFL lighting (CFL bulbs are now being stocked at Aurelio and Felipe's stores; both report that lighting enhances nighttime sales), making it possible to extend activities into the night. Though offset by the monthly user fee, household budgets are freed up by not having to buy candles and batteries (Soluciones Prácticas 2011).

It is clear that Alto Peruanos' main uses of their new electricity serve to break down the feeling of being too rural and disconnected from wider modern society. To have electricity is to be engaged with the broader world: following the 2010 presidential election, calling family, and especially getting their children to study at night, given the close association of education with "development" and "progress" in the minds of most villagers. Some villagers hoped the new electricity would encourage "professionals" from the village to stay put instead of migrating to the city.

What they have not done with these new systems is power machinery that could fundamentally affect their agropastoral livelihood strategy, either due to the amount of electricity available and/or lack of resources (financial, market access, business skills) to launch other income-generating schemes. The initial survey revealed villagers' desire to use this new electricity for a variety of productive uses, including photocopier, computer, woodworking shop (though there is no wood in the area currently, pines have been planted in the lower part so wood production is anticipated, as in neighboring Granja Porcón [cf. Love n.d.]), making clothes with a sewing machine, making some dairy products (none are currently made), using blenders in the store, and irons for clothing.[12] Aurelio's dream of creating a tourism complex at the cold, windy lake hasn't materialized.

Acquiring electricity, whether locally generated or not, requires the technology not only to generate it but also to consume it, and therefore requires cash and other relationships to obtain it.[13] But technology is never neutral; with lights, TV, cell phones, and other trappings of urban life, electrifying the countryside helps rural people the world around overcome the sharp city/country dichotomy imposed on them by the dominant urbanism of late modern cultures—a symbolic if not material victory for rural dwellers.[14]

While this electrification project connected villagers to urban lifestyles, during project development people strategically played their rural identities to their advantage, knowing that the benefits of the project would only be given to people who lived in the community full-time. In an ironic twist, being "local" involved deploying social and cultural capital that could be converted into material benefits (electricity), which in turn enabled people to be more connected to urban lifestyles.

When time came for NGO staff to present their plan at a community meeting, people jockeyed for position to get access to the expected electricity, chiefly by establishing the fact of residency. This was apparent in the seemingly simple task of devising a survey to determine how many people live in Alto Perú as well as in ways people dressed and spoke at community meetings. But who exactly could be included in the list of homeowners, and thus eligible for electricity? What if *mitayos* and other temporary workers stay for a few years and have families? What if the owner also has a house in Cajamarca, where he or she passes a few days of each week or month? What about seasonal residence? And what exactly do people mean by "Alto Perú"? There are households that are spatially in Alto Perú but participate in most community activities in Ingatambo—though now that there is a project, they want to participate up in Alto Perú. And the reverse is also true—there are a few distant houses that identify as being part of Alto Perú, due to family ties and participation in the governance of Alto Perú. And whom do you believe anyway, particularly when everyone is enmeshed in dairying and travel to the city is virtually obligatory? Thus, some people who split their time between

Alto Perú and Cajamarca could quite truthfully say that they lived full-time as agriculturalists in Alto Perú, even if they were absent for long periods.

In sum, rural identity quickly became an asset, and villagers proved adept at using it. For example, a wealthier man who owns a vacant building on the road but lives in the city of Cajamarca attended a meeting about the electrification project; he lauded the engineer's commitment to extending the benefits of the new electricity to *every* resident. The city-dwelling landowner was, by inference, making his claim to "residency" known. At other times, Alto Peruanos would come to the office of Soluciones Prácticas in Cajamarca, wanting to make sure their name was on the list of households to be electrified. In this way, the project "beneficiaries" enacted their rurality, taking agency to access the desired benefits of electricity.

GRID- VERSUS LOCALLY GENERATED ELECTRICITY: MINING, THE STATE, AND COMPETING DEVELOPMENT PATHS

Though Alto Peruanos are clearly pleased with what they are now able to do and enjoy with their own locally generated electricity, they aspire for 24/7 grid power, which since the Velasco era has been in the hands of the national government. Rural livelihood patterns keyed to the agricultural cycle are thus generally inapt for electrification. Currently, the only electrical grid in the area is a high-voltage line that soars overhead en route to a large gold mine, reflecting the government's pursuit of development based on large industry and export economies.[15] The irony, indeed insult, of high-voltage towers marching across their unelectrified landscape to mines farther north is not lost on local residents. Residents commented about the injustice of the electricity lines passing them by. They had advocated to be connected but were told that the lines were high voltage and it was too expensive to connect to them.

Villagers of Alto Perú have a complex, ambivalent relationship with the large mines, including the Yanacocha mine just to the south. They, like villagers throughout the Cajamarca highlands and people in town, are conflicted by apparent environmental damage to the regional water supply caused by leakage from the cyanide process used by Yanacocha and the other gold mines in the area. In December 2011, tensions ignited into huge protests against a new mine, Conga (also owned by Newmont[16]), paralyzing the city of Cajamarca, blocking roads, and leading the national government to declare a state of emergency and send in national troops to patrol the city. Plans for Conga have been suspended for now, but the latest conflict is part of an ongoing struggle over resources, environment, autonomy, and development. The tensions have flared at times in Alto Perú as well, including apprehension of mining prospectors by *ronderos*. Pitting national against local governments, the tension has San Pablo provincial authorities engaged in a fierce legal battle for the water rights in the Alto Perú lagoons, to which Newmont already owns mineral rights (cf. Minería Yanacocha 2009). And yet, poverty and relative opportunity often force compromises of ideology. The local Comité de Defensa de San Cirilo platform is that the mine should relinquish all rights to land and water, but if they don't, Alto Peruanos should be the first ones hired for the new work, instead of bussing in outside workers.

While Alto Peruanos express desire for connection to the grid, they also take immense pride in the locally owned, autonomous wind turbines. After the installation of the orange-and-blue wind turbines, Alto Peruanos independently painted one of their main buildings the same colors. People wanted to put names on them as inauguration. During the installation people spoke admiringly of how beautiful the turbines were, calling them *mariposas que bailan* (dancing butterflies). While the new wind energy intersects these cultural and social patterns, reinforcing some and encountering friction with others, it also enables Alto Peruanos to be more self-reliant even as they're "more connected." Many people, even mine employees, have stopped along the road to inquire about and take pictures of the wind turbines. This new attention has put Alto Perú on the map, so to speak, and as one community leader noted, "We have what others want." Aurelio says that the mine company may want to build wind turbines for other communities (as a local mine is currently installing solar panels). Although time will tell, having electricity may strengthen their image of being a settled community and thus provide sufficient symbolic capital to bolster their claims on the land and water resources vis-à-vis the mine.

Thus are rural dwellers caught up in these large cultural processes; rural electrification illuminates these intersections. The force of these wider cultural logics is so strong that it is hard not to fall into the modern/traditional binary, and in the process miss the layered, textured spaces in which rural dwellers socially construct their worlds. Viewed from the lens of local villagers like Alto Peruanos, lines between these competing development paths are thus blurry. Even as electrification is a powerful symbol of "modernization" and "development," in this context, the electrification of Alto Peru is at the same time a statement for an alternative path of development, set in contrast to the mega-infrastructure projects of the industry-oriented state.

As autonomous energy projects develop in isolated settings, even as national electrical grids stretch across rural landscapes, people's participation in and understanding of "modernity" shifts overnight. Ethnography captures these electrifying transitions—the messy, on-the-ground diverging and converging intersections of energy, society, and culture.

NOTES

1. An earlier version of this paper was presented at the panel "Ethnographies of Energy" at the American Anthropological Association Annual Meeting, Philadelphia, PA, December 5, 2009.

We deeply appreciate the warm reception given us by Alto Peruanos and other villagers involved in renewable energy projects. We gratefully acknowledge the help of SP staff in Cajamarca, particularly Wilder Canto, and Green Empowerment intern John Steed for creating the Alto Perú map. TL appreciates the openness of SP and GE leadership and staff in allowing him and Linfield students to engage their work in Cajamarca, as well as to the Linfield College Collaborative Research Program for funding. AG also extends her thanks to colleagues at SP and to Linfield College for their willingness to collaborate and combine practical project implementation with academic research.

2. The organizations have implemented a range of projects, including two community gravity-fed water projects, eleven biodigesters, thirty-nine wind turbines, thirty-one solar panels, and over fifty micro-hydro projects that provide decentralized rural energy services to over five thousand households (twenty-five thousand people).

3. We utilized a combination of participant observation, semi-structured, and structured interviewing and focus groups over thirty total days of fieldwork in Alto Perú between 2009 and 2011, complementing work done by SP staff before and after installation. From 2008 to 2009 AG was living in Cajamarca, based with SP and managing a grant from Toyota Environmental Activities Grant Program, which included the installation of the energy systems in Alto Perú and related training sessions. TL conducted short-term ethnographic fieldwork in 2009 and 2010. We both contributed to the internal *diagnóstico* (case study) of the community prepared by SP staff for the installation (Soluciones Prácticas 2009).

4. http://www.greenempowerment.org/countries/5/project/12 and http://greenempowerment.wordpress.com/2009/10/08/news-coverage-regarding-alumbre/.

5. Descendants of the *hacendado* family still live on twenty hectares in the hacienda house in Ingatambo, which appears to have historical roots as a way station on the main Inka-era road (*Qhapac Ñan*) heading north out of Cajamarca. An elderly Alto Peruano recalled working eighty days a year on the hacienda without remuneration, and talked about how glad he was to get away from the poverty, the *patroncito/indio* servile greetings, and the whole hacendado/peon relationship of that era.

6. *Rondas* first emerged in the 1970s in neighboring Hualgayoc province to control rustling, which had grown commensurate with the growth in importance of cattle (Gitlitz and Rojas 1983). Having first developed in Cajamarca, the *ronda* model is now well known, having been adopted by the state in the 1980s counterinsurgency effort against Sendero Luminoso. Cajamarca was among the earliest actively politicized highland departments (Taylor 2000), and SL had an active presence in southernmost Cajamarca Department (Taylor 2006).

7. This sort of recent repositioning near roads, related to mining developments, appears to be widespread in the region (Bebbington and Bury 2009).

8. The coincidence of Aurelio's Protestant convictions and his entrepreneurial orientation strongly suggests Weberian connections between Protestant orientation and economic success, but we do not explore this here (see Love n.d.).

9. The plan in this province of San Pablo is now being replicated at the level of the Region of Loreto (funded by the regional government there).

10. The placement of the anemometers was also for Universidad Politécnica de Cataluña and Soluciones Prácticas to undertake a study of the adaptation of wind-modeling software to small-scale wind projects in mountainous areas.

11. Elsewhere we describe the management model, detailing its implementation and operation in Alto Perú, and examine the differences among wind, solar, and hydro as energy sources and their implications for such programs of rural electrification in the region (Love and Garwood 2011).

12. It will be interesting to see whether more household appliances are purchased, including such things as hot plates, which could offset the consumption of propane for cooking (three-fourths of families use wood for cooking, the rest use gas purchased and hauled from Cajamarca).

13. In this sense, the work NGOs are doing is unintentionally fostering the spread of urban-industrial consumption patterns. Sustained examination of the complex normative pros and cons of this shift is beyond the scope of this chapter.

14. Likewise, it is beyond the scope of this chapter to discuss the idealization *and* oppression of rural dwellers resulting from ways urban intellectuals' imagine(d) rural worlds as constituting the antithesis of modern ills—simpler, less consumptive, more honest, bucolic, collectivist, tied to a chemical-free earth, etc. (cf. Kristal 1987 for the Andes; Williams 1973)—and without the powerful machinery driving it all.

15. Extending electricity to sparsely settled, often poorer rural areas has always been difficult on strictly economic grounds (cf. Nye 1990:287ff.; Revolo 2009; Sánchez 2010), requiring state-sponsored development.

16. Yanacocha, the largest gold mine in Latin America, is 51 percent owned by Newmont Mining Company, 44 percent by Cia. Condesa Minería SA, and 5 percent by International Finance Corporation (Minería Yanacocha 2009).

REFERENCES

Bebbington, Anthony J. and Jeffrey T. Bury
2009 Institutional challenges for mining and sustainability in Peru. http://www.pnas.org/cgi/doi/10.1073/pnas.0906057106 (accessed February 19, 2012).

Bury, Jeffrey
2004 Livelihoods in transition: Transnational gold mining operations and local change in Cajamarca. *The Geographical Journal* 170(1): 78–91.

Deere, Carmen Diana
1990 *Household and class relations: Peasants and landlords in Northern Peru.* Berkeley: University of California Press.

Ferrer-Martí, Laia, Anna Garwood, José Chiroque, Rafael Escobar, Javier Coello, and Miguel Castro
2010 A community small-scale wind generation project in Peru. *Wind Engineering* 34(3): 277–288.

Gitlitz, John S. and Telmo Rojas
1983 Peasant vigilante committees in Northern Peru. *Journal of Latin American Studies* 15(1): 163–197.

Kristal, Efrain
1987 *The Andes viewed from the city: Literary and political discourse on the Indian in Peru 1848–1930*. New York: Peter Lang.

Love, Thomas
n.d. From SAIS Atahualpa to 'Sois la luz del mundo': A case study of religion, energy and agrarian reform success in northern Peru.

Love, Thomas and Anna Garwood
2011 Wind, sun and water: Complexities of alternative energy development in rural northern Peru. *Rural Society* 20: 294–307.

Minería Yanacocha SRL
2009 *Memoria anual 2009*. http://www.infomine.com/index/pr/Pa914326.PDF (accessed February 21, 2012).

Ministerio de Energía y Minas
2007 Dirección Ejecutiva de Proyectos *Informe de Gestión Anual*. Lima, Peru: Ministerio de Energía y Minas.
2010a *Anuario estadistico electricidad 2009*. Lima, Peru: Ministerio de Energia y Minas (accessed February 23, 2012).
2010b *Plan nacional de electrificación rural. (PNER) Período 2011–2020*. http://dger.minem.gob.pe/Proyectos_pner2011.aspx (accessed May 29, 2011).

Nye, David E.
1990 *Electrifying America: Social meanings of a new technology, 1880–1940*. Cambridge, MA: Massachusetts Institute of Technology Press.

Revolo, Miguel
2009 *Distribution rural electrification experiences in Peru*. http://www.fema africa.net/attachments/103_11.4%20RevoloM_OSINERGMIN_Distribution%20RE_Peru.pdf (accessed May 25, 2011).

Sánchez, Teodoro
2010 *The hidden energy crisis; How policies are failing the world's poor*. Bourton on Dunsmore, Rugby, Warwickshire, UK: Practical Action Publishing.

Soluciones Prácticas
2007 *Taller regional: Planificación energética y desarrollo de capacidades para mejorar el acceso a la energía y promover el desarrollo local* (internal document).
2009 *Diagnóstico socioeconómico caserío Alto Perú*. Programa de energía, infraestructura y servicios básicos. Cajamarca: Soluciones Prácticas.
2011 *Proyecto de energía microred aerogeneradores, Alto Perú: Evaluación de impacto social y económico*. Programa de energía, infraestructura y servicios básicos. Cajamarca: Soluciones Prácticas.

Taylor, Lewis
2000 The origins of APRA in Cajamarca, 1928–1935. *Bulletin of Latin American Research* 19: 437–459.
2006 *Shining path: Guerrilla war in Peru's northern highlands, 1980–1997*. Liverpool: Liverpool University Press.

Williams, Raymond
1973 *The country and the city*. New York: Oxford University Press.

CHAPTER EIGHT

Space, Time, and Sociomaterial Relationships: Moral Aspects of the Arrival of Electricity in Rural Zanzibar

Tanja Winther

INTRODUCTION

Only a few years after Thomas Edison made the first public demonstration of the incandescent light in 1879 in Menlo Park, New Jersey (Nye 1990:3), electricity appeared in Zanzibar Town. One of the Sultan's palaces was lit in 1886. Since then, the ruling classes, from the Sultanate period to various post-independence governments have used electric light to create and express their distinct versions of modernity and their corresponding images of what Zanzibar is and should be (Winther 2011).

A century after the Sultan's palace was lit, electricity reached ordinary people in rural Zanzibar.[1] This chapter examines some of the central, moral aspects of this shift toward electricity. The sociomaterial organization of people's living rooms will be my primary focus. Before entering this particular space in the evenings to observe the shift in question, a brief introduction to central Zanzibari values will be necessary. I will first show how electricity and adhering appliances have become associated with a modern, desirable way of living. Included here is a moral dilemma in that people's drive for light and new appliances must be socially balanced with the importance of showing material modesty. I will then discuss electricity's degree of commensurability with Islamic values and practices. Successively, we shall enter two distinct living rooms to see how the arrival of electric light and television sets has changed the way people socialize in the evening. One of the homes is connected to the grid and the other is without electricity, and both are typical of their kind in the village in which I resided. By drawing on the treatment of central cultural values, I will point to the ways in which various moral dilemmas were solved in the process of reorganizing the living room. More specifically, I will focus on *why* people in the rural areas adopted the new lighting technology and television practices the way they did and *how* these changes were managed socially, morally, and materially. We shall also see how electricity, by its way of conditioning a new kind of light and people's access to television programs, has changed the pace of life in the village.

This chapter seeks to demonstrate the intrinsic social nature of electricity. Electricity does not serve practical ends in a neutral manner. Its introduction, presence, and uses in a given community are loaded with cultural-specific meanings. As I will show, the existing moral and sociomaterial context will partly affect the outcome of the technology's introduction and uses. Likewise important are the scripts (Akrich 1994) of electric light and electric artifacts, which through their materiality may contribute to producing social change. In this sociotechnical field, light is a particular case. I draw on Bille and Sørensen, who contend that "light is used to reveal people, places and things in culturally specific ways" (2007:266). Their notion of "lightscapes" is instructive, as it captures the shifting ways in which light, shadows, people, and objects together constitute the appearance and meaning of a given space. Lightscapes may reflect important cultural and moral issues as well as mirror representations of ideologies and power relations (Moore 1996; Miller 1998). Embedded in the concept of lightscapes is also the possibility, as Bille and Sørensen suggest (2007:280), that light itself, as an untouchable but highly visible material object, should be attributed with agency. Ortner (1999) identifies agency as the capability or power to be the source and originator of acts and thereby change. I consider the notion of "distributed agency" to be particularly useful for my present purposes (Garud and Karnøe 2005; see also Wilhite 2008). Distributed agency implies that agency is "distributed across people and artefacts" (Garud and Karnøe 2005:89). Thus in addition to people themselves, the sources of acts and thereby social change may be located in the script (Akrich 1994) or materiality of a given object, such as electric light or television programs, positioned in a particular context. Other sociomaterial elements that make up a particular lightscape may also produce effects, such as the degree of darkness and the quality of surfaces.

Light will form a central part of this treatment of electricity, moralities, and sociomaterial relationships in rural Zanzibar. However, to obtain a purposeful and emically attuned analytical perspective, broader practices such as religious ones and adhering objects such as television sets will also be included. I wish to show how social spaces that have been reorganized due to electricity's arrival provide arenas for the production, circulation, and negotiation of meaning.

I proceed by accounting for development and religious ideals, which are central cultural categories in Zanzibar and which have relevance for the forthcoming analysis of electricity's effects. Cultural categories, or shared values, can be understood as justifications for behavior or judgment that do not need further justification (Gullestad 1992:140). Such values may nonetheless sometimes be contradictory and pose dilemmas when new technologies are introduced.

ELECTRICITY AS SIGNIFIER OF DEVELOPMENT

Electric light and appliances such as television sets and freezers are objects of desire and signs of a modern way of living in rural Zanzibar. This yearning for modernity was exemplified during the following incident, which occurred in 2001 when I resided in the village of Uroa during fieldwork.[2] At one point, the street lights stopped working and we were without light for several months. On the day the street lights were switched on, a young man exclaimed in joy, "It is like in town here now!" (Swahili: *kama mjini hapa sasa*). Everybody in the village joined in this celebration of the regained light, which transformed the place visually as well as conceptually. The streetlights demonstrated that Uroa is a modern, attractive place. The man's statement also reflected that town is considered as an ideal within the local discourse of "development" (*maendeleo*, literally "to be moving forward"), which is related to the association between town and a range of consumer goods.

People in rural Zanzibar today regard electricity as a condition for development. Light, in addition to transforming the place to becoming more modern, is also considered to render the place safer. Street light is sometimes referred to as "security light" (the English term is used) because evil spirits (*mashetani*) are said to prefer darkness.[3] Furthermore, electric light is associated with purity by way of making it possible to detect bugs and even spirits and for exposing dirt, thus motivating a host to keep a place clean. Electric light's capacity to reveal dirt has affected people's perceptions of purity.

Electricity's other public uses in terms of improving services like water, health care, and education are also significant and highly appreciated but will not be treated in the present discussion (see Winther 2008). As to private consumption in people's homes, electricity is rapidly becoming normalized in two senses. First, it has become common for households to be connected to the electrical grid. Thirty-three percent of private houses in Uroa had obtained a connection in 2001, and by 2006 more than half had become connected. Second, Zanzibaris consider electricity to be something that a husband *should* provide for his family. Those who cannot afford electricity installation, which costs about five months' worth of income to a fisherman, risk being judged as backwards (*hawataki maendeleo*, literally "they do not want development"). As a result of this normalization process, some young men appear to delay getting married while saving money for electricity. Zanzibaris also consider it important that men who are married to several women treat the co-wives equally.[4] Men who marry another wife before providing the first wife with electricity risk being ridiculed, and the high cost of electricity may therefore be a barrier for marrying several wives. Electricity's status as a desirable but also increasingly normalized good is important to the study of its spatial, temporal, social, and moral implications.

Before signifying objects become normalized, however, they may be socially dangerous (Pantzar 1997). In rural Zanzibar the importance of showing material modesty is highly articulated, and this value in many ways stands in opposition to that of development, progress, and change. The introduction of new appliances may be socially risky. For example, one of my closest acquaintances, Khamis, had purchased a television set in 1998. He told me that three months later he had started aching in his arms and legs up to the point that he had difficulty farming and fishing. Upon seeing a witch doctor (*mganga*) his problem was identified as being caused by one of Khamis's relatives, who had become jealous of Khamis and his new television set. The relative was suspected of having conducted evil magic and causing Khamis harm. After Khamis received proper treatment, his problem disappeared. The incident was nonetheless a reminder to him and his social network of the danger associated with triggering envy and jealousy in the village. Social life is marked by a high degree of frugality, and breaking the norm of sharing may be devastating. Occult knowledge has a strong position, and the effects of people's jealousy or sanctions for misbehavior are perceived as a threat that may be fatal. The point of acquisition and display of a new, desired object seems to be the most risky moment, but one will also be judged at the time of consumption and use of an object. One is expected to show solidarity and share what one has with members of one's group.[5] As we shall see, this stress on sharing also goes for consumption of light and electric appliances in people's homes.

Electricity's Degree of Compatibility with Islam and the Challenge of Gender Segregation

As a Muslim society, people in Zanzibar reference and strive to uphold notions of purity, respectability, and adherence to the Koran. In most contexts electricity's uses are fully compatible with Islamic ideals. For example, long fluorescent tubes have been installed in the village mosques. At nighttime the turquoise, radiating lightscape reinforces the image of a pure and sacred place, underlining the omnipotence of Islam. Electric light installed in mosques enhances the practicing of Islam and people's potential to live as good Muslims; formal worship is facilitated after dusk, even as the luster of the religious tenets of Islam is enhanced. Electricity is also used for other religious purposes. Electrified pumps feed water to the taps located outside the mosques, which provide running water for washing feet and hands before prayers. Together with electric light, the provisioning of flowing water metaphorically enhances the purity of body and spirit. The electrification of speakers, which are used for calling people to prayer,[6] and the use of electric cassette players, which are used for listening to famous Islamic scholars, have similarly led to a strengthening of both the media and the message of Islam. Islamic leaders (and teachers) are also more likely than ordinary people to keep television sets at home. They are skeptical of certain series and movies imported from the West, however, which display what they find to be immoral behavior such as improper dressing manners or people drinking alcohol, but imams generally cherish this new access to information, education, and entertainment. Overall, electricity has strengthened Islam's position in the village.

The segregation of space by gender, however, is a cultural-religious ideal that poses particular challenges in the case of electrification in Zanzibar. During ceremonies such as weddings, women and men always occupy distinct spaces. When food is served on such occasions, men are served first, reflecting the hierarchical gender ideology throughout Zanzibar. In everyday life, the gender division of work also contributes to a segregation of space.

However, as we shall see, the ways men and women spend their leisure time in the evenings have dramatically changed with the advent of electricity and the appliances that electricity enables. To illustrate this shift, which implies that social and moral patterns attached to the ideal of gender segregation have had to be revised, I will now take a look at two distinct Zanzibari homes. Each is typical of its kind. Bi Mashaka's house has not been connected to the grid. In Bi Mema and Khamis's home, they have electricity, light, and a television set.

Two Homes in the Evening

As usual, Bi Mashaka was sitting in her narrow living room together with a female friend one evening. The two were seated close to one another on a mat on the floor. A kerosene lamp had been lit and was placed between them. Their faces were glowing in the yellowish, modest, and lively light, which threw shadows of their silhouettes on the wall. Apart from their bodies and faces, little else in the room was illuminated. The two women held a quiet conversation with soft and nearly whispering voices. The light- and soundscapes (Tacchi 1998) of the room seemed to invite a social form of privacy and secrecy, resembling the aesthetics of darkness one finds outdoor at nighttime in rural Zanzibar (Winther 2008:135–138).

From Meja and Khamis's house the loud noise from the television commercials reached me as I approached the house. Inside the evening guests had arrived, as they did every night, and were seated in the expanded living room. The forty-watt ceiling bulb with no shades was lit and Khamis had brought the television set out from the bedroom. Now it was placed on a table in the living room. Fifteen people were present and were glued to the screen and the program that was about to begin. The children were seated closest to the television set. Next to them, the women were seated on the concrete floor, leaning against a semi-wall in the middle of the room. On the other side of this wall, Khamis's older brothers were standing on the earth floor. These men were less inside the living room than the women. The hosts, on their part, were sitting close to one another on the earth floor out in the corridor. However, they were situated so that they could see the screen through the gap in the half wall. In this position they could notably also watch the door when new visitors arrived or someone left.

The contrasts to Bi Mashaka's house are numerous. First, the physically larger and partly concealed living room provides an arena where the extended family can be together. New houses in Zanzibar tend to contain such living rooms, and this is reflected is a shift in the terms people use to denote various rooms in a house. What was previously called a "living room," *ukumbi*, was the narrow room in which Bi Mashaka and her friend were seated. In contrast, in the modified houses this space is mainly used for passing through the house and referred to as the "corridor" (*njia*). An enlarged, separate space is what constitutes a modern living room (*ukumbi*).

Second, electric light from the ceiling bulb renders visible the frames of the room. To the eye, the light here enlarges the space. There are no boundaries between darkness and the light because the whole room with its protecting walls is there to be seen. The source of the light in terms of a radiating ceiling bulb resembles a torch that reveals everything in its path. The Swedish ethnologist Jan Garnert has pointed to the shifting aspects of space according to the use of various types of light sources (1993:77–78; see also Bille and Sørensen 2007). With electricity, people sit farther away from each other but remain able to read each others' body language. Bright light enhances communication. With light and the greater distances, people also use louder voices, which in darker contexts would appear inappropriate in Zanzibar.

The gender of these two types of spaces constitutes a third difference. In few other contexts would men and women in Zanzibar be located in the same private space, as they are when watching TV. In the secular school, at village meetings, and on the bus they occupy the same indoor space, but these settings are highly public, transparent, and therefore morally safe. By contrast, the inside of a household at nighttime is ambiguous and morally dangerous. It would be unthinkable for Bi Mashaka to start entertaining male guests or inviting a mix of men and women in the evening. The traditional "solution" to the moral danger of mixing sexes inside is that men seldom spend time at home unless eating, washing, fixing the house, or sleeping. Before electricity's coming, indoor space was primarily female. With electric light and television, men in Zanzibar have "entered" (cf. Bourdieu) the house to an extent and in a manner that is completely new. A Zanzibari voluntary officer in town expressed it very clearly when asked about the main impact of electricity in everyday life: "Men have come home."[7]

Interestingly, the potential threat of impurity within households that provide television in the evenings seems to be compensated by the particular light that comes along. Bright light exposes and prevents immoral behavior. The bright fluorescent light is also the most preferred kind of light source in rural Zanzibar, admired for "its environment" (*mazingira yake*). In general, the revealing character of electric light reduces the risks and suspicions of improper conduct when women and men are gathered. Occasionally I observed homes where people were watching TV without having the lights on. If there were no guests, Meja and Khamis would sometimes choose to turn off the light to save on costs. However, they would quickly switch on the light as I entered the house, as if somewhat embarrassed or seeking to prevent an uncomfortable situation from occurring. Among the more affluent and, in their own eyes, thoroughly Islamic-oriented families, the lights would always be on. Light is associated with purity in several senses. People's visual focus and continuous attention to the screen and not to each other may also prevent suspicions of improper conduct. And, importantly, as seen, the particular distribution of people, where men and women occupy distinct parts of the room, represents an important strategy for maintaining the moral and social order. Hence two opposing concerns—the desire to watch television within a limited space and the importance of respectability—are solved by creating a gender-specific micro space within the living room. Electric light contributes to making this reestablished order visible, pure, and acceptable.

As a final observation, the status of the hosting couple appears to be elevated compared to that of the visitors in these particular settings. Their status is partly illustrated in their right to comment on programs. Also important, the husband and wife are seated together apart from the visiting guests and they do not form part of the described gender-segregated order in which the visitors are positioned. With the joint control of the situation they possess in this situation, underlined by the physical remote control they both administer, I have therefore suggested that they here temporarily challenge the ideology of male superiority and highlight the value of development when performing as a modern couple (Winther 2008; Winther 2012).

Life Is Speeding Up

The restructuring of social space in people's living rooms also has temporal implications. At about nine P.M. Bi Mashaka would see her guest out to the front door and then retire to the bedroom to sleep. Before television came to the village, she told me, her husband would normally return home at approximately this hour after having spent time with his mates outdoors. Now, in comparison, he may delay returning home until ten P.M. or even later coming home from watching television in other people's homes. She took me inside the bedroom one evening and explained which of the beds was used by her husband and herself and which were used for the children. In passing, she sighed and said with expressed regret that after her husband started watching television they have less time for spousal intimacy. Thus the new practice of watching television in some homes also affects time management and relations in other homes. In 2001, 9 percent of Uroan homes kept a television set, while as many as 74 percent of men and 52 percent of women reported watching television at least three times a week.[8] Homes that can offer "view-time" in the evenings have become social magnets, and most Uroans seek to view television on a regular basis.

In Meja and Khamis's home it is expected that he stays up late, until the last guest has left the house. She would often withdraw to the bedroom at ten P.M. out of fatigue and when the Swahili dramas she enjoys are over. But for Khamis it would be unthinkable to hint at relatives and family that they were not welcome to continue watching. As a result, it might reach midnight before he goes to bed. Thus in this family, too, husband and wife spends less time together in the bedroom than before. Sexual patterns have become modified due to electricity and television.

At five A.M. Khamis is nevertheless awakened by the call for prayers from the closest mosque. As they started to use amplifiers, he said, the sound reaches "all the way up to here" (*mpaka hapa*). Khamis mentioned this with a smile, as he knew I was aware that he does not always attend the morning prayers. But the effect of late television evenings and the (ideally) early mornings illustrates a general trend after electricity's coming beyond the acknowledged effect that many people sleep less than before. The range of choices of what to do when has changed in a fundamental way. In theory the day now has twenty-four usable hours. For example, a fisherman may repair his fishing net at night; thus the evening is no longer only used for relaxing. Women are preoccupied with earning money from farming seaweed during daytime, and they also want to watch television and relax in the evening. As a result, after electricity's coming they have cut down the number of meals they cook daily from three to two.[9] When I asked women about their views on electric stoves (which are very few in the village), what they highlighted was the benefit of being able to cook quickly (*haraka*), at once (*mara mmoja tu*). I heard the term "there is no time" (*hakuna time*)[10] repeatedly during my stay in the village. An important consequence of electrification is that life is speeding up.

CONCLUSION

This chapter has examined the moment when people in rural Zanzibar started using electricity for the very first time. I have focused on the moral, sociomaterial, and temporal negotiations that ensued. By way of concluding on the chapter's main findings in a somewhat reversed order, I have contrasted people's evening practices in two typical, though different, kinds of homes to illustrate what the shift to electricity implied. In homes in which people rely on kerosene for lighting, women socialize in the evenings whereas men spend their time outdoors or in other people's homes. In contrast, in electrified homes that provide television, men and women occupy the same social space after nightfall to watch programs. This poses a new moral dilemma in that space in Zanzibar ideally should be segregated by gender.

The re-creation of the social order and the forming of new light- and soundscapes in the electrified living rooms are the result of negotiations with several underlying and partly contradictory concerns. First, Zanzibaris desire television programs. Electricity, television, and other media technologies comprise the idea of "moving forward," an ideal of modern life. The hosts enjoy an elevated status when displaying their success as a modern couple during the evening shows. Second, when Zanzibaris watch television programs in the context of the extended family group,[11] this communal viewing is linked to the importance of showing material modesty and solidarity. As seen, at the time of acquisition of new appliances, individuals may perceive themselves as being vulnerable to social sanctions. In one way, Khamis, who experienced such sanctions, can be regarded as compensating for his investment every time he turns on the television set for the extended group to watch . As a result of the desire for television and the expectation that hosts provide viewtime, the time when many people go to bed has been delayed and sexual patterns have become modified. More generally, electricity has caused a different sense of time, which does not follow natural cycles but involves a type of time management that includes more choices of what to do when. The effect of this change in temporality is that the pace of time has increased, and people more often feel to be in a hurry.

Third, the religiously informed concern for respectability, purity, and gender segregation was maintained during the restructuring of the living room. Zanzibari hosts balanced the spatial convergence of men and women by creating gender-specific micro spaces within electrified rooms. Also, in contrast with kerosene-lit homes, newly electrified settings provide bright light, allowing for louder voices and laughter. This light- and soundscape invited a freer atmosphere that resembles daytime setting and interactions. Moreover, the television provided moral legitimacy in these gatherings in two senses, both through its conceptual significance in that programs are desired and modern, and physically as well as socially, by keeping people's visual attention directed to the screen rather than to any potentially inappropriate social relations among the visitors.

There is a high degree of continuity in the way central values are maintained despite the reorganization of Zanzibari villages and living rooms. But in the process, the new lightscapes and practices provide arenas for the production, circulation, and negotiation of meaning. Light and appliances signal development, progress, and success for those who control these technologies while leaving other groups in the shadow. The omnipotence of Islam is strengthened while central religious concerns are handled in new ways and potentially also modified. Perceptions of what constitutes a pure space are also changing. The Zanzibari case illustrates the close interconnections among energy, social life, and morality. The changed sociomaterial forms through which values are expressed and the appearance of new criteria for evaluation may ultimately result in the transformation of cultural values. Finally, the case illustrates how the materiality of a light source and the presence of television sets may transform established forms of sociality and be used for balancing moral concerns. The characteristics of electrical lighting provide new channels for social interaction, even as they raise concerns for established values of morality. With the introduction of new forms of energy and corresponding technologies, new forms of individual agency also arise in the context of contemporary Zanzibar.

NOTES

1. The rural electrification project in Zanzibar was financed by the Norwegian government through the Norwegian Agency for Development Cooperation (Norad) from 1986 to 2006. Zanzibar consists of the two main islands, Unguja and Pemba, and is a semiautonomous state that forms part of the Union of the Republic of Tanzania.

2. I did fieldwork in Zanzibar in 1991 (three months) and in 2000–2001 (ten months), providing the main material for the present work. I later made shorter visits to the islands in 2004, 2005, and 2006. The main village under study, Uroa, was electrified in 1990. I account for the shift from pre- to post-electrification also by comparing life in villages with and without electricity.

3. See Hastrup (1998:69–70), where she mentions a possible connection between electric light and the disappearance of the *huldufólk* ("the hidden people") in Iceland.

4. Co-wives tend to live in distinct houses, and the husband alternates between them on a two-day basis.

5. See Winther (2005) for a treatment of the moralities involved in various types of transactions in Zanzibar.

6. Very few women attend the mosque; they pray at home.

7. As noted elsewhere, associations between gender and space are not only culturally conditioned but also tend to be shifting (Carsten and Hugh-Jones 1995). This also has relevance in Zanzibar, even after electricity's coming. But the described shift in the gender of the room is significant. A similar shift in boundaries has occurred in terms of the distinction between private and public space (see Garvey 2005).

8. The figures are based on a survey I conducted in 2001, which covered 23 percent of the 480 households in Uroa. The sample is not fully representative for the whole population because I wanted to include a sufficient number of electrified households. This is reflected in that 43 percent of survey households had obtained electricity, whereas the real figure was 33 percent at that time.

9. This finding is based on people's accounts and also by comparing life in Uroa with neighboring villages, which also had seaweed farming but not electricity.

10. In this context, the English term *time* was used more often than the Swahili term *wakati*.

11. See Bille and Sørensen (2007) for an interesting treatment of the practicing and display of "hospitality" through the use of light in two distinct contexts, a Danish and a Jordanian Bedouin. The Danish notion of *hygge* has a close parallel in the Norwegian adjective *koselig* (cozy) (Wilhite et al. 1996), which seems just as much attuned inwards as outwards. In contrast, the Bedouin and Zanzibari rooms appear more publicly oriented, as the social organization of outsiders' presence and their evaluations are crucial. The relationship between light and (shifting) public-private boundaries deserves further attention.

REFERENCES

Akrich, Madeleine

1994 [1992] The de-scription of technical objects. In *Shaping technology/building society. Studies in sociotechnical change*, edited by Wiebe E. Bijker and John Law, 205–224. Cambridge, MA: Massachusetts Institute of Technology.

Bille, Mikkel and Tim Flohr Sørensen
2007 An anthropology of luminosity: The agency of light. *Journal of Material Culture* 12(3): 263–284.

Carsten, Janet and Stephen Hugh-Jones
1995 Introduction. In *About the house: Lévi-Strauss and beyond*, edited by Janet Carsten and Stephen Hugh-Jones, 1–46. Cambridge, MA: Cambridge University Press.

Fair, Laura
1994 Pastimes and politics: A social history of Zanzibar's Ng'ambo community, 1890–1950. Ph.D. dissertation, University of Minnesota.

Garnert, Jan
1993 *Anden i lampan. Etnologiska perspektiv på ljus och mörker*. Stockholm, Sweden: Carlssons Bokförlag.

Garud, Raghu and Peter Karnøe
2005 Distributed agency and interactive emergence. In *Innovating Strategy Processes*, edited by Steven W. Floyd, Johan Roos, Claus D. Jacobs and Franz W. Kellermanns, 88–96. Oxford: Blackwell Publishing Ltd.

Garvey, Pauline
2005 Domestic boundaries: Privacy, visibility and the Norwegian window. *Journal of Material Culture* 10(2): 57–76.

Gullestad, Marianne
1992 *The art of social relations: Essays on culture, social action and everyday life in modern Norway*. Oslo: Scandinavian University Press.

Hastrup, Kirsten
1998 *A place apart: An anthropological study of the Icelandic world*. Oxford: Clarendon Press.

Miller, Daniel
1998 Why some things matter. In *Material cultures: Why some things matter*, edited by Daniel Miller, 3–21. Chicago: The University of Chicago Press.

Moore, Henrietta L.
1996 [1986] *Space, text and gender: An anthropological study of the Marakwet of Kenya*. New York and London: The Guilford Press.

Nye, David E.
1990 *Electrifying America: Social meanings of a new technology, 1880–1940*. Cambridge, MA: Massachusetts Institute of Technology Press.

Ortner, Sherry B.
1999 Thick resistance: Death and the cultural construction of agency in Himalaya mountaineering. In *The fate of "culture": Geertz and beyond*, edited by Sherry B. Ortner, 136–165. Berkeley: University of California Press.

Pantzar, Mika
1997 Domestication of everyday life technology: Dynamic views on the social histories of artifacts. *Design Issues* 13(3): 52–65.

Tacchi, Jo
1998 Radio texture: Between self and others. In *Material cultures: Why some things matter*, edited by Daniel Miller, 25–45. Chicago: The University of Chicago Press.

Wilhite, Harold
2008 New thinking on the agentive relationship between end-use technologies and energy-using practices. *Journal of Energy Efficiency* 1(2): 121–130.

Wilhite, Harold, Hidetoshi Nakagami, Takashi Masuda, Yukiko Yamaga, and Hiroshi Haneda

1996 A cross-cultural analysis of household energy-use behaviour in Japan and Norway. *Energy Policy* 24(9): 795–803.

Winther, Tanja

2005 Current styles. Introducing electricity in a Zanzibari village. Ph.D. dissertation, University of Oslo, Norway.

2008 *The impact of electricity: Development, desires and dilemmas.* Oxford: Berghahn.

2011 Les rapports entre État et citoyens à Zanzibar. Un récit ethnographique à partir de la fourniture d'électricité. *Politique Africaine* 121: 107–25.

2012 Negotiating energy and gender: Ethnographic illustrations from Zanzibar and Sweden. In *Development and environment: Practices, theories, policies*, edited by Kristian Bjorkdahl and Kenneth Bo Nielsen. Trondheim, Norway: Akademica Publishing.

CHAPTER NINE

Emergency Power: Time, Ethics, and Electricity in Postsocialist Tanzania

Michael Degani

INTRODUCTION

Before Ali passed away in 2011, the lifelong resident of Dar es Salaam had spent thirty years working at a dance and social club, bringing him into contact with all manner of useful civil servants. When, in the early 1990s, he noticed a manager for the state power utility Tanesco eyeing a barmaid, he "made arrangements" (*nilimfanyia mpango*) and discreetly brokered her services. The kindness did not go unappreciated, and a decade-long exchange of gifts and favors followed. Tanesco workers soon routed a cable from the neighboring clinic to the club's sound system and refrigeration. They left the metered street-side line connected to the lights, furnishing the club with cheap power and the appropriate suggestion of legality. Over the years, the manager steered inspection teams away from the club while Ali provided crates of beer on Christmas, soda for funerals, and use of the establishment for weddings and celebrations. When the manager transferred to another district, he made sure to apprise his replacement of the friendship, tacitly institutionalizing it.

Ali lived in a crowded Swahili-style compound, where he was responsible for maintaining the property and collecting tenants' bills. Around 2002, when electricity tariffs doubled, he arranged the services of a *kishoka* (pl. *vishoka*), a freelance electrician with ties to the utility. For a fraction of the actual cost, the *kishoka* would bring the bill to his contacts in the Tanesco office, mark the account as paid, and return with an "official" receipt. This arrangement worked well for a year or so, until a disconnection team arrived at the compound, citing an outstanding debt and aiming to cut the power. With feigned indignation, Ali presented his receipts—only to be told they were counterfeit. The *kishoka*, he soon learned, had been printing out receipts elsewhere and simply pocketing the money. While Ali trusted he was getting authentically forged receipts, the forgeries themselves were fake.

From Kula shells to ritual masks, cattle to cars, anthropologists have long followed the careers of valuable objects to understand social worlds (Malinowski 1961[1922]; Kopytoff 1986; Comaroff and Comaroff 1992; McGovern forthcoming). In urban Tanzania, the circulation of electrical current is no less illuminating. Its unofficial pathways—friendship and sex, cold cash, or false documents—grow more circuitous and more direct, tracing out the laminations of Tanzania's late-socialist and early capitalist regulatory environments. They reveal a city shaped by an ethos that many of my informants identify as *ujanja*—trickery, or cleverness. In this chapter I analyze the deepening links between *ujanja* and electrical power in the metropolis of Dar es Salaam.[1] I show how corrupted reforms rendered electricity less affordable and reliable at the very moment its economic and cultural significance was expanding, and explore the gray-market labor and transactions that have in turn colonized its ailing infrastructure. Ultimately, I suggest, in these clever redistributions of electricity we can discern a more global dimension of the Tanzanian national experience—what Jane Guyer has called an "evacuation of the near future" (2007:409).

The chapter is divided into three sections. Part 1 sketches Tanzania's energy sector in a broader context of neoliberal reform. Ali's experiences took place as Tanzania's once-socialist government attempted to privatize its national power utility. The ensuing corruption scandals highlight the complex technopolitical field that links state and electrical power in the developing world; the forces of seasonal rainfall, ideological conviction, government rent-seeking, and donor pressure never quite aligned, leading to a central "paradox" of Dar es Salaam today: "unprecedented economic growth despite a plainly failing infrastructure" (Brennan and Burton 2007:65). Part 2 explores the structure of time underlying this paradox, in which the sense of progress toward a near future diminishes. Instead, by dint of force or opportunity, actors are oriented to short-term horizons of calculation. Part 3 explores the informal economy of electricity that has emerged in these foreshortened horizons. It describes how the quasi-trickster figure of the *kishoka* brokers unofficial access to the grid and, in the eyes of those who solicit him, reflects the moral and institutional ambiguities of their nation's postsocialist period.

PART 1. HORIZONS OF REFORM

On one level, the story of energy privatization begins in 1992, when the water level in Tanzania's Mtera Dam reservoir plummeted. The dam was the heart of the country's predominantly hydropower grid, accounting for over 50 percent of its electricity generation. Worryingly, the Great Ruaha River,

the dam's main tributary, had briefly dried up, and Dar es Salaam was forced into daily power rationing for most of the next two years.

The outages also coincided with a new phase in Tanzania's liberalizing political economy. Like many postcolonial nations, Tanzania saw positive growth in the first decades of independence, culminating with President Nyerere's 1967 adoption of Ujamaa, a national project of African socialism. By the late '70s, however, Tanzania was contending with regional conflict, oil shocks, declining industrial production and agricultural exports, and a messy program of collective villagization. In 1986 Nyerere submitted to structural adjustment programs driven by bilateral donors and international financial institutions (IFIs) such as the World Bank and IMF. Lending was made available upon conditions of currency devaluation, trade liberalization, constraints on wage increases, and cuts in public services. The reforms precipitated a deep economic crisis that was mitigated by a flourishing informal economy of petty trades and services (Tripp 1997).

By the early 1990s the shock treatments of structural adjustment had deepened into a sustained vision for the developing world. Good governance was meant to "unleash" civil society and private investment (Rakodi 1997; Tripp 1997; Swilling 1997; Ferguson 2002). The energy sector, accordingly, was a major object of intervention (World Bank 1993; Wamukonya 2003). A small but influential community of private consultants and think tanks associated with IFIs maintained that energy sectors in the developing world required private capital to improve their low rates of access, technical loss, and underinvestment (Gratwick and Eberhard 2008:3949). Over the next decade, this idea became codified in a "standard model" of power reform, a sequence of steps designed to unbundle state power monopolies into private generation, transmission, and distribution components. The first step involved contracting independent power producers (IPPs) to generate and sell electricity to state power companies.

While the Tanzanian government embraced the standard model, its execution was soon entangled in the broader transformation of national politics. The year 1993 also coincided with the first run-up to multiparty elections, making the cuts both economically damaging and politically embarrassing. The ruling Party of the Revolution (*Chama cha Mapinduzi*, CCM) blamed the low reservoir levels at Mtera on drought and environmental degradations occurring upstream of the Great Ruaha. In truth the wet season had quite reliably ensured water in the reservoir; the real culprit for the electricity shortages was more likely poor planning (Walsh 2012:306). Nevertheless, since the early 1990s low seasonal rainfall has indeed created complicated variables for hydropower planning, and the system has suffered periodic water shortages up to the present day. As prelude or pretext, insufficient hydropower has been crucial to CCM's incorporation of unexpected emergency measures into grid management. The onset of multiparty elections and the Mtera "drought" pushed government players into a reactive, crisis-driven version of the energy reform process—a process and narrative of crisis open to political exploitation.

A Tale of Two Power Plants

In 1994, spurred by drought and the reform agenda, Tanesco entered into negotiations with a private power company, Songas, for a World Bank–financed project to construct a 190-megawatt natural gas power plant. But negotiations lagged as Tanesco officials diverted their attention to Independent Power Tanzania Ltd. (IPTL), a joint venture between a Tanzanian company and a consortium of Malaysian corporations backed by state banks. In 1995 Tanesco signed a memorandum with IPTL for short-term emergency power, which then curiously morphed into a lucrative twenty-year diesel generation contract. The decision merited suspicion; IPTL, as many critics point out, was both expensive and redundant. In 1998 it was revealed that IPTL commissioned medium- rather than slow-speed diesel turbines—a significantly cheaper option than agreed upon in the tender—while maintaining the original project costs. Factions in the energy ministry, Tanesco, and parliament balked at the inflated costs, and the World Bank temporarily halted Songas negotiations in protest. Pro-IPTL supporters shot back, defending IPTL on the grounds that it represented a "South-South cooperation" and that its opponents were stooges to Northern donor interest (Cooksey 2002). President Benjamin Mkapa was forced to refer the case to international arbitration, where systematic bribery surrounding the tender was alleged but never proven. Ultimately, the court ruled in favor of Tanesco and lowered the project cost. Nevertheless, for IPTL's opponents it was a Pyrrhic victory. In 2001, after more than half a decade of expensive delays, Tanesco announced that both plants would be coming on line.

Thus in the 1990s hydro shortages and political maneuvering led to the unexpected incorporation of not one but two private thermal generation plants, a situation that both emerged from and undermined the standard model of energy privatization. By 2005, the two power plants represented a significant transformation of the national grid. Thermal sources (natural gas and diesel) outsupply hydropower, comprising over 50 percent of all power produced. Financially, Tanesco now routinely operates at a loss. IPTL was one of the most costly plants on the continent to construct and operate, even after the court-ordered reduction. Moreover, Tanesco pays an average of $13 million per month for the plants' fuel and capacity charges—notably including the expensive diesel fuel needed to run IPTL turbines—well over half of its revenue (Gratwick et al. 2006:46). The financial strain negatively affected the commercialization of electricity supply. This interaction will be key to understanding the difficult conditions in which electricity consumers in Dar now find themselves.

The Problems of NetGroup Solutions

With Songas and IPTL online, Tanesco turned to the next step on the standard model checklist: commercializing utility operations. In 2002 the government hired a private South African management company, NetGroup Solutions, to directly run Tanesco. NetGroup's performance was "mixed" (Ghanadan and Eberhard 2007:30), and its contract was terminated in 2006. NetGroup managers focused heavily on revenue collection, partly because it was the clearest mandate of their contract, and partly because it was the basis on which their lucrative bonus structures were calculated. Tariffs were increased by a total of 28 percent while socialist-era cross-subsidies were reversed. NetGroup increased tariffs for residential and light commercial consumers by 39 percent while lowering industrial tariffs by as much as 28 percent. The residential "lifeline" subsidy for poor households was cut from 100 to 50 kilowatt-hours per month (Ghanadan and Eberhard 2007:22). Overall residential electricity bills tripled by 2005, bringing considerable strain to consumers like Ali.

These practices, as one Tanesco manager explained to me, were part of an attempt to instill financial discipline in a culture still "hungover" from a socialist era of subsidized services and soft budget constraints. Tanesco workers were offered voluntary retrenchment packages, cutting down the workforce by 21 percent. NetGroup promoted prepaid meters among middle-class consumers, selling vouchers in commercial spaces such as petrol stations and supermarkets. It also embarked on a widespread disconnection campaign of poorer, indebted consumers, cutting 3 percent of the customer base each month (Ghanadan 2008:91).

Yet even as revenue collection improved, most other metrics of the company's health were poor. Transmission and commercial losses did not change over the contract period (about 15 and 10 percent, respectively), nor did the rates of new service lines, growing at a sluggish 6 percent (Ghanadan and Eberhard 2007:28). NetGroup had planned to improve technical and customer service issues. Instead, the recuperated revenues were shuffled right back out to pay for unusually expensive Songas and IPTL plants coming on line, which due to a series of mid-decade droughts were soon running at full capacity. In 2005, payments to IPTL and Songas accounted for 69 percent of

Tanesco's yearly revenues (Gratwick and Eberhard 2006:46). In short, despite squeezing customer revenue through raised tariffs, disconnection, and prepaid meters, Tanesco had barely enough revenue to cover generation costs, leaving its service provision to deteriorate.

In 2005 the Tanzanian Parastatal Reform Commission tabled Tanesco's privatization, citing a lack of private-sector interest. Similar denouements have occurred across much of Africa's power sectors; after the "demise" of the standard model, the question is just what, exactly, has taken its place (Gratwick et al. 2008). For Dar es Salaam's growing number of electricity users, the answer is a series of austerity measures that have done little to reduce blackouts or persistent barriers to access. In addition to growing expense, consumers wait weeks for repairs on downed transmission lines and blown transformers or months for new connections. Those with conventional meters are subject to arbitrary or predatory billing practices, and frequent power cuts have continued up to the present day.

PART 2. NO [NEAR] FUTURE

Part 1 described the major dynamics of Tanzania's stalled-out energy sector privatization and its unintended consequences for urban residents in Dar es Salaam. This section considers how such dynamics alter the experience of time and possibility in Tanzania's postsocialist era. It argues they illustrate Jane Guyer's notion of "the evacuation of the near future": the practice of "reasoning toward . . . immediate situations and a [simultaneous] orientation to a very long term horizon" (2007:409). Guyer suspects that the evacuation of the near future is *the* temporal logic of the last thirty years, common to such ascendant world historical forces as evangelical Christianity and economic monetarism. Both replace a cultural confidence in the human/institutional capability to plan with faith in the enigmatic shocks of the market or salvation. To this series we might add political liberalization and oil and, in their interplay within Tanzania's energy sector, note some elective affinities in the way they organize time.

Temporalities of the Air Conditioner and the Veranda

What Guyer calls the "evacuation of the near future" is most evident in the short-term calculations of Tanzania's national leadership. In many ways, "emergency power" was not only a useful pretext for rent-seeking but an apt summation of political strategy in the multiparty era. Cooksey and Kelsall write:

> Tolerance of damaging rent-seeking deals like IPTL . . . can be partly explained by electoral pressures. As political competition within the ruling party and between the ruling and opposition parties increases, so short-term considerations come to dictate political rent-seeking strategies. One interviewee stated: "[These days] nobody is thinking longer-term." Panic rent-seeking prior to elections allows any initiative to proceed, however shoddy. The same interviewee said: "The political apex cannot stop a dubious deal because nobody is asking and nobody knows what is dubious." (2011:34)

To dismiss (though certainly not to excuse) such scandals as rapacious corruption is to ignore how they emerge from the reality of a political system adapting to the pressures of electoral competition. CCM has managed to remain in power since the days of single-party rule, but its hold is tenuous. Economic liberalization undermines its centralized structure by importing new opportunities for enrichment and influence by ad hoc back-room alliances with local businessmen and global corporations—what Emanuel Terray terms a "politics of the verandah" (in Kelsall 2002). In order to remain competitive, politicians must secure sources of "emergency power"; they must court and control these new alliances while maintaining a façade of democratic accountability.

Deals struck out on the veranda undercut the kind of technocratic planning championed, even fetishized, by an opposing "politics of the air conditioner" (in Kelsall 2002)—transnational donor projects exemplified by the World Bank's involvement in the energy sector. The question of Songas or IPTL, of a country powered on local natural gas or imported diesel fuel, speaks to this uneasy coexistence between a donor-guided "politics of the air conditioner" and a national elite "politics of the veranda" occurring right behind it. It is not yet clear how these two modes will be reconciled—whether, say, government rent-seeking can be consolidated and channeled to developmentalist ends—but the outlook is troubling. In 2006, high-level government figures, including Prime Minister Edward Lowassa, staged a farcical repetition of IPTL. They awarded an emergency power contract to a fake "briefcase company," Richmond Ltd, leaving the country "in the dark" for over three months and costing the country millions of dollars in stolen funds and economic paralysis (Brewin 2011).

Time Heats Up

In ways that seem to reinforce the short-term horizons of statecraft engendered by panic rent-seeking, the materiality of oil exerts its own effects on the national timescape. By way of contrast, let us first consider hydropower. For hydropower to provide reliable energy, the wet and dry season ecology central to so much of East African life (see Evans-Pritchard 1940) must be accommodated in countercyclical infrastructures—reservoir storage, spinning reserve capacity—that extend to the "near future." This foresight is especially necessary in Tanzania, where poor rainfall and rapidly growing demand are already taxing an undercapitalized grid. Oil, by contrast, runs fast and hot. A highly concentrated source of energy, it can be transported swiftly and in large volumes, a testament to the fast capitalism and global supply chains in which the human condition is now suspended (Watts 1992; Tsing 2009). But oil's fluid materiality makes for slippery politics. It moves through a capital-intensive infrastructure that labor, civil societies, or national governments find difficult to regulate (Mitchell 2011; Ferguson 2005). In the OPEC era its price and availability are volatile, contributing to short-sighted geopolitical conflicts. Even the nascent postcarbon imagination is fuzzy on the intermediate mechanics of transition, fixated on visions of green abundance or dystopian resource wars (Boyer 2011; Harbach 2007). In its own modest way, Tanzania is part of this global timescape in which short-term interest and end-time faith fold in upon each other. Chained to the IPTL contract, Tanesco is importing oil, and all of its temporal entailments, as a supplement to poor hydrological conditions. Its expense helps foreclose investment in long-term grid improvement, locking in a rhythm of punctuated crisis. As I describe below, the costs of thermal generation are passed on to consumers, further drawing them into its temporal logic. Paraphrasing Weber (2002 [1930]), we might say that Tanzania's light cloak of emergency power has inadvertently thickened to a carbon shell.

I have suggested that the politics of the veranda and air conditioner collapse the national timescape into an emergency "now" and a faith in the "yet to come" (see Simone 2004:5), while actions oriented to midrange near futures are difficult to sustain. We have seen how actors on the veranda operate rent-seeking schemes according to short-term goals of preserving political power in an increasingly competitive environment. The temporal sensibility of air-conditioned actors, with their ultimate goal of unleashing free market growth and near-religious faith in its power (Comarroff and Comaroff 2001), is correspondingly long-term. Though my "tale of two power plants" is a story of their conflict, they are two sides of the same temporal logic. Whether long-term reliance on natural gas or short-term importation of diesel, both displace hydropower, a resource whose temporality is that of the midterm, the renewable (cf. Ferry and Limbert 2008).

At the same time, it is important to contextualize the "evacuation of the near future" within a larger pattern of cultural displacements and reconfigurations. Just as hydropower, though diminished, still shapes the logics of grid management, Tanzanian politics is shaped by the moral and institutional afterlife of socialism. In some ways air-conditioner elites, with their emphasis on anticorruption, channel Nyerere's socialist spirit of moral rectitude, while veranda elites are fighting hard to preserve his institutional body—the rule of CCM. If in new guises, Tanzanians are still contending with the gap between ideology and politics that was characteristic of socialism—indeed, that has been characteristic of aggressively modernizing states in general (Hyden 1980; Scott 1998; see also Ferguson 1990). The gray market for electricity displays this mix of rupture and continuity, of a near future that is not quite the socialist past and not yet the capitalist promise.

PART 3. THE CITY ELECTRIC

Faced with shortage, expense, and unreliability, urban residents have improvised clever arrangements to caulk the gaps in Tanesco's service delivery. At first glance they simply belong to a more general system of "decentralized rent-scraping"—petty bribes to grease state machinery. But in the wake of NetGroup's reforms, many Tanesco workers are not quite state agents. They are retained on short-term contracts, creating a pool of underpaid, part-time, retrenched, or amateur technicians looking to supplement their incomes. Somewhat pejoratively, they are known as *vishoka*, or hatchets.

Hatchets and Trickery

Vishoka is a fluid designation. Tanesco's Company Secretariat defines *vishoka* as "staff or non-staff people" who "commit unethical behaviors" (Tanesco n.d.). Newspaper articles describe them as "unofficial agents" or "hordes of self-trained technicians" (Lazaro 2008; *The Guardian* 2011). Informants have characterized them as conmen or sometimes simply emphasize their status as unlicensed. This terminological imprecision reflects Tanesco's new system of flexible labor and the material properties of the grid, in which the distribution of uniforms, cables, meters, and seals and paperwork creates distinct possibilities for arbitrage. An outline of this sociotechnical ecosystem follows.

At bottom are *mafundi ya mtaani*—street technicians with no official affiliation with Tanesco. These technicians might scrape by a living with small-scale repairs of frayed cables or broken electrical appliances, or find wiring or installation work in the big hotels and compounds that are popping up with Dar's construction boom. Yet they might also be persuaded to expedite a service line application to Tanesco or illegally reconnect an indebted household to the grid. They are often associated with conmen, partly because they are seen as fakes or amateurs lacking technical education, and partly because they are sporadically self-employed and therefore, like many other hustlers in Dar es Salaam's informal economy, more inclined to risk scamming customers. They are known to don the signature green Tanesco coat or ID card and impersonate salaried Tanesco employees in order to extort customers.

Either through informal apprenticeships or friendship, street technicians partially overlap with Tanesco *vibarua* (sing. *kibarua*)—a generic term for day laborer. Tanesco hires *vibarua* on short-term renewable contracts for about three dollars a day, working as line men, meter men, or drivers. *Vibarua* enjoy marginally steadier employment than street technicians but often supplement their incomes with the odd repair job or side trade in Tanesco uniforms, meter seals (the main device for detecting meter tampering), or "cut out" fuses that are removed from indebted households by disconnection teams or revenue protection units. *Vibarua*, finally, with their uniforms and access to Tanesco vehicles and ladders, sometimes partner with street electricians to tamper with meters, providing a measure of discretion against the prying eyes of neighbors or the police. If he is too nervous or inexperienced to do it himself, a street electrician might broker a meter-tampering job to his *kibarua* friend, either overtly or covertly pocketing about 20 percent.

The top tier of electrical work concerns salaried workers with secure employment and benefits. While mostly confined to the office in the wake of NetGroup reforms, some salaried workers do remain out in the field, comprising the regional disconnection teams and revenue protection units. Compared to *vibarua* or street technicians, most remaining salaried workers predate the NetGroup era and are older, more experienced, and in some ways bolder about pursuing sideline revenues. Inspectors may be negotiated with on the spot not to disconnect an indebted or tampered meter, which sometimes evolves, as informants like Ali have recounted, into an ongoing quasi-protection-racket relationship. Foremen in service line construction may sell expedited installation, sharing a small portion of the premium with their *vibarua* crew, while surveyors may accept payment from street electricians to approve applications for houses the latter have wired. For their part office workers may collude in expedited arrangements or engage in varieties of underbilling.

I have given a rather schematic description of how Tanesco's constellation of labor and technical objects contours the pathways of electrical current. In reality street electricians, *vibarua*, and staff may impersonate, denounce, or cooperate with one another, or combine these strategies over the course of a transaction. Nevertheless, for all their empirical mercuriality, we might think of *vishoka* as occupying a distinct structural position between utility and consumer, a "thirdness" that gives the analytically flat notion of an "informal economy" of electricity its historical dimensions (see McGovern 2010:69–71). To borrow the older language of Africanist social anthropology, dealings with *vishoka* cut across the consumer-utility relationship, triangulating a series of segmented oppositions (Fortes and Evans-Pritchard 1940); depending on the circumstances they may act as subsets of citizen or state—or as the point of common contention. Tanesco management repudiates *vishoka* as unskilled and untrustworthy while relying on them for cheap labor. Residents turn to them for bureaucratic shortcuts while fearing legal repercussions, extortion, or scams, sometimes reporting them to Tanesco management if it is judged to serve their interests.

This kaleidoscopic bureaucratic structure means that distributing electrical power encourages an ethic of *ujanja*, an entrepreneurial citizenship or skill set not specified in the rights and duties of abstract legal subjects (cf. Mbembe and Roitman 1995). In a relationship of neither kinship nor contract (Hart 2000), *vishoka*, consumers, and bureaucrats must proceed by trust while, as Ali learned, anticipating its abuse. The ability to negotiate, spot a scam, exploit the ambiguity of official signs and symbols, cultivate social ties, and tacitly communicate intentions all work to partially incorporate *vishoka* as unofficial state agents. Indeed, *vishoka* form a parallel bureaucracy precisely in their capacity to cut pathways through the official one.

Like the connected temporalities of thermal power generation and multiparty politics, the near future falls out of this everyday urban ethic. Power is consumed in anticipation of blackouts, breakdowns, sudden inspections, new rounds of negotiations over payments, or scams. It is punctuated by delays in service, by sudden tariff increases, by the pointillist economic brushwork of incessant "top-ups" on prepaid meter accounts hovering just over zero. These practices are the material counterpart to a jury-rigged sociality of payments and provisional trust, of dealings with "hatchets" whose sharp edges make them both useful and tricky to handle.

Socialist Afterlives

As part of a more general involution of street-level bureaucracy, *vishoka* are symptoms of the contemporary moment—of a thinned-out, post-adjustment African state (Blundo 2006). However, brokerage is also a well-developed social logic in many African societies (de Sardan 1999), and as such represents an old solution to the new problem of electricity shortage. Rebecca

Ghanadan has shown that NetGroup reforms such as prepaid meters and commercial tariffs transform electrical power from a basic service to a partible commodity, like sugar, soap, or petrol (Ghanadan 2008; 2009). Yet in socialist Tanzania, such state-controlled commodities were routinely smuggled, adulterated, or otherwise transferred to unregulated markets (Bagachwa 1995). *Vishoka* recall the (openly) secret networks of illegal distributors and street vendors that allowed the formal economy to function. Like the capitalist "parasites" or "idle thugs" of state contempt (Campbell 2008:172), they are subject to ritual cycles of harassment and denunciation, even as they are an ineradicable, in some ways useful, part of the city's public life. Although such hustlers had an especially negative ideological charge during Tanzania's socialist period, they have roots in a British colonial anti-urbanism and are woven into the fabric of Dar's popular culture (Burton 2007; Lewinson 1998; 2003). They are *wajanja*—clever town men who install themselves in the city in search of not honest productive labor but easy spending money and superfluous pleasures.

Thus, while clever town men may be familiar, their increased presence within Tanesco highlights the ambiguity of the historical moment. Insofar as *vishoka* recall socialist-era fixers trafficking in controlled commodities, they stand as evidence of CCM's inability to reform an economic sector key to capitalist development. Pointing to the success of the privatized telecommunications sector and their ubiquitous mobile phones, residents often suggest to me that, indeed, they welcome private competition when it means reliable service. At the same time, *vishoka* are evidence of the antisocial greed wrought by capitalism, symptoms of a once-egalitarian socialist party embracing a "moral and political laissez faire state posture" (McGovern n.d.; cf. Askew and Pitcher 2006). Many technicians and consumers rhetorically assert that Tanzania's premier *vishoka* are government leaders themselves, and that citizens involved in electricity theft are merely following by example. *Mtoto wa nyoka*, they observe ambivalently, *ni nyoka*—the child of a snake is a snake.

CONCLUSION

When I asked Ali if he was angry about the "fake" forgeries, he shrugged and said there wasn't much he could do in retaliation. Instead, he contacted another friend at the Tanesco office. For about ninety dollars he managed to wipe the debt off his account and obtain an unregistered, prepaid meter with a slowed recording mechanism. It is not clear how long he expected this new round of arrangements to last, but as I hope to have demonstrated, there are good reasons for such provisional pathways of electrical current. They emerge through a distinct assemblage of objects, actors, and energies in postsocialist Tanzania: drought and donor pressure, the technologies of emergency power contracts and thermal generation, the emboldened schemes of a fraying party contending with democratic elections, the austerities of neoliberal reform, and African urban residents' long history of managing everyday economic crises. These processes have suffused electricity with an ambiguous ethic of *ujanja*, allowing residents, technicians, and bureaucrats to improvise an economic order in the foreshortened horizons of urban Africa.

NOTE

1. Research for this chapter was carried out as part of dissertation fieldwork in Dar es Salaam from July 2011 to December 2012. I wish to gratefully acknowledge support from the National Science Foundation, the Social Science Research Council, and the Wenner Gren Foundation. In Tanzania, the Committee of Science and Technology, the University of Dar es Salaam, and Tanesco provided welcoming institutional homes. Finally, while writing this chapter I benefited from rich and insightful conversations with Mike McGovern, James Brennan, Douglas Rogers, Brenda Chalfin, Mohammed Yunus Rafiq, Kulwa Msonga, Rashid Salum Ali, Kamari Clarke, Tanja Winther, and Stephanie Rupp.

REFERENCES

Askew, Kelly M. and Pitcher, Ann M.
2006 African socialisms and postsocialisms. *Africa: The Journal of the International African Institute* 76(1): 1–14.

Bagachwa, Mboya S. D. and A. Naho
1995 Estimating the second economy in Tanzania. *World Development* 23(8): 1387–1399.

Blundo, Giorgio
2006 Dealing with the local state: The informal privatization of street-level bureaucracies in Senegal. *Development and Change* 37(4): 799–819.

Boyer, Dominic
2011 Energopolitics and the anthropology of energy. *Anthropology News* 52(5): 5–7.

Brennan, James R. and Andrew Burton
2007 *Dar es Salaam: Histories from an emerging African metropolis*. Dar es Salaam: Mkuki na Nyota Publishers.

Brewin, David
2011 Corruption—the latest. *Tanzania Affairs* 99: 15–19.

Burton, Andrew
2007 The Haven of Peace Purged: Tackling the undersirable and unproductive poor of Dar es Salaam, c. 1950s–1980s. *The International Journal of African Historical Studies*. 40(1): 119-151.

Campbell, John R.
2008 Corruption and the one-party state in Tanzania: The view from Dar es Salaam, 1964–2000. In *Enduring socialism: Explorations of revolution and transformation*, edited by Harry G. West and Parvathi Raman. Ithaca, NY: Cornell University Press.

Comaroff, Jean and John L. Comaroff
1992 *Ethnography and the historical imagination*. Boulder, CO: Westview Press.
2001 *Millennial capitalism and the culture of neoliberalism*. Durham, NC: Duke University Press.

Tanesco Company Secretariat
n.d. *Tanesco*, http://www.tanesco.co.tz/index.php?option=com_content&view=article&id=75&Itemid=216 (accessed March 11, 2012).

Cooksey, Brian
2002 The power and the vainglory: Anatomy of a $100 million Malaysian IPP. In *Ugly Malaysians? South-South investments abused*, edited by Jomo Kwame Sundaram. Durban, South Africa: Institute for Black Research.

Cooksey, Brian and Tim Kelsall
2011 *The political economy of the investment climate in Tanzania*. London: Africa Power and Politics Programme.

De Sardan, J. P. Olivier
1999 A moral economy of corruption in Africa? *Journal of Modern African Studies* 37(1): 25–52.

Evans-Pritchard, Edward Evan
1940 *The Nuer, a description of the modes of livelihood and political institutions of a Nilotic people*. Oxford: Clarendon Press.

Ferguson, James
1990 *The anti-politics machine: "Development," depoliticization, and bureaucratic power in Lesotho*. Cambridge: Cambridge University Press.
2002 Spatializing states: Toward an ethnography of neoliberal governmentality. *American Ethnologist* 29(4): 981–1002.
2005 Seeing like an oil company: Space, security, and global capital in neoliberal Africa. *American Anthropologist* 107(3): 377–382.

Ferry, Elizabeth and Mandana E. Limbert
2008 *Timely assets*. Santa Fe, NM: School for Advanced Research Press.

Fortes, Meyer and Edward E. Evans-Pritchard
1987 (1940) *African political systems*. London: Kegan Paul International.

Ghanadan, Rebecca H.
2008 Public service or commodity goods? Electricity reforms, access and the politics of development in Tanzania. Ph.D. dissertation, Energy and Resources Group, University of California, Berkeley.
2009 Connected geographies and struggles over access: Electricity commercialisation in Tanzania. In *Electric capitalism: Recolonising Africa on the power grid*, edited by David A. McDonald, 400–436. Cape Town, South Africa: HSRC Press.

Ghanadan, Rebecca and Anton Eberhard
2007 Electrical utility management contracts in Africa: Lessons and experience from the Tanesco-Netgroup Solutions Management Contract in Tanzania, 2002–2006. MIR Working Paper. Cape Town, South Africa: University of Cape Town, Management Program in Infrastructure Reform and Regulation.

Gratwick, Katharine Nawaal and Anton Eberhard
2008 Demise of the standard model for power sector reform and the emergence of hybrid power markets. *Energy Policy* 36(10): 3948–3960.

Gratwick, Katharine Nawaal, Rebecca Ghanadan, and Anton Eberhard
2006 Generating power and controversy: Understanding Tanzania's independent power projects. *Journal of Energy in Southern Africa* 17(4): 39–56.

Guyer, Jane
2007 Prophecy and the near future: Thoughts on macroeconomic, evangelical, and punctuated time. *American Ethnologist* 34(3): 409–421.

Harbach, Chad
2007 The end, the end, the end. *n+1*, December 2. http://nplusonemag.*com/the-end-the-end-the-end* (accessed March 25, 2012).

Hart, Keith
2000 Kinship, contract, and trust: The economic organization of migrants in an African city slum. In *Trust: Making and breaking cooperative relations*, edited by Diego Gambetta, 176–193. Oxford: University of Oxford.

Hyden, Goran
1980 *Beyond* ujamaa *in Tanzania: Underdevelopment and an uncaptured peasantry*. Berkeley: University of California Press.

Kelsall, Tim
2002 Shop windows and smoke-filled rooms: Governance and the re-politicisation of Tanzania. *Journal of Modern African Studies* 40(4): 597–619.

Kopytoff, Igor
1986 The cultural biography of things: Commoditization as process. In *The social life of things: Commodities in cultural perspective*, edited by Arjun Appadurai, 64–91. Cambridge: Cambridge University Press.

Lazaro, Happy
2008 Tanesco devastated by illegal connections. *Arusha Times*, December 12.

Lewinson, Anne
1998 Reading modernity in urban space: Politics, geography and the informal sector of downtown Dar es Salaam, Tanzania. *City & Society* 10(1): 205–222.
2003 Imagining the metropolis, globalizing the nation: Dar es Salaam and national culture in Tanzanian cartoons. *City & Society* 15(1): 9–30.

Malinowski, Bronislaw
1961 [1922] *Argonauts of the Western Pacific: An account of Native enterprise and adventure in the Archipelagoes of Melanesian New Guinea*. New York: Dutton.

Mbembe, Achille and Janet Roitman
1995 Figures of the subject in times of crisis. *Public Culture* 7(2): 323–352.

McGovern, Mike
2010 *Making war in Côte d'Ivoire*. Chicago: The University of Chicago Press.
n.d. The morality of liberty: Socioeconomic transition and bodily comportment in Guinea.
Forthcoming *Unmasking the state: Making Guinea modern*. Chicago: University of Chicago Press.

Mitchell, Timothy
2011 *Carbon democracy: Political power in the age of oil*. New York: Verso.

Rakodi, Carole
1997 *The urban challenge in Africa: Growth and management of its large cities*. Tokyo: United Nations University Press.

Scott, James C.
1998 *Seeing like a state: How certain schemes to improve the human condition have failed.* New Haven, CT: Yale University Press.

Simone, AbdouMaliq
2004 *For the city yet to come: Changing African life in four cities*. Durham, NC: Duke University Press.

Swilling, Mark
1997 *Governing Africa's cities*. Johannesburg, South Africa: University of Witwatersand Press.

The Guardian
2011 Tanesco Must Heed Warning by Ngejela, January 10.

Tripp, Aili Mari
1997 *Changing the rules: The politics of liberalization and the urban informal economy in Tanzania*. Berkeley: University of California Press.

Tsing, Anna
2009 Supply chains and the human condition. *Rethinking Marxism* 21(2): 148–176.

Walsh, Martin
2012 The not-so-great Ruaha and hidden histories of an environmental panic in Tanzania. *Journal of Eastern African Studies* 6(2): 303–335.

Wamukonya, Njeri
2003 African power sector reforms: Some emerging lessons. *Energy* 2(1): 7–15.

Watts, Michael
1992 *Reworking modernity: Capitalisms and symbolic discontent*. Piscataway, NJ: Rutgers University Press.

Weber, Max, Peter Baehr, and Gordon C. Wells
2002 [1930] *The protestant ethic and the "spirit" of capitalism and other writings*. New York: Penguin Classics.

World Bank
1993 *The World Bank's role in the electric power sector: Policies for effective institutional, regulatory, and financial reform*. Policy Paper 11676. Washington, DC: The World Bank.

CONVERSATION 3
Electrification and Transformation

Michael Degani, Anna Garwood,
Thomas Love, Stephanie Rupp, Tanja Winther

The chapters in the third section of *Cultures of Energy* address the social and cultural impacts of changing structures of energy, in particular social change in the wake of the introduction of electricity. In the distinct cultural contexts examined by Thomas Love, Anna Garwood, Tanja Winther, and Michael Degani, it is apparent that changes in the technical circuitry of energy result in changes in social and moral circuitry within newly "electrified" communities. While energy circulates as an invisible force, it nevertheless becomes visible in the form of light and in the dynamics of people's social relationships. The structures and maintenance of energy systems are entangled with political ideologies, practices, and relationships; are implicated in changing moral values and practices; and shape interpersonal relationships. In their chapters and in conversations among the authors, it is evident that energy technologies are socially embedded; change to the technical system affects change in the social system. Within these chapters and the ensuing discussions, authors responded to the following questions and engaged with each other's responses. What impact does being on-grid or off-grid have on the social relationships of energy users? What is the relationship between different kinds of energy technologies and different kinds of social relationships? What is the relationship between political ideologies and energy technologies? In different political contexts, is energy viewed differently (e.g., as a public good, a state utility, or as a consumer good)? What is the relationship between the structure(s) of energy system(s) and the structures of sociality in communities that are recently "electrified"?

ANNA GARWOOD: Most of the rural communities where I've worked become "electrified" with distributed energy sources (such as community-scale micro-hydro or wind power). In these cases the electricity infrastructure was built by the community, and among other intended and unintended consequences, there seems to be an effect of tightening the social structures because community members now jointly manage the system. I am curious, however, how the social effects of becoming electrified vary between communities that are connected to the national grid and those that have electricity through their own sources but are still "off the grid" in a larger sense. What social effects from electrification are caused by the simple access to electricity (change in daily patterns, access to information, communication, television, and so on) and which effects are related to the source and ownership of that electricity?

TANJA WINTHER: Interesting questions! I have had the opportunity to compare systems that are centralized (such as in Zanzibar) and decentralized (as in Sunderbans, India). Although contextual factors obviously count, I have found some marked social and political differences based on the physical organization of the infrastructure structure itself. In a centralized system or grid, which is usually managed by a utility owned by a strong, centralized (or even authoritarian) government, the standardized technology, regulations, and routines tend to be forced on customers who desire electrical service twenty-four hours per day. But these electricity customers may for the first time find themselves in a precarious relationship with the state. They accumulate debt to the state-owned company, do not understand the bills (issued in English rather than more widely spoken languages such as KiSwahili), and find themselves in a relation of dependency vis-à-vis the state-owned utility company. By providing electricity, the state has entered people's living rooms, and people who oppose the government politically become particularly exposed to increased control and sanctions. For example, meter readers are known to report any antigovernment activity that they observe to state authorities. The relationship between utility and customers becomes deeply antagonistic; people who have the courage to engage in illegal connections to the electrical grid will do so, fostering a black market for electrical services and supply, fostering a climate of further suspicion and duplicity. Importantly, however, in villages where culturally valued authorities such as spirit mediums were involved in the changing energy framework, the process of electrification went smoothly after a period of negotiation and moral consideration. In cases where the electricity structure was forced on people within a short time, frame protests erupted, and in one place the project was even stopped. I think it is crucial to consider how actors on the interfaces between technical system and local context meet and interact.

Decentralized electrical systems in Zanzibar are run at least partly by people themselves, including control of consumption in a system without electrical meters but with fixed tariffs according to the agreed number of light points. This system of providing electricity in Zanzibar used to run well, and the level of trust in the technology itself and among the participants in the energy system was high. But because of the lack of technical capacity to expand the system as demand kept growing, supply became limited and those who had access to electricity tended to overconsume, as if wanting to ensure a stock of energy for their households. In this case, the technical system was socially embedded from an early stage, but the lack of financing mechanisms for expansion started the problems. Another limitation was that electricity was supplied for only four hours in the evening, while people gradually wanted access to more machines and more energy demanding technologies. The crux of the issue was how to make existing customers start thinking about expansion of services.

Structurally, I would highlight that in both cases there are excluded groups, whose position became further marginalized with the arrival of electricity. I am referring in particular to women's degree of involvement and benefits, which were limited because of existing structures of discrimination (male ownership of land and houses, inheritance rules, and a high divorce rate in Zanzibar, with a divorcée forced to leave her former husband's household). Electricity becomes extensions of houses, and groups who control houses also decide on matters related to electricity. Short-term benefits (immediate consumption) tend to overshadow the more long-term structural effects (less financial security). I have also seen that elders' power becomes reduced as new criteria for success (such as television and other prestige appliances) replace older ones (such several wives and numerous children).

TOM LOVE: For alternative energy technologies to be effective, their circuitry must not only be technically reliable but also mesh with social circuitry. In Alto Perú, Anna and I found that micro-grids worked well with households that were not only already clustered but also inhabited by people related to each other. While the idea has appeal, tying in as it does with traditional anthropological concerns with kinship and social space, and seems to me more than just a nice metaphor, do people find this idea of "social circuitry" useful?

TANJA WINTHER: The idea of "social circuitry" appeals to me, and even more if you consider the "circuit" aspect as the *momentary*, rather than permanent, structures and dynamics in which a given technology becomes enmeshed. However, in thinking of social circuitry as something enduring, one risks creating an image of a static social reality. Second, I would prefer thinking of circuits in sociotechnical terms, thus how technology and the social context together produce a particular system, a circuit. No doubt the more attuned a technology and its organization is toward existing social realities (socially embedded technology), the more likely its implementation is to be effective. Many factors may account for why you found mini-grids successfully implemented when people were related to each other than when they were. For example, power relations and who has decision-making power in the process is just one such possible factor.

A third reflection concerns the notion of effective technologies, which relates to the developers' perspective and reflects a given image of social change (technology has some direct effects, and success is measured according to the degree that outcomes become as prescribed). In my work I have also tried to look at unintended consequences and also what more profound transformations that new technology may trigger. In turn, more multifaceted goals could be attached to renewable energy interventions, such as achieving social justice, gender equality, and so on.

MIKE DEGANI: I think I can sign on to Tanja's concern that "social circuitry" has something a little soldered about it, a too-rigid sense of social relationships. On the other hand, that is an empirical question, isn't it? In any given time and place the rights and obligations between actors are more or less clearly defined. In Alto Perú the consequences of wronging a mother or uncle or cousin might be obvious enough to make the relationships necessary for using a micro-grid seem as functional and integrated as the technology itself. (Maybe Durkheim got it wrong and it is mechanical solidarity that is the really complex form of organization!)

A negative illustration that possibly supports your point, Tom: I've spent the last few months in Dar es Salaam riding around with the state power disconnection team. In certain older and poorer neighborhoods, there is "Swahili"-style housing, which consists of multiple individuals or families renting out a single room and sharing cooking space, bathrooms—and a single meter for the compound. This arrangement is a recipe for disconnection. One of the tenants is inevitably late on contributing to the bill payment, accusations proliferate about who is using more power than they should, and the conflicts roll on. Now, the big problem is simply that tenants aren't generating enough income to cover all their bills each month. And conflicts with landlords and the utility play a role too. But, pertinent to our discussion, tenants are often strangers—or, at best, friends—and do not feel particularly beholden to each other over the long term. The provisional, short-term, and often deferred nature of money and relationships in Swahili housing is incongruous with a power monopoly that is oriented to long-term continuous service.

Often the result is a kind of alternative shadow system based on theft, debt, and negotiations with particular utility workers. And personally I'm always tempted to see a metaphorical link between a dust-choked, tampered meter and erratic power supply on one hand and this jury-rigged system of payments and provisional trust on the other. Not exactly social circuitry, but the parallels are there. I only have brief impressions about renewable energy but maybe they are worth mentioning. The market for household solar power is just beginning to get going in Dar es Salaam. Solar energy is for wealthier private homeowners as it requires a lot of expensive upfront investment, and at the moment the technology isn't efficient enough to power many electrical objects. But I think conceptually it's sort of slotted next to the gas-powered generator. In true privatized neoliberal fashion it gives one autonomy from all the hassle of the state grid, but there is a bewildering array of makes and models and unless you really shell out for quality you can easily end up with a lemon.

ANNA GARWOOD: I wanted to follow up on Mike's comment about the household solar market. Household solar (and especially solar micro-credit schemes) are the big buzz in the world of international development these days. The organization I work for (Green Empowerment) and most of our NGO partners have been working on community-scale renewable energy projects—often integrating these into the state plan for rural electrification—instead of selling solar to relatively wealthier private homeowners. It has been interesting to watch the boom in small solar home systems (and solar lanterns) over the last ten years. Instead of conceptualizing electricity as a public good or state utility, the supporters of the big solar markets emphasize electricity as a consumer good. Of course the development banks endorse this privatizing strategy. Yes, in many of the places where we work in Latin America there is still an idea that the government should have a role in providing electricity, and thus a resistance to the neoliberal notion that electricity should be distributed through a private market of suppliers and consumers. One could trace the political trends in general from the left to the right by the history of the provision of electrification.

PART 4
Energy Contested: Culture and Power

CHAPTER TEN

Eco-risk and the Case of Fracking

Elizabeth Cartwright

INTRODUCTION

How do we, as anthropologists working on the concepts of energy, energy extraction, and the insatiable consumptive behaviors of humans of the present era, contribute to the transdisciplinary discourse needed to construct intellectual models that can account for both extant distributions of environmental illness and the ever-changing processual nature of disease realities? This is really the anthropology of the unknown, the invisible, the just beyond the senses in many cases. The effects of environmental pollution on human health and well-being are ambiguous, long term, and embedded in multiple, sometimes contradictory, scientific and political discourses. Much is not known about how the human body will react to exposures to various chemicals. The ambiguous nature of the effects of chemical exposures provides ample emotionally charged fuel for the political fires surrounding various extractive processes. In this chapter, I will focus on the issue of hydraulic fracturing, or fracking, as it is more commonly known.

My interest is in conceptualizing a way for anthropologists to move forward in both theorizing the details of the interactions between environment, health, and culture while simultaneously rearticulating ourselves with the thick description and ethnographic richness of lives lived inserted into terrains ever more modified by human creation and waste. This process is rooted simultaneously in biological and cultural realities.

During the process of fracking, large amounts of water, sand, and chemicals are blasted deep into the underground formations where natural gas is found.[1] The chemicals used are proprietary in nature and are increasingly being implicated in water pollution litigation in the western United States and elsewhere. According to the U.S. Environmental Protection Agency's website, the known environmental impacts are:

1. Contamination of underground sources of drinking water and surface waters resulting from spills, faulty well construction, or by other means;

2. Adverse impacts from discharges into surface waters or from disposal into underground injection wells; and
3. Air pollution resulting from the release of volatile organic compounds, hazardous air pollutants, and greenhouse gases.[2]

Energy extraction impacts environmental and bodily functioning at many levels. I argue in this chapter that we need to anthropologically attend to the geologic processes that result in biological exposures to the harmful chemicals in this process. Building upon that level of understanding, I argue for contextualizing these exposures in complex, local biologies. I trace the flow of subsurface water and of subsurface biopower (*pace* Latour and Foucault) as they both drain or are perceived to drain into our bodies via the polluted tapwater that we drink in our homes and communities. I question the genetic reshaping of our future that is implicit in this process, proposing instead an anthropologically oriented conceptual model that I call "eco-risk" for the study of these subterraneanly toxic biological and social processes. The model of eco-risk that I propose here can be extended to other environmental health issues as well.[3]

Fracking is not just something that is happening in the United States. It is a technique of oil and gas extraction that is being used all over the world. Individuals living next to the wells that are being fracked are in remote areas of Latin America, Africa, and Asia. These groups have very different ideas when it comes to human relationships with the environment (Johnston et al. 2007). How will the effects of these extractive processes be viewed amongst indigenous groups in the Andes and in the outback of Australia? Where will blame be placed? Where will points of contestation erupt? As scientists are just beginning to grapple with detecting the health effects of large-scale environmental changes, epidemiological models are insufficient. As McMichael and Woodruff (2005:1) state:

> Scientists fluent in ecology and the earth sciences understand that the current scale of human-induced changes to the biosphere entails risks of systemic dysfunction. Ecosystem processes, being complex and often nonlinear, are somewhat unpredictable in their responses to major external stressors. These issues are not yet prominent or well understood within population health research circles. Yet it is a reasonable expectation that this ongoing impairment of Earth's life-support functions poses substantial risks to human health.

That there is a "reasonable expectation" that large-scale changes in the earth's ecosystem will result in risks to human health seems to be putting it mildly.

LYMPHOLOGIES OF DECEIT

Within the human body the lymph system is a hydraulic mechanism that cleans, drains, protects, and pollutes. As the lymphatic liquids circulate through the body they mount immune responses against intruders, and when those defenses don't work the lymph system transports mutagenic forms of cells that create cancerous metastasis in the forms of malignant tumors throughout the body via the interior canal system of lymphatics. These lymphatic tides are the ebbs and flows of good and bad within the body, the pulsing system that brings life and potentiates death.

I use this description of the lymph system as a metaphor to open up the process of thinking about the consequences of fracking. The geological formations that hold natural gas and the underground aquifers of water are articulated and interconnected at various levels. The natural pulsing of water within the earth is akin to the lymph system. The millions of gallons of fracking water used in each well are laden with chemicals called proppants that serve to facilitate the entrance of the water into smaller and smaller spaces with the goal of pushing out the last remaining pockets of natural gas and oil. The overall result of this is the possible pollution of underground aquifers with unknown quantities of chemicals that may be carcinogenic or otherwise harmful to the health of living things. This process, in turn, creates a situation that may be polluted enough to set off mutagenic processes in living creatures through consumption of water, through breathing in fumes near the drilling sites, and through the consumption of contaminated food grown in the region. This scenario demonstrates geolymphatic flows of possibly enormous magnitude.

The hydraulic systems of the earth transmit not only the water necessary for life but also chemicals that can cause illness and death. Water, in this case, and in many other situations, is a Janus-faced entity.

I use deceit as a heuristic device in my argument to highlight the political nature of fracking. Deceit comes into play in the various representations of the process of fracking. Downplaying and overstating the dangers of the fracking chemicals is evident in local, national, and international media. The manipulations engaged in by the energy industry, citizen interest groups, and the media to (mis)represent the possible dangers show how both sides create discourses of justification. The flexibility of the topic of fracking lies in its ambiguous nature. There is much we do not know yet about how the chemicals used in this process will affect human, animal, and plant life. The scale of the possible effects of fracking is enormous; existing monitoring technologies may well be inadequate to describe the effects of this process on biological systems. It is difficult to assess the risk of fracking.

Risk is not only a "culturally constructed" concept (Douglas and Wildavsky 1983). It is more precisely a particularly lived understanding of, in this case, the dangers of fracking. It is a culturally perceived danger at the crossroads between technologies of diagnosis/quantification and the legal standards that are established to protect citizens from harm. There is a three-way interaction that creates risk in general and eco-risk in this case; it is the process of tacking back and forth between fear (or lack thereof), ability to make visible/quantify/diagnose, and the legalistic structure that constructs codes of behavior and sanctions infractions against given codes that result in harm (Cartwright and Thomas 2000). Risk is not solely a cultural construct of possible danger, it is also quantified and made real by the extant technologies of visualization and quantification and the use of the (often numerical) results to justify actions. Fracking provides an excellent example of how these three interactions take the forms of 1) heightened awareness of possible dangers through media involvement, 2) incomplete scientific quantification data being used to either support or reject its use, and 3) changing legalities surrounding the use of fracking technology.

I turn now to a discussion of the antecedent models that, taken together and specifically focused on the environment, allow us to flesh out the concept of eco-risk.

PATHOCOENOSIS

In the 1960s the French medical historian Mirko Grmek began using the concept of pathocoenosis to describe illnesses as they exist in particular places and bodies at particular times in history. His emphasis was on the way in which *places* create particular configurations of factors that lead to bodies that malfunction in particular ways, and the way those malfunctions are locally perceived and categorized via culturally constructed notions of pathophysiology (Grmek 1991). Grmek defines pathocoenosis in this way:

1. Pathocoenosis is the ensemble of pathological states present in a specific population at a given moment in time. It consists of a system with precise structural properties that should be studied so as to determine its nosological parameters in qualitative and quantitative terms.
2. The frequency and overall distribution of each disease depends on the frequency and distribution of all the other diseases within a given population (in addition to various endogenous and ecological factors).

3. A pathocoenosis tends toward a state of equilibrium expressible in relatively simple mathematical expressions; that state is especially perceptible under stable ecological conditions. (Grmek 1991:3)

Grmek makes two particularly important contributions here. First he addresses the need to include ensembles of pathophysiological states that are present at given moments in time in the understanding of why bodies break down as they do. The concepts of comorbidity and multimorbidity are still not being considered in the vast majority of medical research (Caughey and Roughead 2011). This scant attention to multimorbidities is puzzling in a world where chronic infections of such things as viruses, parasites, and communicable diseases are being suffered simultaneously with a plethora of diseases of development such as obesity and its attendant type 2 diabetes, hypertension, etc. in lived environments heavily polluted with chemicals from industry, agriculture, and combustible engines, among other things. It is the particular combinations of morbidities that are present simultaneously in a living being that can be so difficult to endure and that can so influence how pathology unfolds.

Grmek attends to the environmental component as well, if parenthetically. He states in his second point that frequencies and overall distributions of diseases are dependent on those factors in all the other co-present diseases *plus* various endogenous and ecological factors. The complexity that this brings into the equation is significant, but without it one or two coexisting disease states will continue to be tested at a time in an overly simplistic fashion. We are missing the full ensemble of existing disease states that may influence one another (especially in contexts with high rates of multiple chronic conditions that may be present in subclinical forms) and the ecological (environmental) context that is increasingly befouled with a variety of chemicals that may potentiate existing diseases or create entirely new pathological processes that in turn will influence preexisting conditions.

Grmek's contribution here is to recognize the importance of multimorbidities occurring within specific environments. As we build more complexity in to the proposed model of eco-risk, what does Grmek's contribution look like? Recognizing that multimorbidities create additional complexities in the generation of pathological situations creates a different way of investigating what might be dangerous. If the chemicals used in fracking come into contact with humans, how do they interact with the already existing diseases in that individual and or that population of individuals? One chemical often cited as being present in the environment in larger than natural quantities as a result of fracking is methane. How does extra methane then interact with individuals suffering from type 2 diabetes? What if that individual who is exposed to methane, or any other fracking chemical, has breast cancer, or prostate cancer, or lung cancer? What if that individual or population of individuals has both type 2 diabetes and cancer and is exposed to fracking chemicals?

Eco-risks are terrifyingly complex when viewed in this light. If the scientific purview doesn't take this biological complexity into account, is it creating an oversimplified causal model for the most common of human pathologies? Are we fooling ourselves with too-simple science? Are we missing the real dangers? Should we be more afraid or less afraid? Our sense of danger is rooted in fears both immediate and imagined.

Our abilities to make visible/diagnose/quantify are far behind the questions that we are faced with in this situation. Biopower is just too low resolution for this one. We don't have the ability to see deep into the levels of shale to see where the water and proppants are going, nor can we discern what their ultimate effects will be on various ecosystems, our own bodies included. We just don't know.

And so as we begin to grapple with the laws surrounding the issue of the regulation of fracking, we are constructing our judicial arguments on shaky ground indeed.

Now I turn to adding in Margaret Lock's conceptualization of local biologies. Lock's contribution is to flesh out the importance of perception, interpretation of sensations, and the social aspects of scientific knowledge production. Lock's local biologies give us the intellectual purchase to articulate where eco-risk becomes firmly embedded in local categories of knowledge and perception.

LOCAL BIOLOGIES

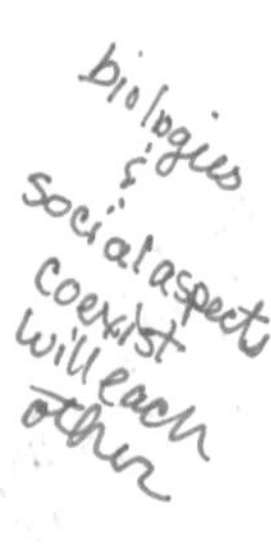

> [L]ocal biologies refers to the way in which the embodied experience of physical sensations, including those of well-being, health, illness, and so on, is in part informed by the material body, itself contingent on evolutionary, environmental, and individual variables. Embodiment is also constituted by the way in which self and others represent the body, drawing on local categories of knowledge and experience. If embodiment is to be made social, then history, politics, language, and local knowledge, including scientific knowledge to the extent that it is available, must inevitably be implicated. *This means in practice that, inevitably, knowledge about biology is informed by the social, and the social is in turn informed by the reality of the material. In other words, the biological and the*

> *social are coproduced and dialectically reproduced, and the primary site where this engagement takes place is the subjectively experienced, socialized body.* The material body cannot stand, as has so often been the case, as an entity that is black-boxed and assumed to be universal, with so much sociocultural flotsam layered over it. The material and the social are both contingent—both local. (Lock 2001; emphasis mine)

I include Lock's full definition of local biologies as the term has tended to be simplified in anthropological discussions. I want to return to the complexity that Lock describes and especially focus on the way in which physical sensations are experienced in particular ways in particular environments and the way in which the sensations themselves are at least partially produced by the environments. Meaning is then given to the sensations in particular cultural contexts that are, sometimes and partially, informed by our notions of biomedicine/science, local dangers, and concepts of illness causation.

In a recent article in *International Business News,* John Fenton, chairman of the Pavillion Area Concerned Citizens and a resident of Pavillion, Wyoming, stated, when asked about the water used in fracking, "that will put sores on your head" (Bertrand 2012). Evident in this quote is the idea that the body becomes the site where the reality of the danger is verified within the logic of local knowledge. It becomes as important to scientifically assess whether fracking chemicals do indeed cause sores on the head (or elsewhere, one might add) as it is to describe how the discourse about those sores is used to challenge the practice of fracking in that little town in Wyoming. The original EPA study carried out in Pavillion found in two deep monitoring wells synthetic chemicals, like glycols and alcohols consistent with gas production and hydraulic fracturing fluids, benzene concentrations well above Safe Drinking Water Act standards and high methane levels. Given the area's complex geology and the proximity of drinking water wells to ground water contamination, EPA is concerned about the movement of contaminants within the aquifer and the safety of drinking water wells over time.

The EPA also has been monitoring drinking water wells around Pavillion. Their findings to date suggest contamination of the drinking water from the fracking processes:

> EPA also updated its sampling of Pavillion area drinking water wells. Chemicals detected in the most recent samples are consistent with those identified in earlier EPA samples and include methane, other petroleum hydrocarbons and other chemical compounds. The presence of these compounds is consistent with migration from areas of gas production.

> Detections in drinking water wells are generally below established health and safety standards. (EPA 2011)

Even though the levels found so far are below established dangerous levels, the EPA has recommended that residents "use alternate sources of water for drinking and cooking, and ventilation when showering." The politics of science need be attended to as well, but the economic side of the scale is illustrative.

Wyoming county health profiles[4] show that residents of the counties that have the most natural gas extraction (Sublette, Johnson, Sweetwater) are high in smoking and heavy drinking, are about average on most health risk factors, but have significantly better salaries than most Wyomingites while attaining low levels of college education. For the moment, the higher-paid jobs in the petroleum industry are assuaging the local economies with $1.9 billion in state taxes in 2010 according to the website of the Petroleum Association of Wyoming (http://www.pawyo.org). This is for a state with about 550,000 people. While the number of jobs being created may be low with respect to the entire population of the United States, the tax benefits at the state level are truly significant.

From an anthropological perspective, this is an example of another layer of the process of creating the idea of an eco-risk. The media has presented and instantiated this example of a locally identified bodily evidence of the effects of the fracking chemicals on human health. Undoubtedly, there are many more locally noticed, deeply felt instances of the bodily effects of fracking. Ethnographic details and systematic, descriptive data on this process are the next step in documenting the evolving local perceptions of diseases coming from exposures to fracking chemicals. The sensorial meanings attributed to a place, how it looks, and how it smells play into the process of giving meaning to how it has been polluted (Reno 2011). As with research that has been carried out on indigenous perceptions of pesticide poisonings, illnesses that are attributed to ambiguous chemical exposures are inserted into preexisting logics of ethnophysiology (Cartwright 2003). Determining who is susceptible to these chemicals, and why, will index themes of how vulnerable or impervious an individual is, or thinks she is, in a particular situation. Personal strength is seen, especially among the working classes, as manifest in the ability to work. Salaries and jobs in an economy of relative downturn ameliorate a lot of concerns, at least in some locales and for those "lucky" enough to be working in the industry. How themes of resistance, illness, strength, and susceptibility play out among residents of the regions experiencing this form of energy extraction and amongst the energy workers will reflect local values, concerns, and fears in the future.

I now turn to a couple of examples of how the concept of fracking is being used in contemporary U.S. media and some of the larger social issues that it is being tied to.

USING THE DISCOURSE OF AN ECO-RISK

The term *fracking* has been inserted into contemporary U.S. media in a way that indexes current economic fears, battles between "Liberals" and "Conservatives," Tea Party politics, environmental exploitation, unemployment numbers, ecofreaks, social class differences, and the price of tea (or was it oil?) in China. In this election year, 2012 discourses of the positive and negative effects of fracking fracture along party lines. *New York Times* columnist Paul Krugman wrote on March 15, 2012:

> Why the hydrocarbon boom? It's all about the fracking. The combination of horizontal drilling with hydraulic fracturing of shale and other low-permeability rocks has opened up large reserves of oil and natural gas to production. As a result, U.S. oil production has risen significantly over the past three years, reversing a decline over decades, while natural gas production has exploded. . . . Put it this way: Employment in oil and gas extraction has risen more than fifty percent since the middle of the last decade, but that amounts to only 70,000 jobs, around one-twentieth of one percent of total U.S. employment. So the idea that drill, baby, drill can cure our jobs deficit is basically a joke. . . . Why, then, are Republicans pretending otherwise? Part of the answer is that the party is rewarding its benefactors: the oil and gas industry doesn't create many jobs, but it does spend a lot of money on lobbying and campaign contributions. The rest of the answer is simply the fact that conservatives have no other job-creation ideas to offer. (Krugman 2012)

Also on March 15, 2012, Rush Limbaugh discussed the topic of fracking in relation to the Democratic governor of North Carolina recently changing her position and coming out in favor of using fracking in her state:

> Now, why would Governor Perdue be all for it now and just months ago totally opposed to it [fracking]? Well, I'll tell you why. She's not trying to hold her base for reelection any longer. She's not running for reelection so she doesn't have to do things to keep the fringe kook Democrat base together. If she were running for reelection she would not have changed her mind on fracking. She would still oppose it. But now that that's not

> a concern and now that she doesn't care what the Democrat Party base thinks, she's going to do for her legacy, for her reputation, for what's written about her after she's no longer governor, she's now gonna do the right thing. She's now gonna allow fracking for natural gas, because, by her own admission, it's jobs. It's a fuel source produced in this country. It's something that can help America and North Carolina be globally competitive. (Limbaugh 2012)

Fracking is about jobs or it's not about jobs. It's a clean source of fuel or it's a potential polluter of massive proportions. It's something that Democrats don't support (especially, according to Limbaugh, if they are going up for reelection). It's something that Republicans use to create pretend jobs. It's a good legacy, it's a bad legacy. Whatever it is, it is certainly emotionally charged and highly ambiguous. Its understanding is embedded in the cultural present and will change as we move into the future.

INTERNALIZED ECOLOGIES: TAKING ECO-RISK ONE STEP FURTHER

> The intense socialization, reeducation, and reconfiguration of plants and animals—so intense that they change shape, function, and often genetic makeup—is what I mean by the term "internalized ecology." (Latour 1999)

When Latour talks about his concept of "internalized ecology," it is in reference to the changes brought about by the agricultural revolution. He points out that there were both physical (mutagenic) changes and social reconfigurations of concepts during that time period. Physical kinds of changes are occurring that are changing biological games in a permanent way. While discourses regarding the meanings of fracking change over time and in different places on the globe, also changing are the baseline physiologies of different creatures, including humans, on our earth. The process of mutations occurs rapidly and in concert with other living entities that reciprocally influence the process. Viruses that are co-present share genetic materials resulting in new viruses, viruses exchange genetic materials with bacteria, mutations in living beings occur at amazing rates and a few are preserved in individuals and passed on to offspring and populations of offspring—all these processes leaving us with a shifting baseline of what we think of as biological reality.

Fracking provides a highly interesting example of an energy extraction process that could result in possible exposures on huge scales, perhaps through underground aquifers, perhaps through surface water contamination, but it is not the only large-scale environmental source of pollution. The number of sources of environmental contamination that are proliferating within our postindustrial societies is staggering. Some are intentional and some are not. The long-term time element that is associated with developing cancers, neurotoxicities, developmental delays, and reproductive problems precludes any immediate actions. Shifts are subtle; exposures are difficult to quantify.

CONCLUSIONS

This book proposes to query how we culturally understand concepts related to energy. This chapter focuses on the particular embodied part of that inquiry as I go through the levels of what I call eco-risk. Spiraling outwards from biological spaces where multiple chemical pollutants interact with humans who have a variety of pathological states, I start to question how we conceptualize how the environment interacts with human health in the presence of multiple diseases and chronic conditions. Grmek's concept of pathocoenosis provides a way to conceptualize this interaction in its complexity.

Risk, as my colleague Jan Thomas and I have defined elsewhere (Cartwright and Thomas 2000), is produced through the process of tacking back and forth between locally perceived dangers, the technologies of perceptions used to quantify/make visible/treat those dangers, and the laws used to regulate practices surrounding what is perceived to be risky. Eco-risk is this process played out against an environmental background. The eco-risks associated with fracking are perceived and understood based on local knowledge and they are embedded in current political discourses. The productivity of the term *eco-risk* as I've applied it to fracking is to keep the levels of cause/effect/perception/manipulation together so that the interactions between the various levels can be conceptualized and problematized.

NOTES

1. This mixture aids in making remote oil and gas deposits accessible for extraction. The process used for fracking in the majority of natural gas wells employ highly carcinogenic and neurotoxic chemicals.

2. http://www.epa.gov/hydraulicfracturing/ (accessed April 14, 2012).

3. I leave the term *eco-risk* in a hyphenated state to set it off from the shorthand version of "ecological risk" that is written "ecorisk" that exists in some governmental literature.

4. See http://www.health.wyo.gov/phsd/brfss/index.html.

REFERENCES

Bertrand, Pierre
2012 Tiny Wyoming town plays big role in fracking fight. *International Business Times*, March 15.

Cartwright, Elizabeth
2003 *Spaces of illness and curing: The Amuzgos of Oaxaca between the Sierra Sur and the agricultural camps of Sonora* (in Spanish). Hermosillo, Mexico: El Colegio de Sonora Press.

Cartwright, Elizabeth and Jan Thomas
2000 Risk, technology and malpractice in maternity care in the United States, Sweden, Canada and the Netherlands. In *Birth by design: The social shaping of maternity care in northern Europe and North America*, edited by Rayomond DeVries, Edwin van Teijlingen, and Sirpa Wrede, 218–228. New York: Routledge.

Caughey, Gillian E. and Elizabeth E. Roughead
2011 Multimorbidity research challenges: Where to go from here? *Journal of Comorbidity* 1: 8–10.

Douglas, Mary and Aaron Wildavsky
1983 *Risk and culture: An essay on the selection of technological dangers.* Berkeley: University of California Press.

Environmental Protection Agency (EPA)
2011 Draft Findings of Pavillion, Wyoming Ground Water Investigation for Public Comment and Independent Scientific Review. Released December 8. http://yosemite.epa.gov/opa/admpress.nsf/20ed1dfa1751192c8525735900400c30/ef35bd26a80d6ce3852579600065c94e!OpenDocument (accessed March 5, 2012).

Grmek, Mirko Drazen
1991 *Diseases in the ancient Greek world.* Baltimore, MD: The Johns Hopkins University Press.

Johnston, Fay H., Susan P. Jacups, Amy J. Vickery, and David M. J. S. Bowman
2007 Ecohealth and aboriginal testimony of the nexus between human health and place. *EcoHealth* 4: 489–499.

Krugman, Paul
2012 Natural born drillers. *New York Times*, March 15.

Latour, Bruno
199 *Pandora's hope: Essays on the reality of science studies*. Cambridge, MA: Harvard University Press.

Limbaugh, Rush
2012 Dumplin' flips on frackin'. March 15. http://www.rushlimbaugh.com/daily/2012/03/15/dumplin_flips_on_frackin (accessed March 16, 2012).

Lock, Margaret
2001 The tempering of medical anthropology: Troubling natural categories. *Medical Anthropology Quarterly* 15(4): 478–492.

McMichael, Anthony J. and Rosalie E. Woodruff
2005 Detecting the health effects of environmental change: Scientific and political challenge. *EcoHealth* 2: 1–3.

Reno, Joshua
2011 Beyond risk: Emplacement and the production of environmental evidence. *American Ethnologist* 38(3): 516–530.

CHAPTER ELEVEN

Specters of Syndromes and the Everyday Lives of Wyoming Energy Workers

Jessica Smith Rolston

Since the 1970s, social scientists and journalists have chronicled the social upheavals accompanying the rapid expansion of the energy industry in the American West. With rare exceptions, they attribute the social ills they observed—from drinking and divorce to depression and delinquency—to the workers who move to rural communities and stress infrastructure, exhaust social services, and upset cultural norms. The kindest accounts draw attention to the socioeconomic and environmental factors that impinge on workers' abilities to participate in community building and to care for themselves and their families. The least generous ascribe these stresses to seemingly innate pathological personality traits that compel people to a never-ending nomadic search for the next big payout.

This chapter turns to Gillette, Wyoming, to reexamine the received wisdom about the social dimensions of energy development. Three major booms—oil in the 1960s, coal in the 1970s, and coalbed methane (a natural gas) in the 2000s—made the town and the surrounding Powder River Basin a focal point in these debates. Graphic accounts of the "Gillette Syndrome" framed the first wave of academic and popular interest in energy boomtowns sparked by the 1973 oil crisis. Forty years later, allusions to the syndrome continue to circulate throughout news media, fiction, and the academy as a resilient "matter of local, regional, and national folklore" (Limerick et al. 2003:18). In fact, the Gillette Syndrome has been used to contextualize everything from oil development in Nigeria (Kashi and Watts 2010:37) to meatpacking in Canada and Kansas (Broadway 2007) to planning for the 1994 Winter Olympic Games in Lillehammer (Leonardsen 2007).

Though Gillette continues to loom as a specter of disastrous development in both academic and popular discourse, scholars have not returned there since the original studies to reassess the syndrome framework for analyzing rural industrial development throughout the West. Drawing on twenty-one months of research in the Powder River Basin and a lifetime of experience growing up there as the daughter of a mine mechanic, I argue that getting to know workers and their families in Gillette challenges the boomtown-syndrome literature's emphasis on the supposed transient lifestyles of energy workers. Their life histories and work experiences draw attention to the less sensationalistic aspects of everyday life that are left out of the literature on energy boomtowns, as well as to the specific features of industries that contour their ability to pursue careers and to raise their families. The ethnographic materials suggest that attributing a suite of social ills to energy or even fossil fuel development writ large rather than to specific commodities hinders our ability to understand such monumental social, economic, and environmental transformations, much less improve the well-being of the people who experience them.

AN INFAMOUS SYNDROME

Gillette has undoubtedly experienced dramatic growth as a result of energy development. Incorporated in 1892 as a railroad stop and small supply center for the area's sheep and cattle ranchers, the town grew steadily until it reached 3,850 residents in 1960. That decade's oil boom drew thousands of workers to the area so that by 1970, the population had doubled to 7,194. Even as the oil boom began to wane in the 1970s, the enormous new surface coal mines attracted another wave of workers and their families. By 1980, 13,617 people lived within city limits, a figure almost double the 1970 census and more than triple the 1960 census. In the 1980s and 1990s, the population leveled off as the energy industries stabilized. The 1990 and 2000 censuses reported 17,635 and 19,646 residents, respectively. The coalbed methane boom that began in the mid-2000s once again prompted rapid expansion: the 2010 census counted 29,087 residents, a 48 percent increase from the 2000 figure, though population estimates during the throes of the boom reached 34,000 people.

Social science research on booms has grown and changed since the first studies in the 1970s. Social psychologist ElDean Kohrs (1974) originally proposed the term "Gillette Syndrome" in a 1974 conference paper that rested on a simple statistical analysis showing increases in rates of divorce, crime, alcoholism, and depression accompanying the population expansion. As scholars turned their attention to other communities, the social disruption hypothesis quickly became conventional wisdom for analyzing energy boomtowns around the West (Davenport and Davenport 1980; England and Albrecht 1984; Smith et al. 2001:429; Summers and Branch 1984:154). Researchers documented the disintegration of rural neighborliness among longtime residents and ranchers (Gold 1974; Tauxe 1993); the harms suffered by workers and their spouses forced to move frequently in search of work (Feldman 1980; Massey 1980; Moen 1981; Walsh and Simonelli 1986); and the possibility that social stresses could hinder economic growth (Gilmore 1976).

Later research challenged the ubiquity of the severe and negative effects from energy development posited by the first wave of studies. Scholars called for caution in extrapolating from statistics based on small numbers—since a 500 percent increase in crime could reflect an increase from two to ten cases—and raised the possibility that shocking statistics reflect double counting, when multiple people receive treatment from multiple institutions for a single incident (Freudenburg 1986:68; Summers and Branch 1984:155). Critiquing the romantic view of pre-boom life characteristic of the first wave of research, they documented cases in which considerable disruption occurred before development began (Nelson 2001; Summers and Branch 1984; Tauxe 1993; Wilkinson et al. 1982).

The later generation of researchers overall found that "the assumption that energy development causes social disruption in Western communities is based on undocumented assertions, questionable interpretations of evidence and superficial analyses" (Wilkinson et al. 1982:275). With a larger data set, researchers found two of Kohrs's three major assertions to be factually incorrect: divorce rates were not higher in rapidly growing Wyoming communities, welfare payments did not increase as a result of population growth, and criminal activity increased only in areas associated with property (Wilkinson et al. 1982:35; see also Summers and Branch 1984:155; Thompson 1979). Further studies in other communities persuasively showed that the impacts of energy development were both positive and negative (see Smith et al. 2001:431–432 for a summary) and that boomtown residents themselves recognized both aspects of this growth (England and Albrecht 1984; Freudenburg 1986; Nelson 2001). Moreover, longitudinal studies showed that the negative elements of the booms waned as social services caught up with growth and workers became established in communities (Freudenburg 1986; Smith et al. 2001:446).

This critical reconsideration of the tropes used to analyze energy development did not extend to journalistic coverage in prominent outlets such as the *New York Times*, *Wall Street Journal*, and *National Geographic*, all of which graphically portrayed Gillette as a dangerous, hedonistic boomtown populated by pathological criminals. Even more alarming, some current research about energy development uncritically reproduces the flaws of the original misleading studies. Conservation biologists Joel Berger and Jon Beckmann blame both ecological decay and community disturbance on itinerant workforces associated with energy rather than agrarian or recreation development (Berger and Beckmann 2010:895). They characterize them as transient migrants who disturb communities—even though they did not interview any workers (2010:894)—and recurrently cite a writer famous for documenting the life of a *local* Wyoming man who died working on the natural gas rigs (Fuller 2008). Their main argument rests on their finding that the absolute and relative frequency of registered sex offenders (RSOs) "grew approximately two to three times faster in areas reliant on energy extraction" (2010:891). But their data actually shows per capita changes increasing from .75 to 2 per 1,000 residents in energy extraction counties (2010:894), illustrating the argument made twenty years earlier that "spectacular increases in percentages generally need to be interpreted with caution, particularly if the 'base' for computations is quite small" (Summers and Branch 1984:155). Berger and Beckmann explain this increase by suggesting that RSOs are attracted to energy boomtowns because they can more easily find high-paying jobs, without considering the possibility that increased RSO frequencies could result from spatial autocorrelation rather than energy extraction—in other words, the counties with energy development all also happen to be adjacent to one another. Moreover, their conclusions fail to factor in their observation that the one nearby county not exhibiting increased RSOs (Teton County, home to Jackson) is "considered a magnet for celebrities and political leaders" and has the "highest mean per capita standard of living in the United States" (2010:893). In other words, RSOs might be less prevalent there because its residents have the cultural and financial capital to navigate the legal system.

Perhaps most problematic, Berger and Beckman erroneously cite the boomtown literature to ground their argument. They argue that "the evidence that social ills accrue in energy boomtowns has been well chronicled . . . primarily through analyses of changes within individual communities over time" (2010:895). But they cite two studies (Wilkinson et al. 1982 and Summers and Branch 1984) that actually critique overstated statistical analyses, showed that energy development could not be reduced to simply negative social consequences, and critiqued the dearth of research extending beyond the initial boom times.[1]

Understanding the social impacts of energy development requires less research insinuating that energy workers are delinquent sexual predators and more research understanding the structural factors that create inequalities and compel certain groups of people to move in order to earn a paycheck. It also calls for more thorough analysis of the perceptions and experiences of workers and long-term residents alike in order to interpret statistical measures. Unlike the survey work that dominates the boomtown literature, ethnographic data does not channel people's thoughts and experiences into preset categories but maintains narratives in a way that opens up the possibility of challenge to scholarly and popular expectations (Nelson 2001:398).

LONG-LASTING RELATIONSHIPS, NOT ISOLATION

No one who lived in Gillette during the booms denies that the town went through difficult adjustments in providing adequate housing and social services, and people who experienced the booms draw on the same Western imagery in describing those years as "wild." Yet they do so selectively and differentiate their characterizations of that period from prominent representations.

The first key difference is that energy workers remember booms for the long-standing relationships they helped foster, not for the isolation or conflict posited by journalists whose informants were interviewed almost exclusively in bars.[2] Roy moved to Gillette with his brother in the early 1970s and remembered Gillette being "nuts, a boomtown." Unlike many others, he and his brother easily found housing since many of their hometown friends had recently moved there as well. They both began working on the oil patch, but Roy became frustrated because it would "get dead and there would be nothing to do sometimes." After four years, he was hired at a mine to do ranchwork on reclaimed land and eventually became an equipment operator. At the time of our interview, he had been at the same mine for seventeen years and with his first and only wife for just a few years less. They planned to stay in Gillette at least until their son graduated from high school. They remained close with Roy's brothers, who all stayed in the area and continued to work in the industry.

Patty, who worked her way up to management after starting as general labor, also qualified her initial description of Gillette as wild by pointing to the friendships she maintained:

> I hate using the word *wild*, but it was wild. People worked hard and people played hard. You knew everybody either from, I hate to say, either from work or the bar. . . . Anywhere you went, you knew basically everybody that was there. . . . I think the best way to sum it up is that people worked hard, because you were either basically in mining or the oil field. You played hard, you all went out, you all went to the bar, you all went out to eat. You partied together. It was crazy.

Patty describes many of the same things covered by the journalists, but notes the stereotypes and moves beyond them in highlighting the close social connections crafted by workers. She emphasizes camaraderie among friends in marked contrast to the loneliness and alienation highlighted by the reporters. In the rest of our conversation, she spoke about how she eventually got married and raised two daughters.

Another group of friends remembered moving to Gillette, making friends, building a church, and raising a family—activities that are common throughout the community but scarce in the dominant portrayals of it. One couple started out living in a mobile home but eventually bought a house and raised their children along with a close-knit group of friends on their street. One of the neighbors explained, "We raised those kids just like brothers and sisters. They ran between our houses. One day they'd put together and perform a play for everyone at our place, and the next day they'd have a sleepover at the other." The couples also fondly remembered Friday night get-togethers at the local pizza parlor and Saturdays at the bowling alleys. They enjoyed going to high school sports games, movies, and out to eat at the Chinese restaurant, and during the summer they went camping in the Big Horns. One husband described the bar scene as "rowdy . . . loud . . . a lot of young people in their twenties drinking, dancing, and smoking cigarettes." But he and his wife could only name a few bars, despite all of the sensational accounts of the town's drinking scene, and remembered "not going out much." The wife said, "We weren't real crazy people." In an interesting twist on the partying stereotype, he remembered holding worship services in the American Legion before the church was built. "Going in there on Sunday morning after a night of drinking, it stunk like beer and cigarettes! And the beer cans were still there, and the beer bottles were still on some of the tables. Yeah, you just moved it aside and had church. I remember that." These couples' memories are typical for people I came to know, who remember that period for the friendships they established and maintained.

The themes of family and place continue to appear in contemporary criticisms of national portrayals of energy booms. During the spring of 2007 I worked with two classes of high school students in Gillette.[3] Together we read the articles analyzed in this chapter. Students wrote short responses to the texts and then discussed them as a group. The students, especially those with parents working in the mining or oil or natural gas industry, criticized the stereotypes that emerged from the readings. Not only did they critique the authors for making it seem as if the town was a "giant dirtball" where only two groups of people—ranchers and miners—lived, they pointed out the lack of attention to family life. They enjoyed recounting stories of family trips to the local reservoir and mountains during their parents' days off from the mine. In their writing and class discussions, they emphasized stability by drawing attention to how long their families had lived in Gillette and how long their parents had worked with the same crews. Because the biggest cohort of miners is now in their early and mid-fifties, many of their children are just finishing high school so that many of the students in the class I worked with had been born and raised in Gillette.

These firsthand accounts of the booms all draw attention to the importance of social ties formed during periods of upheaval and their aftermath. This emphasis on family and friends stands in marked contrast to the Gillette Syndrome literature that treats energy workers as isolated individuals who drown their stresses in alcohol and violence.

STRATEGIES OF MOBILITY, NOT PATHOLOGICAL WANDERLUST

The second difference is that whereas the syndrome writers suggest that people flocked to Gillette to party, escape social conventions, and make a quick buck, most of the people I met moved there to start families or find a steadier source of financial support for them. They migrated because their work fluctuated along with mineral markets, but they did so reluctantly and only after thoughtful consideration, in what can be considered "strategies of mobility" (Andrews 2008:90). These findings correspond with recent research showing that wanderlust was a weaker predictor of migration among boomtown construction workers than time in the community, job security, age, housing integration, and dissatisfaction with facilities and services (Fahys-Smith 2011).

All of the families I met dealt with a bust on an industry-wide scale. Copper market downturns pushed miners from the famed copper mines in Montana to find work in Gillette. Stan had been working at one of the large southern coal mines for twenty-two years when I climbed up on an electric shovel to spend the day with him. He was born and raised in Butte, and most of his male relatives worked in the mines. Once out of the military, Stan sought work there as well because "that was the only game in town that paid anything." A few years later he was laid off along with everyone else. "I had two small kids, so I had to leave the area and find work somewhere. I couldn't take care of my family in Butte." Stan remembers telling his uncle, who disapproved of the move, "I dearly respect you, but I have to eat." When they reopened the copper mines a decade later, he was offered his old job back. Although he and his wife did not originally intend to stay in Wyoming, they passed up the offer in Montana, explaining, "Coal was pretty secure. Everybody had adjusted down here, and the kids all had friends and everything." In this case, Stan and his wife made a choice they believed would afford them and their children the most stability.

Kelly, a well-respected equipment operator, also grew up in a Montana copper-mining family. When the mine shut down, her father transferred to a Nevada molybdenum mine. During summer breaks from college she worked in the same department with her father and tried to get full-time work, but the company did not hire her because they wanted to avoid encouraging students to drop out of school. After working as a waitress, she went to work at the mine full-time when the company started hiring again. When it shut down in the mid-1980s, she transferred to a mine in Gillette, where she was still working when I met her in 2006. She relished talking about the "wild and crazy boom times" when she did whatever she wanted to do, such as a buying an expensive new sports car and partying with her friends. But she also remembered, "People eventually grew up. You couldn't go out partying at all times of the night and still get up for your shift in the morning." Laughing, she added, "We all got old and worn out."

The frequency with which mines open and close created social networks around the West. A mechanic named Al also grew up in a mining family. His father worked in phosphate, silver, gold, and uranium mines. Dissatisfied with college and his first job, Al moved back to his hometown in central Wyoming in the late 1970s to work in the booming uranium industry. "It was a bustling time," he said, but it did not prevent him from making social ties: "I made a lot of really good friends there when I was there." He eventually started working for the same company as his dad, but left to return to school. After finishing a degree in diesel mechanics, Al and his wife moved to Gillette in the mid-1980s when they realized that the uranium industry was not going to emerge from the bust that began a few years before. He was hired on at a new mine by virtue of a connection from his previous work in uranium. The person who recommended him, he said, "knew my work history, he knew what kind of mechanic I was, and knew that I took a lot of pride in my work and that I'd been working on machines and that I'd be reliable. And he got me the interview, and I ended up getting the job. I remember it was pretty exciting for my wife and me."

Al's brother also moved to Gillette years later to work as a mechanic for the same company. Their mother believes it is no accident that her sons followed the same career path as their father. She speculated that it takes "the same mind" to be a mechanic, but also noted the scarce job opportunities around town: "When you stop and think about it, the only work that was around here was mining." Al agreed with her, citing the lack of other appealing job opportunities as his primary reason for working in the industry. In fact, almost every single miner I came to know originally sought work in the industry not for an inherent love of it but because it allowed them to stay close to family. In other words, for Al's family—as well as for Kelly, Stan, and Roy—mining seemed to offer an opportunity to *stay* in a place where they felt they belonged rather than a ticket for a lifetime of geographic mobility as the Gillette Syndrome writers suggest.

Placing popular portrayals of the Gillette Syndrome in dialogue with local appraisals of the booms raises critical questions about attachment to place and the socioeconomic pressures that imperil it. The people who have made Gillette their home, from the large cohort of coal miners to the most recent group of Michigan autoworkers, are neither the first nor last to move in search of stable jobs that would allow them to lead lives and raise families as they deem fit. In fact, the history of coal mining in Wyoming—as elsewhere—was a history of migration from its very beginning, as the first mines drew British, German, Scandinavian, Slavic, Austrian, French, Italian, and Greek laborers. Getting to know Wyoming mining families reveals that far from being inherent transients, they create places and become attached to them for as long as it makes sense given their families' trajectories within an ever-shifting extractive economy.

LIMITS OF BOOMTOWN BIFURCATION

The boomtown literature overstates the impermeability of social divisions between "workers" employed by the company and communities impacted by mining (see also Smith and Helfgott 2010). An exclusive focus on heightened social divisions during the early years of booms—usually between longtime residents (especially ranchers) and new energy workers—obscures the ways in which people not only form friendships across categories but move among them. As Steve showed me how he operated the dragline, he looked over the horizon and pointed to a stretch of land. He grew up on a nearby sheep ranch, where money was tight until his father started building houses during the oil boom. One of the first employees at the mine, he recalled that he and his "crew family" felt empowered because they wanted to "do things right. We wanted to make it the safest place possible." Looking back, he described Gillette in the 1970s as being full of young twentysomethings, a "wild and woolly boomtown. Everybody was drunk, lots of parties." He then added, "But we've all aged and matured. The town has always followed us. Now we're all geezers, ready to retire."

A significant number of local ranchers like Steve end up working in the mines and on the patch because making a living by agriculture alone is nearly impossible for all but the wealthiest ranching families. "I need a steadier paycheck to feed my ranching habit," as one miner who still lived on his family's land explained to me. A coworker of his who wears steel-toed cowboy boots to work as an equipment operator appreciates that the multiple days off in a row allow him to care for his cattle ranch. Conversely, many people who moved to Gillette for mine employment eventually buy ranches of their own, oftentimes to replicate their own agricultural childhoods. Jo grew up on a farm in the Midwest and started working at the Wyoming mines when they first opened. She and her husband enjoy the solitude of living on a few acres out in the country. Carrie similarly enjoys living on a ranch after having grown up on one in the central part of the state. Her children help care for the animals and participate in the same extracurricular activities like 4-H and Future Farmers of America along with their classmates whose parents are full-time ranchers.

The years miners and their families have lived in Gillette opens up the category of "established residents" to include energy workers. The main cohort moved there in the 1970s or 1980s. Many workers like Roy started off on the oil patch but eventually took mine jobs to provide more stability for the families they were growing. Their children spent most of their lives living in the same town, if not in the same house. The workers and their families distinguish themselves from the few ranching families who can trace their history in the area back to the late 1800s and early 1900s but lay claim to their own status as long-term residents, enmeshed in social relationships and invested in the town's lasting well-being. Taking a longitudinal perspective on boomtowns—and following workers into middle age and retirement—shows the increasing malleability of social categories: people invoke categories such as rancher, miner, or roughneck, but they also acknowledge movement among them.

CONCLUSION: POSSIBILITIES FOR STABILITY

Tracking the career trajectories of energy workers reveals concrete reasons why they move among towns and industries. Viewing mining as a strategy for staying close to their families, they move when market downturns close mines and shut down entire industries. They value mine jobs precisely because they are more stable than those in construction, oil, or natural gas. In the Powder River Basin, the 1960s oil boom eventually dried up and even the more recent coalbed methane boom was beginning to wane by 2010, spurring many of the workers to move their families to western Wyoming and Colorado. The coal mines, on the other hand, remained a steady source of employment. When smaller mines were shut down during the market downturns of the early 1980s and 1990s, employees were transferred to larger mines in the basin. On a daily basis, mine employees report to the same place to work with the same crews, whereas people working in oil or natural gas move among rigs. Comparing accident rates reveals that mines are also much safer than oil or natural gas rigs. And while rotating mine shift-work schedules are demanding, they are more regular than those in endured by oil and gas workers, making it easier for miners to spend time with their families and participate in community events (Frosch 2012; Fuller 2008; Ring 2007; Rolston 2010). Thus specific workplace features, combined with differing market vagaries, makes it comparatively easier for mining families to integrate themselves into the Gillette community.

Despite these salient differences, the boomtown literature rarely distinguishes among specific industries. The communities forming the worst-case scenarios for the Gillette Syndrome literature were undergoing oil and natural gas development; in fact, the later articles tack on coal miners to roughnecks or hot shots without considering the differences among them. This fact has not been subjected to sufficient scrutiny, even though scholars long ago raised the possibility that coal would engender very different kinds of development (Summers and Branch 1984:156). Future research about the social impacts of energy development should analyze the specific working conditions of different industries to better understand these monumental transformations and improve the well-being of the people who experience them.

No resident would argue that booming Gillette escaped the social pressures accompanying rapid expansion. But by focusing their attention on the booms, journalists and researchers did not see the creative ways in which local residents ameliorated the stresses of boomtown life, formed community organizations, and built infrastructure. Nor did they observe the miners' gradual shift away from the "work hard, party hard" attitude that accompanied their increasing responsibilities for taking care of their children and the introduction of twelve-hour shifts that left them precious little free time during their workdays.

The purpose of this chapter is not to argue that issues such as drinking, depression, delinquency, and divorce were not present in Gillette during the three energy booms. For some residents, they certainly were. The more interesting question is why certain moments of Gillette's history have come to dominate popular and academic portrayals of the town and the energy industry at large. Portrayals of residents as miners and roughnecks, and of miners and roughnecks as rowdy drinkers and troublemakers, are not necessarily untrue. Many residents do work in the energy industry, and some of these workers enjoy drinking and carousing every now and then. The problem is that this delimited set of images constitutes most of what nonresidents believe they know about the area and the people who live and work there. For observers, Gillette's history is equated with the troubles of oil and coalbed methane booms, and accounts of these booms caricature and denigrate the workers who produce the energy upon which consumers throughout the United States rely.

NOTES

1. Berger and Beckmann did not engage in research showing that social ties "were neither disrupted nor did they give way to formal ties" (England and Albrecht 1984) or documenting the waning of differences between newcomers and old-timers after a period of adjustment (Nelson 2001:398).

2. Bars are a central research site for studies of miners and other working people, but scholars should critically consider how the social conventions of bar interactions influence how people talk about which topics.

3. Thank you to John Bayles for facilitating the classroom project.

REFERENCES

Andrews, Thomas
2008 *Killing for coal: America's deadliest labor war*. Cambridge, MA: Harvard University Press.

Berger, Joel and Jon Beckmann
2010 Sexual predators, energy development, and conservation in greater Yellowstone. *Conservation Biology* 24(3): 891–896.

Broadway, Michael
2007 Meatpacking and the transformation of rural communities: A comparison of Brooks, Alberta and Garden City, Kansas. *Rural Sociology* 72(4): 560–582.

Davenport, Joseph III and Judith A. Davenport
1980 Grits and other preventive measures for boom town bifurcation. In *The boom town: Problems and promises in the energy vortex*, edited by Joseph Davenport and Judith A. Davenport. Laramie: Department of Social Work, University of Wyoming.

England, J. Lynn and Stan L. Albrecht
1984 Boomtowns and social disruption. *Rural Sociology* 49(2): 230–246.

Fahys-Smith, Virginia
2011 Migration of boom-town construction workers: The development of an analytic framework. *Environmental Geochemistry and Health* 5(4): 104–112.

Feldman, Dede
1980 Boomtown women. *Environmental Action* (May): 16–20.

Freudenburg, William R.
1986 The effects of rapid population growth on the well-being of boomtown residents. In *New directions in urban geography*, edited by Chiraji Singh Yadav, 65–104. New Delhi, India: Concept Publishing Company.

Frosch, Dan
2012 Report blames safety lapses for an epidemic of deaths at Wyoming job sites. *New York Times*, January 12.

Fuller, Alexandra
2008 *The legend of Colton H. Bryant*. New York: Penguin.

Gilmore, John S.
1976 Boomtowns may hinder energy resource development. *Science* 191: 535–540.

Gold, Raymond L.
1974 *Social impacts of strip mining and other industrializations of coal resources*. Missoula: Institute for Social Science Research, University of Montana.

Kashi, Ed and Michael Watts
2010 *Curse of the black gold: 50 years of oil in the Niger delta*. Brooklyn, NY: PowerHouse Books.

Kohrs, ElDean V.
1974 "Social consequences of boom growth in Wyoming." Paper presented at the Rocky Mountain American Association of the Advancement of Science Meeting. Laramie, Wyoming.

Leonardsen, Dag
2007 Planning of mega events: Experiences and lessons. *Planning Theory & Practice* 8(1): 11–30.

Limerick, Patricia Nelson, Claudia Puska, Andrwe Hildner, and Eric Skovsted
2003 *What every westerner should know about energy*. Boulder: Center of the American West, University of Colorado at Boulder.

Massey, Garth
1980 Critical dimensions in urban life: Energy extraction and community collapse in Wyoming. *Urban Life* 9(2): 187–199.

Moen, Elizabeth
1981 Women in energy boom towns. *Psychology of Women Quarterly* 6(1): 99–112.

Nelson, Peter B.
2001 Rural restructuring in the American West: Land use, family and class discourses. *Journal of Rural Studies* 17(4): 395–407.

Ring, Ray
2007 Disposable workers of the oil and gas fields. *High Country News* 39: 7–21. Paonia, CO.

Rolston, Jessica Smith
2010 Risky business: Neoliberalism and workplace safety in Wyoming coal mines. *Human Organization* 45(1): 43–52.

Smith, Jessica and Federico Helfgott
2010 Corporate social responsibility and the perils of universalization. *Anthropology Today* 26(3): 20–23.

Smith, Michael, Richard S. Krannich, and Lori M. Hunter
2001 Growth, decline, stability, and disruption: A longitudinal analysis of social well-being in four western rural communities. *Rural Sociology* 66(3): 425–450.

Summers, Gene F. and Kristi Branch
1984 Economic development and community social change. *Annual Review of Sociology* 10: 141–166.

Tauxe, Caroline
1993 *Farms, mines and main streets: Uneven development in a Dakota County*. Philadelphia: Temple University Press.

Thompson, James G.
1979 The Gillette syndrome: A myth revisited? *Wyoming Issues* 2(2): 30–35.

Walsh, Anna C. and Jeanne Simonelli
1986 Migrant women in the oil field: The functions of social networks. *Human Organization* 45(1): 43–52.

Wilkinson, Kenneth P., James G. Thompson, Robert R. Reynolds, and Lawrence M. Ostresh
1982 Local social disruption and Western energy development. *Pacific Sociological Review* 25(3): 275–296.

CHAPTER TWELVE

Energy Affects: Proximity and Distance in the Production of Expert Knowledge About Biofuel Sustainability

Derek Newberry[1]

INTRODUCTION

Fueled by widespread concerns about climate change and fossil fuel depletion, global markets for clean energies have grown into the hundreds of billions of dollars. As these markets continue to expand, the question for sugarcane ethanol manufacturers in Brazil is a complicated one: how to remake an energy source that has been produced domestically for a century into a modern "sustainable" biofuel in order to capitalize on this demand?

Ethanol was first produced on a large scale in Brazil as a result of the Pro-Alcool subsidy program of the military dictatorship during the 1970s. While production dropped off in the late 1980s and 1990s amid declining oil prices and public backlash against the policy, the introduction of flex-fuel cars to the market spurred a recovery for this alternative energy in the early 2000s (Martines-Filho et al. 2006). Beginning around 2005, amid a resurgent domestic market, ethanol producers began looking abroad as global concerns about the impending decline of fossil fuels and the perils of climate change meant new potential demand for alternative energies in lucrative U.S. and EU export markets (Nass et al. 2007). Despite being initially heralded as a viable transition fuel to a green energy future, biofuels quickly came under fire over criticisms originating primarily among NGOs and academics who feared that the negative effects of large-scale biofuel production would far outweigh any potential environmental benefits. Sugarcane ethanol in Brazil has been singled out for the potential deforestation that could result from massive expansion of sugarcane production in the interior to meet new demand (Martinelli and Filoso 2008) and for claims that harvesting this crop requires dangerous, low-wage migrant labor (Alves 2006; Silva 2006).

These biofuel controversies are emblematic of a broader discourse of energy crisis that has reemerged in recent years—one that links long-standing fears about the implications of growing fossil fuel consumption to contemporary concerns about climate change, food security, and resource scarcity, particularly for vulnerable populations. As the contributions to this volume demonstrate, this discourse has sparked a resurgence of scholarly interest in energy as environmental anthropologists, historians, and geographers in particular have begun to ask: How can we make sense of these complex networks of energy circulation? And how do we remake these chains to meet future consumption needs in ways that are more equitable, stable, and resource-efficient—that is, "sustainable"?

Due to the technoscientific nature of governing sustainable energy commodity chains, it turns out that many of our informants in energy business, research, and policy circles are asking the same questions. Since 2009, I have been studying one such group at a government-funded ethanol think tank in Brazil which I will call the National Biofuels Institute (NBI) to understand how its researchers have responded to the public biofuels controversy. My fieldwork with this group explores the politics of producing socioenvironmental impact studies of biofuels—research that is used by the industry to convince other scientists and policy makers in the European Union and United States to adjust their energy regulations in ways that facilitate importation of Brazilian sugarcane ethanol. As I detail in the following pages, these scientists view energy as a problem of global proportions and, much like many of the environmental anthropologists, historians, and political ecologists giving renewed attention to this problem, they are concerned with conceptually encapsulating energy's flows in ways that render legible their myriad effects. The stakes are high for the outcomes of these studies, as they will help to determine whether and how much Brazilian ethanol should be exported to immense markets for alternative fuels. As such, the role of scientific debate in the burgeoning international biofuels market is increasingly politicized.

A number of scholars across the social sciences are discovering that as contemporary governance itself has become highly technoscientific in its workings, it is crucial to study the experts who make up the epistemic or transnational communities that are driving these new modes of governance (e.g., Haas 1996; Keck and Sikkink 1998; Djelic and Quack 2010) as well as explore how to grapple with the uncomfortable closeness of their knowledge-production practices and social contexts to our own (Riles 2000; Boyer 2008). I argue that studying and denaturalizing the all-too-familiar work practices and professional discourses of these experts is an especially important component of an agenda for resurgent anthropologies of energy. In doing so, we are better able to understand the significance of the assumptions and contradictions inherent in these experts' practices and cosmologies—phenomena that play a central role in how energy assemblages are formed. This perspective is also especially useful in enabling us to denaturalize and contextualize our own assumptions about the scale and characteristics of energy flows, given that we share many of these assumptions with our informants.

To demonstrate this, I will first describe NBI scientists' understanding of their own role in rendering biofuel production risks legible by exploring one of their primary tools of inquiry—the life-cycle analysis (LCA)—as a "technology of distance" (Porter 1996). By this, I mean that LCAs are imagined to be highly credible because they give a holistic perspective on energy impacts that are detached from the individual sentiments and social contexts of their creators—a moral economy shared by social scientists and historians developing their own critical analyses of energy circulation. I then describe some of the proximate, intuitive practices of knowledge production that are in fact crucial to how NBI scientists create particular representations of "sustainable" ethanol. Last, I turn this critique back on my own discipline to posit that because our theoretical frameworks for energy stem from similar historical roots and everyday practices, contemporary energy debates in anthropology and political ecology are at risk of a similar lack of reflexivity, obscuring the contextual circumstances that shape our ethical understandings of how these energies should be produced and consumed.

THE MORAL ECONOMY OF THE LIFE-CYCLE ANALYSIS: CONSTRUCTING "TRUST AT A DISTANCE"

One of the tasks of NBI researchers is to engage other scientists and policy makers in the European Union and the United States in order to convince them to adjust their own calculations of Brazilian ethanol's impacts, in terms of greenhouse gas (GHG) emissions, water use, and other indicators. This seemingly arcane theoretical exercise will have far-reaching implications as these calculations are used by the U.S. Environmental Protection Agency (EPA) and European Commission to define quotas for different types of transport fuels. The result will determine whether and how much Brazilian ethanol should be imported into these potentially immense markets for alternative energies. It is in this context that NBI scientists work to produce numerous studies showing the social and environmental effects of Brazilian sugarcane ethanol, especially LCAs of the ethanol production process. LCA is a methodology aimed at measuring the net impacts of an energy commodity from "well to wheels"—that is, from the moment when petroleum is pumped from a well, or sugarcane is grown on a farm, to processing at a refinery to distribution and eventually combustion in a vehicle. As an example, GHG emissions of sugarcane ethanol can be compared to those of petroleum by calculating the net CO_2 emissions generated by the production and consumption of each energy source to determine which is lower. These calculations can be done in Excel, but modelers usually use specialized software for entering predefined "inputs" of resource use (such as fertilizer, water, and type of machinery) into the various stages of the production process.

The structure of the LCA heavily reflects the managerial logic of the institutionalized global environmental movement, which is also responsible for creating the sustainability of biofuels as a particular object of concern. The LCA has its earliest roots in U.S. Department of Energy studies in the 1960s as a holistic quantitative technique to measure the total fuel requirements of various production technologies. It was quickly adopted by the private sector as essentially a public relations tool to assuage consumer fears about resource waste in corporate commodity chains (Curran 1996; Hunt and Franklin 1996; Gabathuler 2006). These studies emerged from the influence of—and were later standardized by—U.S.- and EU-centered transnational networks of nongovernmental organizations, intergovernmental organizations, and activists that began forming in the 1960s and 1970s around three common beliefs: first, that there is a global environment that needs to be protected from a new proliferation of large-scale risks; second, that these risks require scientific research to render them visible; and third, that they require transnational governance initiatives to mitigate them (Jasanoff and Martello 2004; Bartley 2007; cf. SETAC 1994).

The first proposition arose from the ecosystems framework of the life sciences, which displaced the image of "nature" as fixed spaces to be conserved with the concept of the "environment" as a complex set of interrelated material flows and processes (Kline 2007). This new global environment was one that was constantly threatened by the widespread, often invisible risks of industrial capitalism—be it through small traces of cancer-causing chemicals in one's water supply or invisible noxious emissions in the atmosphere (Hays 2000). As for the second and third propositions, the combination of public pressure on government and business to manage these new risks became channeled through a neoliberal logic of decentralization, institutional efficiency, and accountability, giving rise to the audit as a mode of governance. "Audit societies" (Power 1999) or "audit cultures" (Strathern 2000) describe the now-pervasive practices of managing programs through quantitative metrics. These practices have become popular as a response to heightened public scrutiny of policy-making; audits are intended to address this public skepticism through performances of monitoring that rely upon the perceived transparency of numerical data as a means of projecting objectivity and neutrality (Jasanoff 1986; Espeland and Vannebo 2007).

As a product of these intertwined paradigms, the LCA represents a particular moral economy of managerial environmentalism and audit cultures in the sense Daston uses the term—as "a web of affect-saturated values that stand and function in well-defined relationship to one another" (Daston 1995:4). This moral economy frames nature as numerous interconnected, quantifiable material flows that should be holistically managed through transparent, objective forms of socioenvironmental risk assessment. Illustrating this point, the actors involved in standardization of the LCA in an early-stage workshop cited the need for a "quantitative approach based on science and engineering, thereby minimizing subjective biases" (World Wildlife Fund 1991:86) to combat the waste of resources, which they define as a problem of distinct product streams that must be made visible and whose outputs must be managed.

The NBI has multiple lines of research that utilize the LCA model to estimate the net GHG emissions and, in the near future, other environmental impacts of standard Brazilian ethanol production processes, such as total water use throughout the production process. NBI researchers use these results to engage international scientific debates over how desirable sugarcane ethanol is as an energy source as compared to conventional fossil fuels and other biofuels such as corn ethanol (cf. Shapouri et al. 2002; Wang 2005; Blottnitz and Curran 2007). The channels for engaging these debates are publications in prestigious peer-reviewed journals and a steady stream of conferences and workshops held across the globe to share the most recent research on issues of energy and sustainability.

In adopting the practices of LCA methodologies created in the United States and the European Union, NBI researchers also share the life-cycle moral economy as a way of judging ethanol production and its effects. When asked what "sustainability" means for energy, they point to a number of environmental and social impacts or indicators through which sustainability of a total production and consumption cycle should be understood and calculated. In terms of the political controversies over the moral acceptability of biofuels, NBI researchers, like their interlocutors abroad, see themselves as being the neutral, objective arbiters of what has become an intensely politicized and emotional subject. When I asked Thales, one of the NBI researchers, to describe the sorts of issues they have with engaging public policy, he lamented the unscientific nature of the legislative process: "It is not well based in very solid data . . . it is more about putting out a fire (you make legislation when a disaster happens) than making environmental policy for the long term. This doesn't exist." Thales goes on to lament that the public and policy makers tend to be too quick to assign blame for environmental hazards without systematically determining their real causes, a sentiment echoed by many of his colleagues at NBI. Implicit in Thales's comments is the view that part of this unscientific knowledge-production process is to focus on isolated cases of environmental or social harms that are not representative of the whole of Brazilian ethanol production. This was a point made explicit numerous times to me by NBI researchers emphasizing the importance of painting a portrait of the sector in its entirety, not relying upon isolated "case studies" of bad actors in the industry. As members of the scientific sphere, NBI researchers thus see their duty as informing policy dialogues in the political sphere that are too often driven by subjectivity, emotion, and overly particularized forms of knowledge that fail to capture the reality of the socioenvironmental benefits created by the sector as a whole.

I would argue that the methodological challenge of demystifying and contextualizing this moral economy is all the greater for scholars in the environmental humanities and social sciences as we often operate within the same networks, inhabit the same professional practices, and even share a common intellectual lineage with NBI researchers and their colleagues. The first time energy received sustained treatment within anthropology was in cultural ecology, most famously in the work of White (1943, 1959) and Rappaport (1967, 1984) to study energy flows in societies as a means of determining their developmental stage or their symbiotic relationship to their local environments. These energy frameworks in cultural ecology arose during the same time period and from the same social context as the early LCA studies—in fact, both Rappaport and LCA modelers cite ecologist Howard Odum's (1970) holistic concept of net energy analysis as an influence for their own studies of energy flows (cf. Rappaport 1971:130; Spreng 1988). It is therefore unsurprising that the quantitative, holistic lens through which these anthropologists and the creators of LCAs understood energy would share a great deal in common. Similarly, the risk/audit discourse of LCA modelers also pervades White's and Rappaport's ethics of energy circulation, which revolve around concerns that overconsumption of this omnipresent resource will lead to societal decline (White 1943) and environmental disaster (Rappaport 1971).

While quantification has been largely eschewed by contemporary energy studies in anthropology and the more critical theory–inclined practitioners of political ecology, the holism of cultural ecology—the tendency to view energy as a Big, Global Problem that structures nature and society—remains as firmly planted in its disciplinary offspring within the social sciences and humanities (environmental anthropology, political ecology, and environmental history) as it does in the systems ecology of contemporary LCA modelers. Environmental historians studying paradigms of energy production have constructed grand narrative arcs about the massive increases in global energy flows required by industrial societies and dramatic concerns about their depletion (c.f. Sieferle 1990; Wrigley 1990; Smil 1994; McNeill 2000; Crosby 2006). Many political ecologists and environmental anthropologists also take a large-scale systemic analysis to the structural violence (Watts 2001), inequalities (Hornborg 2001, 2009, and this volume), and international assemblages (Gillon 2010; Hollander 2010) of energy circulation. In terms of professional practices, members of these fields participate in sustainable energy research programs alongside quantitative modelers, particularly in European academic institutions, and take part in meetings to create sustainable biofuel standards; I have been at more than one such conference with NBI researchers where political ecologists have attended as presenters. Given such proximity, how to reconstruct an anthropology of energy that is capable of critically distancing and interrogating the practices and cosmologies of these energy experts, as well as our own?

THE LCA IN PRACTICE: TRUST IN PROXIMITY

Since at least the 1970s, there have been calls for anthropologists to expand their ethnographic gaze beyond our traditional marginalized subjects to include powerful actors in policy, business, and scientific spheres, exemplified by the informants in my own research (e.g., Nader 1974; Nelkin 1975). More recently, a burgeoning body of scholarship has gone further in identifying the tensions that exist when anthropologists study experts who share many of their professional practices (Mosse 2006; Riles 2006; Boyer 2008; Schwegler and Powell 2008) and analytic categories (Riles 2000; Choy 2005). As a means of contextualizing these practices, some argue that anthropologists must shift their ethnographic lens away from the rationalism that is most publicly associated with these experts and toward what Boyer calls "the halo of sentiments, affects, intentions and aspirations, none of which should be reduced to secondary status in expert knowledge-making" (Boyer 2008:45). Similarly, Holmes and Marcus (2005) call for exploration of the "paraethnographic" practices of experts; that is, the intuitive, experiential knowledge practices they draw upon when grappling with uncertainties in their data and decision-making strategies. I argue that studying these paraethnographic practices among the relatively small clusters of policy makers and experts that have a strong role in creating the techno-legal structures of energy assemblages is a profitable direction for a renewed anthropology of energy. This perspective encourages us to explore the everyday practices and affective, embodied knowledge through which these experts experience energy production—

experiences that are influential in how they develop their scientific analyses and thus how they shape the energy assemblage itself.

While at NBI I became highly familiarized with the productive disjunctures between these researchers' moral economy of "trust at a distance" and the "paraethnographic," or intuitive knowledge, practices they apply in their knowledge labor. I found that the former is most prevalent in their public discourse about what they do, as I describe above—in particular when drawing contrasts between the validity of their own research and the politicized, narrow-minded rhetoric of antiethanol activists and politicians. As with other "rituals of purification" (Latour 1993) their ability to imagine a pure separation between objective scientific research and subjective forms of knowledge relies on a preexisting hybridity of fact and value, objective measurement and intuition. While NBI researchers may shun the experience-based expertise of labor and environmental activists, at other moments they are quite reliant on this form of expertise for their own work. This happens in two ways: through data collection visits with *usina* experts whose opinions they value precisely because of the long-term personal experience they have working in the sector, and simultaneously through the firsthand knowledge they gain from visiting *usinas* (ethanol refineries) themselves.

In the first case, the intuitive knowledge of industry experts is crucial to the work of constructing quantitative LCA models where there is uncertainty about how its inputs should be calculated. The goal of the LCA is to create, in the software, a representation of a "standard" ethanol production process—one that is reflective of the ethanol/sugar refineries in a given area, which can be as large as an entire country but, in the case of the analyses done by NBI, is never smaller than a multicity production region. In LCA programs such as SIMAPRO or ASPEN+, the user enters figures for resource inputs at each designated stage of this standardized production process.

In practice, determining what a standardized production process is and collecting the necessary data on every input from diesel used by harvesting equipment to the hourly rate of sugarcane processes poses a number of challenges. Even within a certain state, where climate and soil conditions are roughly similar for different *usinas*, there can be a substantial amount of variation in the technology used from one refinery to the next. There is even more uncertainty in the agricultural phase in which sugarcane is harvested for transport to the *usina*. Here, as Thales explained to me, there is much greater variability in the factors affecting production such as climate and soil type, so there are more uncertainties about modeling impacts from this agricultural phase of production, yet it is also where the greatest magnitude of impacts is concentrated.

In this environment, NBI researchers rely on *usina* experts both to provide data for LCAs and to guide them on where in a particular range of production metrics a "normal" *usina* would lie. In the first case, a researcher will go to an *usina* where access has been established (relatively few, given the sensitive nature of such data collection). They will sit with the managers of a particular aspect of the production process and ask them for data on a series of metrics they need as inputs for the LCA model they will construct. As there are often misalignments between the needs of the researcher and the actual measurements taken by *usinas* for their own purposes, a dialogue often ensues where the director and the researcher will attempt to translate data into a form that is legible by the LCA model, often providing educated guesses where such translation is difficult.

In other cases, an LCA modeler might already have data from an existing source or from their own estimates but must consult an expert to make sure that he or she identifies the correct figure or range for a given input that matches a given region. For instance, a researcher might want to know what range of quantities of fertilizer are used to produce one hectare of sugarcane in southern Goiás. In this case, they will consult with an expert who has firsthand knowledge of ethanol production, having extensively worked in or with *usinas* as an engineer or consultant—someone who has had enough experience working in the sugarcane industry to be able to provide an educated guess on the right figure. The quasi-ethnographic nature of this work of constructing LCA representations extends to the ways these experts are identified—frequently they are among the friends and social networks of the researchers themselves, people NBI employees have come to know over the years through conferences, social engagements, and personal referrals.

These visits are crucial not only for the researchers to benefit from the experience of *usina* experts but also for the researchers themselves to develop or confirm their own ideas of what ethanol production is like. In conversations, NBI researchers acknowledged the importance of their own experience of "being there" in the field (or at the production site, in this case) in gaining an accurate and authoritative understanding of the socioenvironmental impacts of ethanol. A frequent complaint I heard about American and European critics of Brazilian ethanol is that these skeptics have not been to the country and seen ethanol production there, so they hold onto false assumptions about it, such as that it requires a large portion of available land or that refineries use outdated, inefficient technologies. The explanation for this given by one foreign researcher is telling; when I asked him why he needed to come to Brazil to conduct his research when he could get the same numbers by making phone calls from abroad, he replied: "To realize in the field . . . it's very important rather than just having a conversation on the phone, because I have visited sugar fields, sugar mills, distilleries—that is real feeling, you know? It increases the level of confidence in what you are doing." Compare his statement with what one labor activist told me about the value of his own work: "When you put a foot in the mud—you go to the field, you're there—you'll feel what the *boia fria* [field laborer] feels, you'll eat the food the *boia fria* receives from the company, see what taste it has, understand? This gives you something extremely precious. . . . You will see, you will be there, you will hear."

While NBI researchers may draw sharp divisions between their moral economy of objectivity and transparency-in-quantification to antiethanol activists' and policy makers' reliance on the intuitive and experiential, in their own work they tend to valorize these ways of knowing to get a "real feeling" for the nature of ethanol production in Brazil. They utilize these ways of knowing when they consult industry experts to get a sense of what figures or range of input values they should be including in their LCA models, as well as when they use the firsthand experience of these *usina* visits to form their own impressions of the production process. These knowledges are legitimized not because they come from precise, reproducible numerical calculations but rather because they are products of their own lived experience as well as that of informants who have been embedded in ethanol production for years or decades.

RELOCATING ETHNOGRAPHIC PARTICULARITY IN CONTEMPORARY ANTHROPOLOGIES OF ENERGY

Exploring the disjunctures between the discourse and everyday practice of sustainability science is a critical goal for future anthropologies of energy because it sheds light upon the ethical commitments and imaginaries of experts that affect how they craft energy policy. These researchers' primary firsthand encounters with ethanol production take place with mid-to-upper-level technicians and administrators at highly professionalized ethanol refineries. This contrasts with the engagements of anti-ethanol activists in Brazil, who tend to focus on understanding the lived experience of lower-wage laborers at sugarcane farms that may not even be directly owned by the ethanol companies they supply. These are the very different sorts of experiences that produce for NBI researchers an affect of sustainability for ethanol that evokes modernity, cleanliness, professionalism, and efficiency as compared to an affect of unsustainability for activists that evokes squalor and exploitation. As global sustainable production standards for biofuels are finalized and implemented, these experts' intuitive knowledges of the sector will shape how they approach issues such as labor practices and community engagement in ways that may be obscured to them by their particular moral economy. In describing these disjunctures, my goal is not simply to deconstruct the LCA models NBI researchers create for being based in part on situated, qualitative experiences and therefore "just their opinion." To do so would be as counterproductive as merely dismissing the work activist groups do to study labor conditions as being political and therefore nonfactual. Rather, I am arguing that by understanding and productively engaging experts about the ways in which their basic intuitions about the nature of ethanol production are shaped, we can make energy policy more inclusive and responsive to the needs of workers, local residents, and others who risk being rendered invisible in knowledge-production processes about energy's impacts.

Studying expert practices is not only a useful means for developing anthropological theory and applied interventions. It also serves to remind those of us in the social sciences who are studying energy issues of our own cultural cosmologies of energy circulation—especially as they are so close to those of our interlocutors. During this moment of resurgent interest in energy, there is a familiar divergence of approaches between theoretical frameworks that take the large-scale, structuring effects of energy as a starting point for a big-picture analysis and more ethnographically grounded critiques that emphasize the phenomenological, everyday experiences people have with energy (compare Hornborg to Wilhite in this volume). The question I hope to have raised here is: If LCA modelers' holistic, arm's-length analyses are in fact generated from experiences marked by embodied knowledge and intimate social relationships, what naturalized assumptions are embedded in our own theoretical understandings of energy as anthropologists, given our shared intellectual genealogies and professional practices?

This is not to say that anthropologies of energy should avoid structural analyses altogether, but in terms of the general direction of this growing subfield in anthropology, we need more, not fewer, ethnographic descriptions of the sorts of assumptions and ethical commitments held by the makers of these assemblages that become reflected in their techno-legal structures. We must keep our sights on the forms of affective experience and embodied knowledge that animate and are in turn reconfigured by these networks of energy circulation. At the same time, viewing energy uncritically as a Big Problem, we risk making the same mistake with regards to understanding our own positionality within these networks. A critical, practice-based approach to energy can enable us to remain reflexively attentive to the specific intellectual genealogies, professional relationships, and affective experiences of energy that subtly structure our own understanding of this phenomenon.

New anthropologies of energy should ask: What sorts of assumptions do we make about the natural scale of energy, and about the societal problems it poses? What are our ethical commitments in making these assumptions? What and whose alternative cosmologies do we elide in constructing these conceptual frameworks?

NOTE

1. I am grateful for the many productive dialogues with contributors to this volume as well as members of the Penn Knowledge Production working group from which this chapter benefited greatly. Funding for this research was generously provided by grants from the University of Pennsylvania School of Arts and Sciences, the Wenner-Gren Foundation, and the National Science Foundation.

REFERENCES

Alves, Francisco
2006 Por que morrem os cortadores de cana? *Saude e Sociedade* 15(3) (December): 90–98.

Bartley, Tim
2007 Institutional emergence in an era of globalization: The rise of transnational private regulation of labor and environmental conditions. *American Journal of Sociology* 113(2): 297–351.

Blottnitz, Harro von and Mary Ann Curran
2007 A review of assessments conducted on bio-ethanol as a transportation fuel from a net energy, greenhouse gas, and environmental life cycle perspective. *Journal of Cleaner Production* 15(7): 607–619.

Boyer, Dominic
2008 Thinking through the anthropology of experts. *Anthropology in Action* 15: 38–46.

Choy, Timothy K.
2005 Articulated knowledges: Environmental forms after universality's demise. *American Anthropologist* 107(1) (March 1): 5–18.

Crosby, Alfred W.
2006 *Children of the sun: A history of humanity's unappeasable appetite for energy*. New York: Norton.

Curran, Mary Ann
1996 *Environmental life-cycle assessment*. New York: McGraw-Hill Professional Publishing.

Daston, Lorraine
1995 The moral economy of science. *Osiris* 10 (2nd Series, January 1): 3–24.

Djelic, Marie-Laure and Sigrid Quack
2010 *Transnational communities: Shaping global economic governance*. Cambridge: Cambridge University Press.

Espeland, Wendy Nelson and Berit Irene Vannebo
2007 Accountability, quantification, and law. *Annual Review of Law and Social Science* 3(1) (December): 21–43.

Gabathuler, H.
2006 The CML story—how environmental sciences entered the debate on LCA. *International Journal of Life Cycle Assessment* 11 (April): 127–132.

Gillon, Sean
2010 Fields of dreams: Negotiating an ethanol agenda in the Midwest United States. *Journal of Peasant Studies* 723.

Haas, Peter M.
1996 Introduction: Epsitemic communities and international policy coordination. In *Knowledge, power, and international policy coordination*, edited by Peter M. Haas, 1–36. Columbia: University of South Carolina Press.

Hays, Samuel P.
2000 *A history of environmental politics since 1945*. Pittsburgh: University of Pittsburgh Press.

Hollander, Gail
2010 Power is sweet: Sugarcane in the global ethanol assemblage. *Journal of Peasant Studies* 37(4): 699.

Holmes, Douglas and George E. Marcus
2005 Cultures of expertise and the management of globalization: Toward the re-functioning of ethnography. In *Global assemblages: Technology, politics, and ethics as anthropological problems*, edited by Aihwa Ong and Stephen J. Collier, 235–252. Malden, MA: Blackwell.

Hornborg, Alf
2001 *The power of the machine: Global inequalities of economy, technology, and environment*. Walnut Creek, CA: AltaMira Press.
2009 Zero-sum world: Challenges in conceptualizing environmental load displacement and ecologically unequal exchange in the world-system. *International Journal of Comparative Sociology* 50(3–4) (June 1): 237–262.

Hunt, R. G. and W. E. Franklin
1996 LCA—How it came about—personal reflections on the origin and the development of LCA in the USA. *International Journal of Life Cycle Assessment* 1(1): 4–7.

Jasanoff, Sheila
1986 *Risk management and political culture: A comparative analysis of science in the policy context*. New York: Russell Sage Foundation.

Jasanoff, Sheila and Marybeth Long Martello, eds.
2004 *Earthly politics: Local and global in environmental governance*. Politics, Science, and the Environment. Cambridge, MA: MIT Press.

Keck, Margaret E. and Kathryn Sikkink
1998 *Activists beyond borders: Advocacy networks in international politics*. Ithaca, NY: Cornell University Press.

Kline, Benjamin
2007 *First along the river: A brief history of the U.S. environmental movement*. 3rd ed. Latham, MD: Rowman & Littlefield Publishers, Inc.

Latour, Bruno
1993 *We have never been modern*. Cambridge, MA: Harvard University Press.

Martinelli, Luiz A. and Solange Filoso
2008 Expansion of sugarcane ethanol production in Brazil: Environmental and social challenges. *Ecological Applications* 18(4) (June): 885–898.

Martines-Filho, Joao, Heloisa L. Burnquist, and Carlos E. F. Vian
2006 Bioenergy and the rise of sugarcane-based ethanol in Brazil. *Choices* 21(2): 91–96.

McNeill, J. R.

2000 *Something new under the sun: An environmental history of the twentieth-century world.* Global Century Series. New York: Norton.

Mosse, David

2006 Anti-social anthropology? Objectivity, objection, and the ethnography of public policy and professional communities. *Journal of the Royal Anthropological Institute* 12(4) (December): 935–956.

Nader, Laura

1974 Up the anthropologist: Perspectives gained from studying up. In *Reinventing anthropology*, edited by Dell Hymes, 284–311. New York: Vintage Books.

Nass, Luciano, Pedro Pereira, and David Ellis

2007 Biofuels in Brazil: An overview. *Crop Science* 47(6): 2228–2237.

Nelkin, Dorothy

1975 The political impact of technical expertise. *Social Studies of Science* 5(1) (February 1): 35–54.

Odum, Howard T.

1970 *Environment, power, and society*. New York: Wiley-Interscience.

Porter, Theodore M.

1996 *Trust in numbers*. Princeton, NJ: Princeton University Press.

Power, Michael

1999 *The audit society: Rituals of verification*. Oxford: Oxford University Press.

Rappaport, Roy A.

1967 Ritual regulation of environmental relations among a New Guinea people, in environment and cultural behavior. In *Ecological Studies in Cultural Anthropology*, edited by Andrew P. Vayda. Garden City, NJ: The Natural History Press.

1971 The flow of energy in an agricultural society. *Scientific American* 225(3) (September): 117–122.

1984 *Pigs for the ancestors: Ritual in the ecology of a New Guinea people*. New Haven, CT: Yale University Press.

Riles, Annelise

2000 *The network inside out*. Ann Arbor: University of Michigan Press.

2006 *Documents: Artifacts of modern knowledge*. Ann Arbor: University of Michigan Press.

Schwegler, Tara and Michael G. Powell

2008 Unruly experts: Methods and forms of collaboration in the anthropology of public policy. *Anthropology in Action* 15: 1–9.

SETAC

1994 Guidelines for life-cycle assessment: *Environmental Science and Pollution Research* 1(1) (January 1): 55.

Shapouri, Hosein, James A. Duffield, and Michael Wang

2002 *The energy balance of corn ethanol: An update*. Agricultural Economic Report. United States Department of Agriculture. http://www.transportation.anl.gov/pdfs/AF/265.pdf (accessed October 24, 2008).

Sieferle, R. P.

1990 The energy system—a basic concept of environmental history. In *The silent countdown: essays in European environmental history*, edited by Peter Brimblecombe and C. Pfister, 9–20. Germany: Springer-Verlag.

Silva, Maria Aparecida de Moraes

2006 A morte ronda os canaviais paulistas. *Revisa Abra* 33(2) (December): 11–143.

Smil, Vaclav
1994 *Energy in world history*. Boulder, CO: Westview Press.

Spreng, Daniel T.
1988 *Net-energy analysis and the energy requirements of energy systems*. Westport, CT: Praeger.

Strathern, Marilyn
2000 *Audit cultures: Anthropological studies in accountability, ethics, and the academy*. London: Psychology Press.

Wang, Michael
2005 Updated energy and greenhouse gas emission results of fuel ethanol. San Diego, CA: Argonne National Laboratory. http://www.transportation.anl.gov/pdfs/TA/375.pdf.

Watts, Michael
2001 Petro-violence: Community, extraction, and political ecology of a mythic commodity. In *Violent Environments*, edited by Nancy Lee Peluso and Michael Watts, 189–212. Ithaca, NY: Cornell University Press.

White, Leslie A.
1943 Energy and the evolution of culture. *American Anthropologist* 45(3) n.s. (September): 335–356.
1959 *The evolution of culture: The development of civilization to the fall of Rome*. New York: McGraw-Hill.

World Wildlife Fund
1991 *Getting at the source*. Washington, DC: Island Press.

Wrigley, Edward Anthony
1990 *Continuity, chance and change: The character of the industrial revolution in England*. Cambridge: Cambridge University Press.

CHAPTER THIRTEEN

Local Power: Harnessing NIMBYism for Sustainable Suburban Energy Production

Scott Vandehey

"I don't mean to sound NIMBY, but I don't want those powerlines ruining our community!" a member of the Rancho Peñasquitos town council exclaimed one warm Southern California summer evening in 2006 as the group discussed a proposal by San Diego Gas & Electric Company (SDG&E) to construct high-tension electric transmission lines through the suburban neighborhood. The woman's comments echoed a growing sentiment throughout the community that the neighborhood was under threat and needed to respond accordingly. Later that summer a group of over five hundred community members packed the small local library and spoke out against SDG&E's planned powerline project. Over the next few months pressure built as community members organized to challenge SDG&E's plans. A public awareness campaign was launched to bring the issue to the attention of the neighborhood as a whole, public meetings swarmed with individuals from the community expressing their discontent with the powerline proposal, and official legal challenges were made by members of the community to the California Public Utilities Commission (CPUC).

I spent over sixteen months during the height of the powerline controversy conducting participant observation in Rancho Peñasquitos through 2006 and 2007. During that time I conducted in-depth interviews and attended countless public meetings to investigate the civic participation and engagement of the residents. In many ways Rancho Peñasquitos is a nondescript suburban community. Located approximately fifteen miles north of downtown San Diego, the community wraps around the west, south, and east sides of a large hill referred to as Black Mountain, which marks the northern extent of the community. To the south, the Los Peñasquitos Canyon denotes the southern border of the community. Rancho Peñasquitos is a fairly average neighborhood for the San Diego region. It spans approximately 6,500 acres and as of the 2000 census had a population of 47,588 individuals. The median age is about thirty-five and each household averages just over three people. Ethnically, 59 percent of residents in the community identify as white, 25 percent as Asian, and 8 percent as Hispanic. Of the adult population over the age of twenty-five, 94 percent have completed high school. Additionally, 61 percent have attained a college degree at some level, a bachelor's degree being the most common at 34 percent. Approximately 78 percent of the housing units in Rancho Peñasquitos are single-family detached homes (San Diego Association of Regional Governments 2006).

This chapter focuses on the contestation over power between residents of Rancho Peñasquitos and San Diego Gas & Electric utility company. *Power* usefully connotes both electricity and the ability to have one's way, both central to issues present in the Rancho Peñasquitos powerline dispute. I begin by describing the struggle over power (in both meanings) in the community. The proposal to run electric lines was centered on a fight over authorities' political power to construct electric power transmission lines. I then show how the flow of power (in both meanings) is a social process that is structured through an interplay of local and regional interactions. Finally, I argue that the social power of NIMBYism might be transformed in productive ways to provide sustainable electrical power to our communities on the local level.

HIGH TENSION OVER LOCAL POWER

San Diego Gas & Electric called its Sunrise Powerlink Transmission Line Project, first proposed in 2005, "Our Connection to a Clean Energy Future." The ongoing project will culminate in a 117-mile, 500-kilovolt electric "superhighway" from Imperial County, California, to San Diego with 1,000-megawatt capacity—approximately enough energy for 650,000 homes (SDG&E Whitepaper 2010). The project was designed to meet the San Diego region's growing energy demand and develop the infrastructure upon which to base a "renewable energy" future. According to the Renewable Energy Transmission Initiative, eastern San Diego County, along with neighboring Imperial County and Northern Baja California, has a potential aggregate generating capacity of roughly 6,870 megawatts of solar energy, 3,495 megawatts of wind power, and 2,000 megawatts of geothermal energy. Together, these renewable resources could potentially produce enough electricity to power eight million homes (State of California 2009). SDG&E has publicly stated that the Sunrise Powerlink transmission line is intended to carry this untapped renewable energy generated by these sun, wind, and geothermal sources (California Independent Systems Operator 2006). In its 4–1 decision to approve the line, the California Public Utilities Commission stated that construction of the Sunrise Powerlink was crucial to the development of Imperial Valley renewable energy projects (California Public Utilities Commission 2008b).

High-voltage transmission lines, such as those proposed in the Sunrise Powerlink, carry electricity over long distances and are the essential backbone for transferring electricity from power-generating plants to final destinations. Once generated at a power plant, electricity is fed through a step-up transformer, which produces the low-current, extremely high voltages (typically 110,000 volts) required for long-distance electricity transmission. Voltage is analogous to water pressure in pipes. The higher the voltage the more "pressure" the electricity is under. The higher the voltage, the more efficiently electricity can flow over the lines, but the more dangerous it becomes. Before this electricity can be used by an end consumer, therefore, it must be reduced in voltage. The voltage of a standard electrical outlet in the United States is 120. In order to reduce the voltage from tens of thousands to 120, long-distance transmission lines feed into power substations where transformers reduce the voltage to a level acceptable for smaller distribution lines to handle. This reduction in voltage may happen multiple times before the electricity reaches the end consumer. Each reduction makes transmission less efficient, so it is done as close to the destination as possible.

While necessary for long-distance energy transmission from the Imperial Valley to San Diego, the Sunrise Powerlink is not without controversy, even among policy makers. In a dissenting opinion, a CPUC commissioner stated that the application should be denied based on inefficiency, ratepayer cost, and negative environmental impacts and that other less costly projects should be pursued instead (California Public Utilities Commission 2008a). These same concerns were voiced by many diverse groups and individuals throughout the San Diego and Imperial Valley regions. San Diego County Supervisor Dianne Jacob actively opposed the route and testified before the CPUC in opposition to the project. Utilities watchdog groups such as the Utilities Consumer Action Network also objected to the Sunrise proposal. The Center for Biological Diversity challenged the CPUC's approval in the California Supreme Court. In addition, the project was being litigated against in federal court by a coalition of three community groups: the Protect Our Communities Foundation, East County Community Action Coalition, and Back Country Against Dumps. These community groups challenged the Bureau of Land Management's approval of the project on BLM land, on the grounds that the BLM failed to perform its legally required environmental reviews of the project. Many other grassroots organizations formed and concerned individuals spoke out to oppose SDG&E's Sunrise Powerlink. Despite such vocal and broad-based opposition, however, the project was approved. The California Public Utilities Commission gave its nod in 2008, and the Bureau of Land Management and U.S. Forest Service both gave the go-ahead in 2010. Construction started in December of 2010 with an expected completion date in late 2012, which as of this writing is still on schedule.

The original Sunrise Powerlink plan proposed by SDG&E in the fall of 2005 routed the transmission lines right through the middle of Rancho Peñasquitos in an easement granted to the company before the community was built. The proposed route ran from the Imperial Valley west to San Diego, connecting the line to an existing electrical substation just west of Rancho Peñasquitos. In order to get to the substation, the lines had to run through the middle of the neighborhood. Residents of the suburban neighborhood were not pleased with the plan, and shortly after the announcement of the proposed project a number of concerned community members in Rancho Peñasquitos joined together to form the Alarmed Citizens of Peñasquitos (ACP).[1] The ACP had two goals. Primarily the group worked to prevent the powerlines from running through Rancho Peñasquitos. Their secondary goal was to educate members of the community about SDG&E's plans and to mobilize community members in opposition to the project. The ACP opposed the Powerlink project for a number of reasons. Of chief importance, the group felt that the project seriously threatened the suburban character of the neighborhood, would lower property values, and posed potential health risks to the families in the community. The chairman of the organization informed me that he had spent hundreds of hours and thousands of dollars fighting the powerline project. When asked why he was so dedicated, he responded, "This threatens everything. I'd give as much as I can to protect my family, home, and community. I'll do whatever it takes and thankfully I'm not the only one." The ACP also argued that the additional powerlines were not necessary and the benefits did not outweigh the substantial costs and risks—both monetary and otherwise. The ACP was extremely successful in raising local opposition to the Powerlink project. During an initial public hearing, required by the CPUC, over five hundred community members packed the local library to express their opposition to the Powerlink proposal and they made their concerns quite clear. Over a period of two years the ACP continued its pressure on SDG&E. Public meetings were packed with individuals opposing the Powerlink, legal challenges were made, and complaints were lodged with the CPUC. Through their efforts, the ACP won significant concessions from SDG&E, which are explored in the following section.

The conflict between SDG&E and the residents of Rancho Peñasquitos illustrates a tension over energy as a resource, but it also highlights a tension over the fundamental meaning of energy. To SDG&E energy was a resource and commodity. As a producer and purveyor of energy, SDG&E desires the ability to transmit that resource to its customers, and the Sunrise Powerlink would fill that need. To the residents of Rancho Peñasquitos—especially those involved in the ACP—energy was conceived of very differently. Less a resource (although it was that too), energy was thought of as "dirty," "dangerous," and "ugly." Community members were especially concerned about the "radiation" that powerlines produce and the potential health effects long-term exposure might have on their children. Furthermore, residents were concerned that the towers and lines would destroy the natural views from their homes and detract from the beauty of the neighborhood, thereby lowering property values. To the residents of Rancho Peñasquitos, then, energy is more than just a resource and commodity. It is also a danger to health, lifestyle, investment, and even nature itself. These differing images of energy underpinned the dispute between SDG&E and the residents of the community, one that pitted a large corporation against a small but tenacious group of concerned community members.

POWER FLOWS

SDG&E's original plan was to run aboveground transmission lines through the community. After considerable and concerted opposition was raised, SDG&E changed its plan and offered to run the lines underground. Though this is much more expensive, SDG&E felt that this would alleviate concerns regarding property values, as there would be no powerlines or transmission towers to mar the landscape. While this did address one concern raised by Rancho Peñasquitos residents, it heightened another. Residents were also concerned that the electromagnetic radiation associated with high-tension powerlines might be detrimental to the health of their children. The new proposal, which would place the lines six to ten feet underground, would bring the lines closer to people living in the neighborhood and thereby increase exposure to any possible risk factors. SDG&E attempted to alleviate residents' fears by explaining that the lines would be shielded by concrete and that there were no foreseeable health risks, but the residents and ACP members were not convinced and kept up their strong opposition.

In response to the continued pressure, SDG&E proposed a third option. Rather than run the lines through the easement, they would route the project underneath a major roadway that ran through the neighborhood. This suggestion would have moved the lines out of the easement abutting residential lots and away from houses and under a four-lane road with a center median. This option would have kept the lines underground and out of sight, and also placed them far from homes and locations frequented by children. Although this proposal seemed to mitigate all of the concerns raised by residents in Rancho Peñasquitos, it managed to raise an entirely new set of concerns.

Putting the transmission lines under the roadway would have required long-term and disruptive construction on a major traffic artery into, out of, and within the community. Residents regularly complained about the local traffic and this proposal would have only made it worse.

Despite SDG&E's attempts to come up with workable solutions that would have addressed all concerns and allowed the project to move forward, the ACP did not let up the pressure. The community group continued to fight the Sunrise Powerlink proposal in all of its manifestations. The group lodged a formal complaint to the CPUC and due in large part to vocal, constant, and plentiful ACP opposition, the CPUC required SDG&E to propose numerous routing alternatives that would keep the transmission lines outside of Rancho Peñasquitos entirely. SDG&E complied with the request and produced three additional routing options that did not include Rancho Peñasquitos. The final route approved by the CPUC was one of these alternatives, and the Sunrise Powerlink was kept far outside of the community.

The conflict between SDG&E and the residents of Rancho Peñasquitos pit a powerful corporation against the power of a community. Both groups, when organized correctly, can be quite successful in achieving their goals. At the end of the day, the ACP managed to force the powerlines out of their community. They were able to usurp the power of a large company—one that held right-of-way privilege through an easement established when the land the community now sits on was still an undeveloped sprawling ranch—and force SDG&E to change its plans on numerous occasions. Simultaneously, however, SDG&E was eventually able to get its proposal approved. The flow of power in this situation is quite complex. Social power flowed away from SDG&E and toward the ACP and Rancho Peñasquitos community members. At the same time, electric power (in the form of the Sunrise Powerlink) flowed away from the community. While both power flows seem to be running in a similar direction, the flow is actually quite complex. SDG&E, while forced to alter its proposal, remained a powerful player and was able to gain final approval for the rerouted project. Likewise, even though the power was moved away from the community, Rancho Peñasquitos will reap the benefits of electric power once the project is complete. The electric power grid enables power to flow in multiple directions simultaneously from multiple sources to multiple destinations. Social power, it would seem, follows a similarly complex network of interconnections.

Social and electric power are perhaps more than just metaphorically related. Adams (1975) makes this point quite strongly. Social power—the ability to have one's way—is centrally important in the decisions made in regards to energy distribution. The opposite is true as well. Control of energy often leads to social power. As middle-class citizens, the residents of Rancho

Peñasquitos had access to resources that could be used to garner political/social power and affect the routing of electrical power. In a very real sense, the community members used their social power to take control of the electric power. As we will see, they were able to do so largely because they could rely on a preexisting supply of cheap and reliable energy. Furthermore, the strategies and discourses employed by these middle-class community activists were justified and strengthened by an appeal to class interest.

TRANSFORMATIONS

The conflict in Rancho Peñasquitos highlights important issues regarding energy, its distribution, and the power to make energy-related decisions. While important on a local level, when connected to a larger context, the case described above is representative of important transformations that will need to be harnessed to ensure a long-term, sustainable, and viable electrical distribution network. On one hand, the actions of the ACP and residents of the community might be interpreted as the typical NIMBY (not in my backyard) reaction in which middle-class suburban residents oppose any development that might negatively impact their property values and lifestyle. On the other hand, this case demonstrates the power of local organizations and manifestations of local citizenship. This small community was able to successfully challenge the power of a public utility to distribute electricity to its customers. This type of local grassroots organization centered around one's neighborhood will be quite valuable as we transition into an energy-scarce environment.

Residents in Rancho Peñasquitos shared experiences, values, and concerns that focused attention around the neighborhood itself. The neighborhood was seen as the primary location of action. The ACP and others could have opposed the Powerlink project on numerous grounds that spanned the entire length of the line. Instead, they focused primarily on the part of the proposal that impacted their community directly. The inward-oriented focus of suburban citizenship is not unique to Rancho Peñasquitos. Many suburban communities demonstrate this trend, and have been criticized for acting in self-interested ways that are opposed to the larger public good. Examples abound of suburban residents opposing things like garbage dumps, prisons, and factories being built too close to their communities. The sentiment in Rancho Peñasquitos was no different, as the quote from the community member opening this article illustrates: "I don't mean to sound NIMBY, but I don't want those powerlines ruining our community!" The term *NIMBY* has come to be used pejoratively in description of such behavior. In essence, NIMBY refers to the opposition offered to any development that might lower property values no matter how necessary or desirable that development might be for the greater society.

While it may be tempting to place negative judgment on NIMBY-type behavior, doing so obscures some important insights that might be gained from a more objective analysis. I do not see NIMBYism as positive or negative. Rather, I see it as an attempt to protect a perceived ideal way of life, assert local control over outside forces, and as a refocusing of citizenship on a local level. The example of the resistance to the Sunrise Powerlink project serves as perfect illustration of the NIMBY phenomenon at work. Despite the potential benefits to the larger San Diego community, the proposed project threatened the idealized suburban lifestyle in Rancho Peñasquitos. Community residents stood up in opposition to the project in order to protect what they saw as their right to live according to the suburban ideal and not have that right impinged upon. We could label this behavior as simple NIMBYism and discount it as selfish homeowners acting to preserve their own self-interests, but this criticism misses the fact that NIMBYism is not just about selfish political action. It is also about working to maintain the values and goals of a community. If this local political power can be harnessed and directed into productive channels, it could be quite beneficial for ensuring our energy future. In fact, it may be the primary way to secure our future energy consumption styles at anything near current levels.

As we transition into an energy-scarce world, much will need to change. Life for residents of Rancho Peñasquitos—and for all suburbanites—will be altered greatly. The suburban lifestyle in particular has been signaled out by many as extremely energy-intensive—if not wasteful—and therefore untenable in an energy-scarce environment. Over the past half century—during the peak of plentiful and cheap energy—suburbanization has flourished and suburbs have become the most popular living arrangement in the United States (Nicolaides and Wiese 2006). Even as suburbs grew, they became the target of criticism regarding their supposed inefficiency, unsustainability, and wastefulness. Characterized by low-density building and physically separated land use, the suburbs rely on cheap and abundant energy to maintain the relative inefficiencies of decentralization. Furthermore, the detached freestanding single-family house, which makes up the majority of suburban units, must be individually heated, cooled, lit, and otherwise infrastructured and maintained. Atkinson (2007a, 2007b, 2008) has argued that the suburbs will fail as a living arrangement when fossil fuels are no longer readily available and cheap. Kunstler presents a similar view:

> America finds itself nearing the end of the cheap-oil age having invested its national wealth in a living arrangement—suburban sprawl—that has no future . . . We constructed an armature for daily living that simply won't work without liberal supplies of cheap oil, and very soon we will be without both the oil needed to run it and the wealth needed to replace it. Nor are we likely to come up with a miraculous energy replacement for oil that will allow us to run all this everyday infrastructure even remotely in the same way . . . [T]he tragic truth is that much of suburbia is unreformable. It does not lend itself to being retrofitted into the kind of mixed-use, smaller-scaled, more fine-grained walkable environments we will need to carry on daily life in the coming age of greatly reduced motoring. [T]his suburban real estate, including the chipboard and vinyl McHouses, the strip malls, the office parks, and all the other components, will enter a phase of rapid and cruel devaluation. Many of the suburban subdivisions will become the slums of the future. (2005:17–18)

A post-oil suburbia is perhaps quite bleak. As Leslie White (2007[1959]:56) explains, "culture advances as the amount of energy harnessed per capita per year increases, or as the efficiency or economy of controlling energy is increased, or both." As a cultural product, suburbia emerged with the availability of vast amounts of fossil fuel energy. As that energy is depleted, the culture that is built upon it would seem to be on shaky ground. In his study of social organizational collapse, Diamond (2005) documents many cases in which societies fail based upon how they respond to limited resources. It is quite clear that social failure in the face of resource scarcity is a very real concern, and the scenarios painted by Kunstler and Atkinson could very well await us.

Diamond, however, also documents cases where societies succeed in the face of resource scarcity. The small Pacific island of Tikopia, for example, faces resource scarcity due to its small size, relatively high population density, and isolation from other societies. In spite of these conditions, the Tikopians have maintained continuous occupation of the island for over three thousand years. They have faced numerous challenges including deforestation, species extinction, and overpopulation. Despite the array of factors against them, the Tikopians have managed to survive by developing a combination of sustainable gardening and population control managed by a decentralized political structure. Decentralization, coupled with the small and close-knit community, ensures that problems are dealt with in a bottom-up manner that allows the society to be extremely adaptable and responsive (Diamond 2005). In all cases presented by Diamond, successful societies were able to make radical and effective changes to their society and culture that allowed them to adapt to a resource-poor environment. The NIMBY response in Rancho Peñasquitos to the Sunrise Powerlink conforms in many ways to the Tikopian case. While not a small, isolated, or resource-poor community, the ACP does represent a bottom-up form of social organization that is highly adaptable and powerful.

Kunstler's and Atkinson's portrayal of a failed suburbia relies upon a rather large assumption that the suburbs are static and unchanging places. A brief historical overview of the suburbs illustrates that this is far from the case. The suburban way of life in the United States has been constantly evolving and changing to new situations for the past hundred years or more (Nicolaides and Wiese 2006). The current model of suburbanization from the 1950s and '60s—a middle-class nuclear family living in a freestanding house well outside of a central city—is not the only form suburbia has taken or might take in the future. In the mid-nineteenth century railroads, electric streetcars, cable cars, and other forms of mass transit made transportation cheap enough to be affordable to a significant segment of the urban population. This allowed more people to live farther from their places of employment and opened up the suburbs—which had been restricted to the upper class—to an entirely new class of urban residents. The increased access led to the first great wave of mass suburbanization in the United States, which has been documented by numerous authors, including Warner (1978), Jackson (1985), and Fishman (1987). As suburban areas opened up to increasing numbers of people from differing classes, the way the suburbs were used changed. House sizes and the space between houses decreased in order to make the homes more affordable, boarders and multigeneration families were not uncommon, and working-class suburban neighborhoods were often home to chickens and garden plots. Over the past half century, suburban spaces have changed drastically. Suburbs are not static or fixed entities tied to a particular lifestyle. Instead, suburbs are embedded in a particular context and are constantly changing to meet changes as they occur. Technology, social forces, market condition, cultural values, and governmental policies all have a profound effect on the manifestation of the suburbs in which we live. The real question is the degree to which suburbs are adaptable: Can they adapt fast enough? How will they adapt? And are residents willing to adapt? Fundamentally, is transition to a post-oil suburbia possible?

Ethnography, participant observation, and other methods employed by cultural anthropologists are important and necessary tools to help address these questions. Stereotypes often skew our perceptions about how people live and ought to live. Perhaps this is nowhere more obvious than in how we as a society think about the suburbs. Images of suburbia are all around us and they have a powerful impact on our perception of the suburban experience and lifestyle. Many urban-based authors who may have driven through suburbia but spent little time there themselves have a lot to draw from in their critiques, but a deep and rich analysis needs to draw from up-close and long-term observation. Ethnography can provide the details of the lived experience and illustrate the nuances (and even contradictions) of everyday life. To fully understand the way we think about and consume energy, scholars need to enmesh themselves in the cultural practices of the people they study. Nowhere is this more important than in those places that we take for granted, such as suburbia.

As a society, we have invested unfathomable amounts of money, resources, time, and labor—not to mention energy—into the suburbs. Abandoning them would amount to a colossal waste, and a workable alternative might not be possible in an energy-scarce environment. Rather than replace suburbia, we should be looking for ways to alter its current use. Vail (2008c) provides a number of very thoughtful and valuable insights on how a suburban environment might be adapted and reshaped in an environment of expensive energy. Vail argues that although suburbia is currently more energy-intensive that urban spaces, it has the potential to be less so (2008c). Due to its relative low density, suburbia has a lot of space to work with. Currently taken up by yards, parks, landscaping, and other "natural" areas, these spaces could be relatively easily changed into food growing areas, which would reduce suburban reliance on mass-produced and trucked-in food. The detached homes of suburbia provide a huge amount of roof space per capita that can be used for both rain catchment and solar energy production. According to Vail's calculations, with a bit of frugality and effort, most suburban communities could become self-sufficient in terms of food, water, and electricity (2008c).

While true self-sufficiency of suburbia is certainly debatable and depends largely on the local area, it is undeniable that suburban areas could become much more energy-efficient and self-sustaining. This will become a necessity as energy becomes less readily available. But how do we move toward a sustainable suburbia, and is it even possible? Vail (2008c) suggests that the decentralization of suburbia may well be its single strongest asset in terms not only of self-sustainability but also self-government. As Rappaport (1971) points out, large hierarchical and centralized systems are energy-intensive to maintain and difficult to change. The larger the system, the more energy it takes to keep it functioning. Rappaport suggests that in order to survive with declining energy supply, societies may need to decentralize and return to more local resource dependencies. Far from the urban-biased stereotype, suburbs may well prove to be far more resilient elements in our society's coming transition out of the fossil fuel centralization model. Suburban neighborhoods and communities provide a natural location for small-scale decentralized political units to form. Because they are small-scale and in tune with local issues they would be less energy-expensive, more adaptable, and a ready-made answer to a looming energy crisis. Anderson et al. (1996) support the idea that decentralized spaces can actually be more energy-efficient if local resources are used advantageously. The focus on local solutions are not limited to urban and social planners. In their look at available future energy sources, Armstrong and Blundell (2007) also argue that local energy sources will be a fundamental necessity to maintain current energy levels in the future.

The events in Rancho Peñasquitos and the fight over high-tension powerlines suggests that residents of suburban communities are already well aware of the power and necessity of local action. Superficially, the residents of this neighborhood seemed to be acting against their energy interests. What they displayed, however, was a clear ability to organize effectively when their way of life was threatened. This local, decentralized, and small-scale community organization will be at the center of suburban adaptability.

As energy becomes scarce, the threats to suburbia will become more numerous, and the ability to maintain a suburban lifestyle will be challenged. Conservation will become a necessary reality. Gilbert et al. (2009) propose that people conserve because of a concern for the environment, but that this sentiment is relatively "shallow" and fades rather quickly when their accustomed way of life is challenged. In an environment of energy scarcity, this calculus will change. Conservation will no longer be a threat to an accustomed way of life, but rather it will be the only way to maintain something approaching it. The phenomenon of NIMBYism might well be the suburbanites' best weapon. As we have seen, it is used to stave off potential threats to the ideal suburban lifestyle. If suburbanites can redirect their NIMBY protectionist attitudes and impulses towards changing their communities and making them more self-sufficient, suburbia may just survive in an energy-scarce environment. We are in fact already seeing the initial signs of this throughout the United States and elsewhere. Many suburban neighborhoods have begun growing community gardens. Solar panels and rain catchment systems are becoming ever more popular, and with governmental subsidies and increasing electric and water costs the trend is likely to continue. While the woman opening this chapter may have been embarrassed by her NIMBY attitudes, it is precisely those sentiments and willingness to act that saved her community and may end up doing so again in the not-too-distant future. In so doing, suburbanites like her will be transforming the face of both political and electric power across the nation.

NOTE

1. All names regarding individuals and groups within Rancho Peñasquitos have been changed.

REFERENCES

Adams, Richard Newbold

1975 *Energy and structure: A theory of social power.* Austin: University of Texas Press.

Anderson, William P., Pavlos S. Kanaroglou, and Eric J. Miller

1996 Urban form, energy and the environment. *Urban Studies* 33: 7–35.

Armstrong, Fraser and Katherine Blundell

2007 *Energy . . . Beyond oil.* Oxford: Oxford University Press.

Atkinson, Adrian

2007a Cities after oil—1. Sustainable development and energy futures. *City* 11(2): 201–213.

2007b Cities after oil—2. *City* 11(3): 293–312.

2008 Cities after oil—3. *City* 12(1): 79–106.

2009 Cities after oil (one more time). *City* 13(4): 493–498.

California Independent Systems Operator

2006 California ISO board approves Sunrise/Greenpath transmission project. http://www.caiso.com/1847/1847bb8a57f70.pdf (accessed September 15, 2008).

California Public Utilities Commission

2008a Agenda ID#8065. http://docs.cpuc.ca.gov/efile/ALT/93073.pdf (accessed January 3, 2009).

2008b Agenda ID #8136 (Rev. 1). http://docs.cpuc.ca.gov/PUBLISHED/AGENDA_DECISION/95357.htm#P2638685. (accessed January 3, 2009).

Diamond, Jared

2005 *Collapse: How societies choose to fail or succeed.* New York: Penguin.

Fishman, Robert

1987 *Bourgeois utopias: The rise and fall of suburbia.* New York: Basic Books.

Gilbert, Liette, L. Anders Sandburg, and Gerda R. Wekerle

2009 Building bioregional citizenship: The case of the Oak Ridge Moraine, Ontario, Canada. *Local Environment* 15(5): 387–401.

Heinburg, Richard

2003 *The party's over: Oil, war and the fate of industrial societies.* Gabriola Island, Canada: New Society Publishers.

2004 *Power down: Options and actions for a post-carbon world.* Gabriola Island, Canada: New Society Publishers.

Jackson, Kenneth T.

1985 *Crabgrass frontier: The suburbanization of the United States.* New York: Oxford University Press.

Kunstler, Howard

2005 *The long emergency: Surviving the end of oil, climate change, and other converging catastrophes of the twenty-first century.* New York: Grove Press.

Nicolaides, Becky M. and Andrew Wiese

2006 *The suburb reader.* New York: Routledge.

Rappaport, Roy
1971 The flow of energy in an agricultural Society. *Scientific American* 225(3): 117–132.
San Diego Association of Regional Governments
2006 Data warehouse. http://datawarehouse.sandag.org/ (accessed February 17, 2006).
Scheer, Hermann
1999 *The solar economy: Renewable energy for a sustainable future.* Sterling, VA: Earthscan.
SDG&E Whitepaper
2010 Sunrise Powerlink, Electronic Document. http://regarchive.sdge.com/sunrise-powerlink/docs/srpl_whitepaper.pdf (accessed June 22, 2010).
State of California Renewable Energy Transmission Initiative Steering Committee
2009 Renewable energy transmission initiative. http://www.energy.ca.gov/2009 publications/RETI-1000–2009–001/RETI-1000%202009–001-F-REV2. PDF (accessed November 5, 2009).
Vail, Jeff
2008a A resilient suburbia? 1: Sunk cost and credit markets. http://www.theoildrum.com/node/4720 (accessed June 8, 2009).
2008b A resilient suburbia? 2: Cost of commuting. http://www.theoildrum.com/node/4741 (accessed June 8, 2009).
2008c A resilient suburbia? 3: Weighing the potential for self-sufficiency. http://www.theoildrum.com/node/4774 (accessed June 8, 2009).
2008d A resilient suburbia? 4: Accounting for the value of decentralization. http://www.theoildrum.com/node/4844 (accessed June 8, 2009).
Warner, Jr., Sam Bass
1978 *Streetcar suburbs: The process of growth in Boston (1870–1900).* Cambridge, MA: Harvard University Press.
White, Leslie
2007 [1959] *The evolution of culture.* Walnut Creek, CA: Left Coast Press.

CONVERSATION 4
Energy Contested: Culture and Power

Elizabeth Cartwright, Thomas Love, Derek Newberry, Jessica Smith Rolston, Sarah Strauss, Scott Vandehey

In Part 4, we shift attention to questions of boundaries across scale, moving from individual bodies to wider geopolitical interaction and on to the earth system. The papers in this section, by Elizabeth Cartwright, Jessica Smith Rolston, Derek Newberry, and Scott Vandehey, engage the relationships and points of conflict across these domains. They ask how thinking about scale might be a useful tool for responding to some aspects of energy development and transition, while addressing questions about health and illness, as well as sustainability and renewable energy transitions, at the individual, community and national scales.

As part of their process, the participants in this section queried the conceptual links among political systems, geological systems, and individual biological systems (people's bodies), which resulted in an interesting exchange about "coal famines" and the overall utility of energy metaphors for bridging different scales of interaction. How do people and groups of people—private individuals, private corporations, public communities, public political structures—value energy, and the risks associated with it? Are these values for energy in line with each other? At odds with each other? How are conflicting values of energy resources and risks resolved? The group also focused on the issue of energy transitions, whether across source or scale (or both), responding to questions about the movement from large-scale energy systems to smaller-scale community-based energy generation, transmission, and distribution systems.

JESSICA SMITH ROLSTON: Thinking about the conceptual links among political systems, geological systems, and individual biological systems (people's bodies) is a very interesting question. It's telling that in the late nineteenth and early twentieth centuries in the United States and Europe, coal shortages were widely described as "coal famines." These shortages (or fears of them) were associated with labor actions, especially strikes, that would shut down production of the fuel that had become indispensable to everyday life.

The term *coal famine* likens coal to food, suggesting that people and entire societies cannot function without coal, just as individual human bodies cannot function without food. The most famous coal famine was the result of the United Mine Workers of America's Coal Strike of 1902 in eastern Pennsylvania's anthracite fields. Fear of a similar shortage played a large role in the unfolding of the infamous Ludlow strike and massacre in Colorado in 1914. In his book about Ludlow, historian Thomas Andrews quotes the Denver Chamber of Commerce summing up the region's dependency on coal by simply stating, "We cannot exist without it" (2008:85). Quick searches of news databases reveals that the term *coal famine* ceased almost completely in the 1950s as coal was replaced by other energy sources. Current news favors the more generic term *energy famine*.

People make similar analogies on mine sites. Shovel operators in northeast Wyoming's Powder River Basin describe the coal face, as chunks of it slide toward the shovel's bucket, as "feeding" them. The process by which truck drivers haul the coal to the plant, where it is broken down and loaded onto trains, is also described as "feeding" the plant. Symbolically grounding energy sources in biological systems highlights their importance in our everyday lives. But does it also naturalize (Yanagisako and Delaney 1995) our dependence on them as necessary or inevitable (to think back to Tom Love's comment about energy slaves in Conversation 1)?

SCOTT VANDEHEY: Jessica, your description of coal being conceptualized in biological terms is quite fascinating. In answer to your concluding question, I would have to say that yes, it does serve to naturalize our dependency. The metaphor of "feeding" is telling, I think. Food is a necessary component of life, and here energy resources are being conceptualized in the same manner. Although not conclusive by any means, a recent quote from the *New York Times* illustrates how we talk about (and think about?) energy more generally: "In 1973, oil accounted for 46 percent of the world's total energy consumption; by 2005, its share had declined to 35 percent. But oil remains well ahead of other energy sources: coal meets 25 percent of the world's energy needs, natural gas is next with a market share of 20 percent, and nuclear power meets 6 percent of the planet's energy needs." The words *consumption* and *needs* in association with energy continue the trend of thinking about energy as a type of biological necessity analogous to food.

DEREK NEWBERRY: You both make excellent points, and I think this demonstrates the usefulness of energy as a theoretical tool. The flexibility of this concept allows it to traverse the three spheres mentioned in this question (roughly, the social, the natural, and the individual/biological) in ways that enable us to connect all three. Importantly, as you note, these same metaphors are used by our informants themselves to imagine these linkages.

In the biofuel sector, where there exists a great deal of concern about the "sustainability" and energy efficiency of these alternative energies in relation to fossil fuels, policy makers rely heavily on life-cycle models that measure the total energy input-output ratio over the entire commodity chain of ethanol in order to set policies regulating its production. In these models, energy refers to more than the ethanol commodity itself as it leaves the refinery and is distributed to consumers filling up at a gas station. Energy is measured through a series of transformations as it enters production via sunlight nurturing stalks of sugarcane and a diesel-powered machine harvesting them to its eventual output as an alcoholic liquid after a fermentation process. Here geological and techno-political systems are connected seamlessly as they are reduced to the accumulation of so many energy flows.

It seems that as with "coal famines" and "peak oil," the ways our informants talk about energy tell us a great deal about how they imagine these complex systems of production and consumption as well as how they participate in and reconfigure them. In the example of energy input-output analyses I gave above, certain activist groups have tried to extend these networks of energy flows to include the energy exerted by humans who harvest sugarcane for ethanol. In my conversations with them, they frequently discuss the exhaustive nature of harvesting—how it saps the energy of cane cutters and "uses them up." Extending the energy metaphor to these workers is a way of including them in the discussion about ethanol "sustainability" and challenging the way in which labor for ethanol production is organized. Of course advocates of ethanol I have spoken to claim the industry trend toward mechanized harvesting is eliminating these difficult jobs and that such working conditions do not reflect the average production process. Either way, how these actors conceptualize energy heavily affects how the production chain is shaped, as their advocacy is translated into regulatory policy. Understanding these "energy imaginaries" therefore seems inseparable from analyzing the complex assemblages of energy circulation from a political economy perspective. As anthropologists, studying the ways people experience energy as producers, or in the case of Stephanie's work, as consumers, we can begin to make sense of the large-scale systems through which they circulate, much as Alf is doing in his own research.

SCOTT VANDEHEY: Derek, I think you make a very important connection between the modern conceptualization of energy and the emergence of modernity itself. We could spend all day debating the meaning of the term *modernity*, but regardless, you make a very valid observation. The way we think of energy has a lot to do with the way we use energy, which was fundamentally altered with the emergence of three interrelated developments: capitalism, industrialization, and fossil fuel usage.

As I'm thinking through these linkages, an interesting idea keeps popping into my head. I haven't had a chance to tease this out fully, but it seems to me that energy is playing a mediating role between "nature" (whatever that means) and "technology." In the realm of industrial capitalism, energy is the connection between the natural world and the industrial world that humans have created. As such, it seems to metaphorically be related to "food." Food is also a mediator between humans and "nature." Pushing the metaphor just a bit: humans need food to survive, just as our creations need energy to survive. Perhaps this is why energy has come to be seen as a "need." Just some "food" for thought. (Pardon the pun, I couldn't resist!)

JESSICA SMITH ROLSTON: Thank you both for the great points. It seems that one of the common threads is redefining consumption. Scott, you point out that consumption can be naturalized as a "need" (which is so often forgotten in energy policy debates focused on increasing production instead of curtailing consumption). Derek, you show how the activist groups try to draw attention to the consumption of human labor and energy that is involved with the production of biofuels. This is a topic that resonates with my work. I'm reminded of Janet Finn's (1998) research in Montana and Chile, in which she creatively redefines mining terms such as *reclamation* and *consumption* to capture the everyday experiences of copper miners and their families. I also see this in my work with coal miners, who have to work rotating (day to night) twelve-hour shifts. They all love the seven days off that they get each twenty-eight-day rotation but critique the exhaustion they endure for the other three weeks.

When the miners make statements about the rest of the country not knowing or appreciating where their electricity comes from, they, like the cane cutters, are trying to expand the visible network of energy flows to include them and their labor. I've found Robert Foster's (2008) work on disjunctures and misrecognitions in product networks to helpful in this regard. There are energy networks that link disparate places and people (miners in Wyoming who produce coal, power plants that burn it, people who flip on a light switch, etc.), but these are not always socially recognized (especially when energy sources are located so far away from consumers). It's only after someone connects the dots that they can consider "the moral relations and ethical responsibilities implicit in the movement of products from one set of hands to another" (Foster 2008:240). I'm inspired by Foster's call for "caring from a distance," an ethical engagement that "hinges on the respectful and serious regard given by people, connected to each other as agents in a product network, for each other's concerns: a politics of mutual recognition" (2008:240).

Your larger "linking" comments made me wish we had more space—the Powder River Basin miners do strongly critique specific aspects of the labor process (both explicitly by documenting the bodily and social stresses of rotating shift work and implicitly by raising their children to find other kinds of work), but their relative strength in a tight labor market has also made it possible for them to make the mines generally safe and dignified places to work. I've also found that they are generally very critical and aware and concerned about the larger environmental impacts of their work in the industry. Miners and their families are avid outdoorspeople who spend their days off hunting, fishing, hiking, and so on, and many were born and raised in ranching families that depend on healthy ecosystems—those kinds of engagements ground their concerns about local environmental impacts. On a broader scale, people I met worry very much about their industry's contribution to climate change, so much so that I once walked into a mine shop where there was a debate about global warming taking place on one of the big white boards where they also kept track of assignments and projects—and the consensus was that it existed, not that it was a threat created by greenies to take their jobs. At the same time, they need a job so they keep working there, hoping that the industry will last until they can retire.

My interest has always been in trying to understand how people inhabit potentially alienating places like mines or soften the blows of market vagaries, but you bring up an equally important counter question: what are the implications of feeling "safe, happy, and okay"? What other social and environmental processes does this enable? To go back to the main argument of my chapter, it is important to distinguish among energy workers. Working conditions, attachments to place, possibilities to lead meaningful lives, etc. are very different for coal miners and rig workers of all types, which helps to explain the very different environmental and safety records for the two industries even within a single state such as Wyoming. It seems to me that energy policy debates need to include more attention to the concerns of energy producers. How does our energy consumption engender particular possibilities for people to engage in safe, dignified work—and avoid being totally "consumed" by it?

ELIZABETH CARTWRIGHT: Interesting ideas so far. These posts bring to mind some conversations I had with coal miners in Mexico, years ago. The miners would go in to the local clinic for lung function tests on a fairly regular basis, but as far as I could see they would then just go back to work.

Finally, someone told me that the men all wanted to work until they had only 25 percent of their lung capacity left, because if they were 75 percent disabled they would receive more retirement from the government. This also makes me think of the farmworkers in my dissertation work, who were indigenous Mexicans. They understood the pesticides they were working with caused short-term problems (runny noses, red eyes, skin disease), but they didn't have any idea of the long-term problems of using DDT and Deldrin and Eldrin and being in contact with these whilst pregnant.

I think what is required is a more fierce anthropology of work, and particularly of the biological/social/genetic/economic morasses that people are in and that they accept because they provide Life, when that really might not be the case at all and identifying the levels of bodily repercussions and keeping them intact. . . . I guess what has made me remember this after all this time is not that the notion of getting more disability benefits is new, but rather that this very "scientific" way of measuring, complete with numbers and probable years of life left, seemed to be one of the most direct recognitions of the damage being done to one in this kind of extractive process. Life energy was being directly traded for coal energy.

JESSICA SMITH ROLSTON: Yes, exactly. Your examples here are really rich and point to larger problems about both temporality and scale in thinking about (and ameliorating?) risk in its many forms. All of the miners back home started working in the industry back in the late 1970s when they were in their twenties, so most of them are now contemplating retirement. Listening to them talk about how their experiences of shift work currently compare to what they remember back in their younger years, or listening to them document all of the chronic injuries that are just now appearing after two to three decades, drives this point home as well.

TOM LOVE: The question of whether it is possible to move away from large-scale energy systems to smaller-scale community-based energy generation, transmission, and distribution systems is an important one. At the Association for the Study of Peak Oil conference in November 2011, Richard Heinberg "spoke quite literally about 'the end of growth.' Point by point, he argued that":

1. Cheap energy led to mass production, which led to advertising and credit.
2. Payment of debt requires future growth.
3. Increasing debt for consumers is increasing wealth for banks leading to increased power for financials.

4. We are living at the end of history's greatest credit bubble.
5. Our economic future will have persistent high unemployment, declining income and net worth, and financial instability.
6. We need to build local resilience, but it is at odds with economic efficiency.
7. Rapid economic growth is an artifact of the fossil fuel age.
8. We can have a better quality of life with reduced consumption.

SCOTT VANDEHEY: Jeff Vail (2008a–d) has made very similar arguments in regard to the development of suburbia. In essence, his argument is based on the observation that suburbia was made possible through the mass production, economic boom, and the energy abundance of the oil age. Rather than argue against suburbia as untenable in a post–cheap oil environment (as many have done), Vail points out that we have invested too much into suburbia to go back. Our entire financial system is built on the back of financing and supplying the suburbs and the majority of Americans live in suburban places. To remake the system at this late date is impossible, according to Vail. Even if we wanted to scrap the suburbs and start fresh, we probably couldn't do it, as the energy resources are no longer available.

Instead, Vail suggests retooling the suburbs for more efficient use. He argues, in fact, that the suburbs may actually be better positioned to adapt to a post–cheap oil future than other types of communities. As relatively low-density living arrangements, the suburbs have plenty of land to grow food, collect water, and generate "green" energy. Theoretically, much of this could be self-sustaining.

Vail's argument relies upon a massive shift away from centralization. He offers "open source" as a model of future economic and political organization, as opposed to the hierarchical structures that are dominant today. I find Vail's suggestion of open source quite intriguing. It seems to be a way for him to get around a central problem. For over ten thousand years human societies have been trending toward increased social complexity and hierarchy. This increase in complexity is closely correlated with the more efficient harnessing of energy. Now the human species faces a challenge: what to do in the face of declining energy?

I can't predict how we will address the issue socially, but it is clear that the two are related. I can't see a fully open-source system being implemented anytime soon. I do, however, see signs of things moving in that direction. Individuals are coming together to grow community gardens, collect their own rainwater, install solar power, and so on. Perhaps there is a small-scale movement in the works?

DEREK NEWBERRY: This is something several of the energy and climate scientists I have spoken to have expressed as well. Although they do not go so far as to locate energy circulation of the future on the community scale, they predict that energy sources will at least have to be regionally sourced. While this seems logical, since the viable energy alternatives to fossil fuels are extensive and/or difficult to distribute long distances (with the exception of nuclear power), there is very little understanding of what this would mean for the structure of the global economy. As Scott indicates, the "thingness" of energy heavily influences the structure of the production and consumption systems it fuels—suburbia was made possible by seemingly limitless supplies of cheap, dense, and highly mobile energies. As Tom notes, new energy paradigms will be inextricably linked with new paradigms of production and consumption, but what changes will this entail for the large-scale systems that powerfully mobilize people and nature to produce goods and services at a scale and speed that would be otherwise impossible?

Regarding the point about self-sustaining suburbs, the book *Natural Capitalism* (Hawken et al. 2000) has a brief description of the work that has been done to reconfigure the Washington, D.C. suburb of Arlington, Virginia, in exactly this way—by developing communities around subway stations. I would add that as this energy question is often tied into larger concerns about the global economy it should not be separated from other ethical concerns about the underlying conditions of contemporary commodity chains. For example, efforts by the fair trade movement to make agricultural production chains visible to consumers and to protect small producers against the vagaries of the market are also a result of concerns with the diffuse and sometimes invisible "risks" generated by the structure of the global economy. In these discourses of "sustainability," there is a common preoccupation with protecting peoples' rights, dignity, and self-sufficiency against these unwieldy risks. In my comments above on the separate question of circulations, I mentioned that we should understand why people conflate these risks—frequently using energy metaphors—as a way of examining how they interact with and constitute commodity chains. Here I would suggest that amid these conflations we might find possibilities for developing new ethics of need and consumption that reconfigure (or deconstruct?) those commodity chains and the sometimes exploitative relationships of dependency they can produce.

PART 5
Energy Contested: Borders and Boundaries

CHAPTER FOURTEEN

Oil's Magic: Contestation and Materiality

Gisa Weszkalnys

> [Oil] anesthetizes thought, blurs vision, corrupts . . . [It] kindles extraordinary emotions and hopes, since oil is above all a great temptation.
>
> —From Ryszard Kapuscinski's *Shah of Shahs*, cited in Watts 2009:84

"Oil," Michael Watts asserts, "has always been vested with enormous, often magical, powers. Sometimes I think we are too easily seduced by these deceits." In this chapter I want to rethink this particular rendering of oil as a seductive and cursed substance.[1] What is this magic of oil that has seduced scholars, corporations, and people living in oil-producing countries? What tools does anthropology offer to fathom it? Crude oil is readily the most coveted but also one of the most contested sources of energy in the world today. Despite the expertise and increasingly costly, specialized technology required for its making, oil also remains one of the cheapest forms of energy we have. Its unusually high energy density, its ready transportability at ambient temperatures, and its extreme versatility as a primary resource have made it almost indispensable to the fabric of contemporary societies. The "magical" power of oil *as energy* has been largely conceived of as invisible or immaterial. By contrast, I will highlight in this chapter oil's *material* powers—not just as a raw material for the production of all sorts of consumer goods but as a distributed materiality that spans this substance's physical and chemical constituents as well as the specialist equipment needed for its extraction, the practices of abstraction and valuation that go into its making, and the people doing the extracting, contesting, and transforming of oil. I argue that anthropological insight into the flows of oil across a range of material sites and across the boundaries of nature and culture goes a long way in explaining how oil's magic has come about.

World oil demand, according to the International Energy Agency, is now exceeding ninety million barrels per day[2] and rising, while oil supply hovers at several million barrels below that level. The search for additional sources of oil supply is growing increasingly frantic. While historically countries may have benefited from possessing oil to boost industrial growth, the geographical pattern of oil's distribution has long since departed from the pattern of the distribution of economic development and wealth (Mitchell 2009; Selby 2005). It is becoming evident that the inhabitants of old and new oil-producing countries are demanding their share of the wealth derived from the resource, resulting in new and sharpened episodes of contest. This is nowhere more apparent than where I conducted my research, on the African continent whose onshore and offshore oil fields are widely regarded to constitute one of the most promising and sought-after resource frontiers today (Clarke 2010; Frynas and Paolo 2007). U.S. military forces in the form of the newly established Africa Command (AFRICOM) are deployed to watch over a region characterized, on one hand, by a continuous sense of looming crisis and, on the other, by growing Chinese investment, specifically in the resource sector (Keenan 2008; Klare and Volman 2006; Taylor 2006). African oil, in particular, has entered into the limelight of public debate in the last decade or so due to its apparent capacity to bring about social and political friction and military conflicts. From the Niger Delta to Southern Sudan, oil exploitation appears to be accompanied by protracted violence and civil wars. Oil has been pronounced a uniquely "cursed" substance (e.g., Auty 1993; Ross 1999, 2012; Sachs and Warner 2001).

In other words, the increasingly scarce sticky substance we call oil comes to us not simply with promises of resource wealth. It is also imbued with "fetishistic qualities" (Watts 2004:61) and a rich mysticism of abundance, affluence, human greed, and power. In São Tomé and Príncipe (STP), the African Atlantic island state where I carried out ethnographic fieldwork, such ambivalence is palpable. In this former Portuguese colony, petroleum has long been an object of rumor and speculation. In a recent chronicle of STP's oil, the current director of the country's National Petroleum Agency notes that Santomeans had wondered about the dark liquid oozing out of the muddy ground at obscure places in the islands' dense forests long before their colonizers initiated an official search for oil (dos Prazeres 2008:11). During the last fifteen years the search for STP's oil has resumed, now focused on the country's maritime territory in the oil-rich Gulf of Guinea.[3] STP is a tiny country with barely two hundred thousand inhabitants, about half of whom live at or below the poverty line. For them, the prospect of oil could bring some positive and far-reaching changes to their life and that of their children and grandchildren. Friends told me that when rumors regarding STP's oil wealth first emerged, there were some who expected that soon everyone would have a barrel of oil stashed away in the backyard or a tap from which *petróleo* would flow freely and on demand. Such stories were usually accompanied by laughter and shaking of heads at the patent credulity of their compatriots entertaining such hopes. But these stories were also about questions that have loomed large on people's minds: What will it mean for their country to "have" petroleum? What forms will it take? Who will benefit from it? My Santomean friends, who worked as university teachers, bank clerks, technicians, or lawyers in the island state's capital city, knew that there was not just one answer to these questions. Pointing to their neighbors in the region—wealthy Gabon with its inflated bureaucracy that threatens to collapse as oil is running out; Equatorial Guinea, plagued by extreme poverty despite the unspeakable wealth of its president; or the much-admired Portuguese-speaking "relative" Angola, resurrected from decades of civil war, partly thanks to its vast petroleum wealth—they commented on oil's baffling capacity for causing both happiness and suffering, wealth and corruption, development and social friction, but doubted that the general population would benefit much from the income derived from it.

What would be the anthropological take on the scholarly and popular invocations of oil as a "fetish" or a "cursed" substance with apparently "magical" powers? While the language of the occult has partly been used to cover methodological inadequacies within the context of economic knowledge production (Weszkalnys 2011:356), it is also more than metaphor. Yet it has rarely been subject to scholarly reflection.[4] Here, I want to draw on anthropology's tradition of exploring magic and the occult to develop a fresh understanding of the nature of oil's magical agency or efficacy.[5] The association of natural resources and magical powers is not unfamiliar to anthropologists. Bolivian miners of the late 1960s feared the destructive and destabilizing forces exercised by the Lord of the Mountain, who embodied and controlled the potential riches of the minerals buried in it. They famously made offerings to his various versions as Supay, Tío, or the Devil to ensure both their safety and the prosperity of the community (Nash 1993; Taussig 1980). A well-known invocation of oil's association with magic is found in Fernando Coronil's remarkable study of Venezuela as a "magical state" (Coronil 1997). However, the magic of the resource, in these accounts, has been one imputed by way of either cosmological analogy or a false projection of desires, which fetishizes the products of capitalist production. For example, Taussig's Marxist-inspired reading of Bolivian mining—resting on a fundamental division of subject and object, society and nature, agency and passivity (Maurer 2006)—did not allow for the miners' own resource ontology to be taken seriously as a constituent of local protest and power. In this view, magic is about make-believe and a false, or at least selective, rendition of reality; it is a power wielded never by the resource itself and always by those who control it.[6]

I suggest that in order to grasp fully oil's magic we need to reconsider the question of its resource materiality.[7] When explaining the efficacy of magic, anthropologists have often pointed to the "magical power of words" (Tambiah 1968; see also Favret-Saada 1980). The objects and substances of various kinds that are routinely employed in magical activities have, from this perspective, mattered largely as props in an orchestrated performance of gestures, words, and actions (Tambiah 1990:73–74). Such human-centered explanations reflect the conceptual and practical problem that magic generally, and the fetish in particular, has posed for us "moderns" derived from the modernist denial that material things can actually "*do* something to us" (Pels 2010:613). More recently, anthropologists have begun to pay rather more attention to the things handled by those who believe in supernatural powers, to the ways that putatively magical objects "captivate" their audience, thus producing a sense of magical causation, and to the material effects of the sensing and doing of practitioners of magic (e.g., Engelke 2005; Gell 1998; Ochoa 2007; Pels 2010). This literature suggests that even in instances where magical objects appear to be merely accessories of its trickery, without those objects there would be no magic (Taussig 2003; cf. Evans-Pritchard 1950; Weiner 1983). In rethinking what I term oil's magic/materiality[8] I want to take up the analytical possibilities afforded by these insights to suggest how we might be able to develop common invocations of oil's magic. I stress the distributed and associative nature of oil's magic—a relational force that is inherently material and would be nothing without the substance we call oil. Yet oil's magic/materiality is more than the substance's chemical or physical properties. It is distributed across things, words, practices, institutions, values, and technologies, which may indeed matter very much in the formation of its apparently "evil" efficacy.[9]

CASTING THE SPELL

Oil makes itself felt in heat and smell as it combusts spontaneously or deliberately; oil powers global transportation; oil pollutes, by oversight or accident, large stretches of the earth's surface; and oil leaves harmful traces on human and nonhuman bodies that mix with it in uncontrolled ways. Toxic waste and sludge are the material ramifications of oil extraction, part of a resource curse that is destructive not only of the immediate resource environment, including people, animals, and plants living in the vicinity of extractive sites, but often also at a distance.[10] Oil infrastructures—the platforms, pipelines, and pump jacks—are the great icons of our extractive age, and both suppress and enable forms of oil's magic and its contestation. Oil's material presence is blatant, overwhelming, awe-inspiring, and terrifying. Yet, as I will demonstrate in this section, this fact has not generally led to a systematic analysis of oil's materiality in the literature. Much of the critical assessment of the so-called oil curse, as Tim Mitchell (2009) points out, has seen the resource in only one particular form—a monetary value (rents, revenues, and so on)—rather than the modes of extraction, circulation, and usage that make up an important part of oil's peculiar nature. As a result, they have overlooked the most basic foundations and contradictions of our present-day carbon-based democracies.

For the last few decades, the notion of the "resource curse" has received almost disproportionate attention among economists and political scientists and at one point had considerable influence on the work of global institutions such as the World Bank. Skepticism regarding the capacity of natural resources to contribute to sustained economic growth is not new in the history of resource economics.[11] It became reinforced during the last three decades by ineffective resource booms, resource-related violence, and booming oil prices that did not translate into development. In a recent study, Michael Ross (2012:xiv) claims that the curse is in many ways a new phenomenon and that oil states, in particular, are growing ever more distinct. Over the decades, economists and political scientists have teamed up in diagnosing more clearly a unique capacity of precious resources (including hydrocarbons) to bring to their proprietor nations not wealth and prosperity but violence, corruption, and—paradoxically—economic stagnation. The hunt is on for the empirical confirmation that the curse "really exists" (Sachs and Warner 2001:828).

For the economists, political scientists, and development specialists who invoke the resource curse and marvel at oil's evil magic, the language of the occult is more than a metaphor. Oil's capacity to "do evil" is considered both economic and political. First, for many experts, it is the presence and not the scarcity of oil that is the problem (cf. Peluso and Watts 2001). This notion—like magic—defies established economic knowledge and "common sense," and begs for explanation. In this view, resource booms—especially in developing countries—are often accompanied by unhealthy economic imbalances and a diminishing of nonexport or manufacturing sectors. Phenomena such as the "Dutch disease" are seen to point to economic "distortions" brought about by the export of large quantities of the highly valuable mineral substances (Auty 2001; Gelb 1988; Sachs and Warner 2001). A parallel literature studies what oil exploitation does to local politics, highlighting the weakening of democratic institutions and the behavior of local elites (Ross 1999). It points to "rent-seeking" as an endemic practice in resource-extractive contexts (Karl 1997); it is also an archetypal form of unproductive wealth—a somewhat immoral way of making one's living from largely opaque sources and, in certain cultural contexts, associated with occult powers (Comaroff and Comaroff 2001; Taussig 1980). Ruling elites' interest in ensuring economic growth and investment is believed to diminish as abundant resource revenues secure their income and position, while public accountability is drastically reduced. Without real development, so-called white elephant projects may "engender collective fantasies of progress" and mesmerize citizens into "a condition or state of being receptive to [the State's] illusions—a magical state" (Coronil 1997:5). Third, analysts have pointed to a correlation between resource exploitation and violent conflict (Collier and Hoeffler 2000; Klare 2001; Le Billon 2005). The ensuing scholarly debate has pivoted on notions of greed or grievance, pitting an ideologically motivated agent against a rational one. In such analyses, conflict is interpreted as the result of the greed of profit-maximizing players rather than of genuine political grievances or a sense of entitlement expressed by minority groups (Collier et al. 2006). Everybody, even if originally motivated by political concerns, is quickly seen as trying to seize the resource trophy.

The resource curse thesis has played a significant role in shaping a contemporary notion of African countries as abundant but perilous and corrupt resource providers. The notion indicates a distinctive phase in the continent's long-standing representation as *l'Afrique utile*, an exploitable resource environment, subject to particular forms of government (cf. Ferguson 2006:39–40). The image of African actors as rationalistic, calculating entrepreneurs seems to have replaced an earlier image of Africans as irrational perpetrators of unspeakable violence, occasionally helped by magic, which crystallized during the violent civil wars affecting the continent in the 1990s (Richards 1996; Shaw 2003). Nonetheless, from the viewpoint of some of the analysts cited above, this makes them no less susceptible to the seduction of oil; depending on the precise lens one chooses, they appear to be victims of oil's magical powers or the perpetrators of its evil.

Some of the premises and conclusions of the resource curse thesis have now come under scrutiny. For example, the contributors to a recent World Bank publication assert, through an empirical and statistical unpicking of vast sets of data, "that natural resources are neither curse nor destiny" (Lederman and Maloney 2007:3). They show, among other things, that it is not the presence of natural resources but the lack of export diversification, investments in education and human capital, technological development, and appropriate economic policy, more generally, that are at issue. Here, the curse seems to make itself manifest in a kind of material vacuum: it becomes visible in the schools that are not constructed, the institutions that are not set up, the investment opportunities left untried, the factories that are shut down, the engineers who are not trained, and the decrees that are not passed.

It is possible to detect in current scholarship a fatigue with environmental determinist explanations, which attribute simple causality to the presence of precious natural resources in a particular site. Anthropologists and human geographers, if not rejecting the notion outright, have moved on to interrogate some of the overly resource-deterministic assumptions that have characterized initial analyses of the resource curse. As Watts (2009:81) notes, in talking about an oil curse we risk mixing up the effects of oil exploitation with the broader adversities and challenges brought on by capitalism and global markets. Thus, he asks: "Is Nigeria cursed by oil or corruption (or corporate capitalism); by petroleum or politics (or ethnicity)?" This scholarship aims to resist glib "materialist" arguments that a specific resource would automatically be either a positive or a negative determining force in resource-producing countries. Instead it illuminates the country-specific factors and historical underpinnings of the curse, emphasizing the role of human motivations and actions (e.g., Behrends et al. 2011; Soares de Oliveira 2007). These very laudable attempts to peer through the curse's "smoke and mirrors" (Obi 2010), perhaps unintentionally, reduce oil's magic to a matter of perception, false consciousness, and essentially human agency. This approach risks turning the question of resource substance into a purely sociological problem. Instead of hastily dismissing the notion of oil's "magic" as also a materially grounded issue, it might be worthwhile to demonstrate just how oil's materiality matters and is "locally" appreciated (in Africa, global policy circuits, or the academy).

OIL IN ITS ABSENCE

My concern with oil's magic/materiality stems from my fieldwork experience in São Tomé and Príncipe—an emergent oil economy from which oil was largely absent. One of the questions Santomean friends and acquaintances would ask me repeatedly was whether I believed that there is oil in STP. It turns out that, to this date, no Santomean has received any oil extracted from national waters, whether or not free of charge. At the same time, oil's presence and effects could be felt, for example, in the international banks that have rushed to STP in anticipation of investments, in the villas constructed to house the Santomean *nouveaux riches* or maybe immigrants from Angola and Nigeria, and in the technical assistance program for STP's oil sector sponsored by the World Bank.

My research has demonstrated that the oil that is anticipated in STP has generated, and is generated by, a specific arrangement of words, practices, and objects, including commercial and scientific exploration techniques as well as forms of self-consciously ethical capitalism. These are indicative of the coupled "technological zones" of measurement and qualification that Andrew

Barry (2006) has identified as typical of the contemporary oil industry. Seismic studies have produced data and colorful maps depicting the geological structures, on the basis of which STP's oil prospectivity is established. During a period of three months, the Deepwater Pathfinder vessel drilled test wells to depths of 4,196 meters, targeting numerous oil sands and yielding estimations of hundreds of millions of barrels of unrisked prospective oil and natural gas, thereby earning its owners a reported U.S. $600,000 per day. It was an attempt to turn possible and probable into proven reserves, but so far with little success. While the commercial viability of STP's oil has not been officially confirmed,[12] in the hearsay and rumors told in Santomean bars, streets, and markets, the sense prevails that international corporations already "have" oil but deliberately conceal their possessions from the country's governors and citizens.

Santomeans have also encountered their oil in a welter of analyses, technical assistance programs, and NGO campaigns aimed at preventing a resource curse by preparing the country for oil (Weszkalnys 2008, 2011). Thus, aside from having the generic resource potential to bring wealth (Ferry and Limbert 2008:8), STP's oil is deemed to have a potential for the evil magic that can be observed in other countries lining the Gulf of Guinea coast. This latter repulsive force of oil's resource potential has significantly shaped STP's economy of expectations. It is embodied in the country's National Petroleum Agency, its new oil legislation, in NGO campaigns, and other "prospective infrastructures" that seek to help the country achieve material, procedural, and behavioral change and thus to escape the curse.

In the same way that STP's oil may be best understood as a distributed resource materiality—composed of an array of data, technical standards, infrastructures, local and international technical staff, national and private profit-seeking ventures, and so on—so is its associated magical/material power. To paraphrase a recent account of magical-spiritual practice, magical "objects" or "substances" (read: oil's magic/materiality) emerge as an effect of the sensual and sensed "forcefulness of matter" (Ochoa 2007:488). Similarly, oil as a contested energy source emerges from the relations of humans and nonhumans, which occasionally congeal into "oil" but always in historically specific ways. To emphasize oil's material aspects is not to propose a neo-Marxist or purely "social" reading of oil's magic but to include in the analysis both its physical and social properties, its social value and natural wealth (Coronil 1997:59–61). It is also not to reveal the substance (oil) as a mere prop in an orchestrated theater of magical trickery, and to unmask its power for what it really is (Taussig 2003), but to reinstate its substantive place in generating the magic in the first place. It may be time to concede that neither the Bolivian tin miners nor the World Bank economists who invoke the magic of the resource (nor the resource itself) have ever been as modern as we might like them to be (cf. Latour 1993).[13]

POINT RESOURCE, ENCLAVE, PLATFORM, AND PIPELINE

Before considering oil's magic/materiality through an analysis of four constitutive elements—the point source, the enclave, the platform, and the pipeline—I want to briefly point to the material and conceptual processes that constitute oil as we know it. Oil is in many senses an easy resource, tractable and versatile; yet it is not easily appropriated. Without access to the required technologies, it is of only limited practical value (Selby 2005:206). Only by removing them from the ground in which they are trapped can the complex hydrocarbons become "crude" and be sold on a world market. Crude is the "not yet" of petroleum, requiring an assortment of technologies that play on oil's affordances and its ability to transform from a raw and largely useless material into refined and multiply consumable substance. Technological and cerebral labor, and processes of standardization and commodification—whether through chemical refinement or through financial calculations—enable oil's potential to unfold. Its specific chemical and physical properties collude with specialized transformative technologies to make the substance a traded good, implying commensurability, which permit its global circulation as a highly priced source of energy.

For some observers, oil's potential to curse starts already in this initial abstraction of the substance from the environment. Oil is a "point-source resource," which, in contrast to timber, agricultural products, or fluvial diamonds, is highly confined and requires capital-intensive technologies of extraction (Le Billon 2005:8). Oil's extraction and management are readily centralized in institutions that articulate with historically shaped environments of colonial and postcolonial governance, private corporations, and state bodies. This complex techno-social-managerial arrangement wrought around oil carries an air of regulation but has become the focal point in an illicit system of distribution of the wealth and influence accruing from the resource (Karl 1997; Watts 2004). Where in the past coal workers were able to ground nations to a halt and achieve political ends by blocking coal's circulation (Mitchell 2009), contemporary administrative and government "oil workers" occupy much sought-after positions in which they can block and divert the flow of oil and revenues into illicit channels. Institutions such as Angola's national petroleum company Sonangol or the authority managing STP's joint development zone with Nigeria can potentially provide fertile playing ground for oil's magic (cf. Soares de Oliveira 2007).

Second, resource curse analysts have noted the "enclaved" character of oil, referring to the ways oil extraction operates largely independently of other economic processes, industrial sectors, and without much need for local labor (Humphreys et al. 2007:4). As James Ferguson (2006) has shown, in contrast to earlier "socially thick" projects of mineral extraction, which included workers' housing, health services, and recreational facilities and were grounded in paternalistic ideologies, oil extraction—where it is not already located offshore—involves the establishment of gated enclaves. These "socially thin" environments are frequently divorced from the cultural, economic, and legal contexts in which they are situated. They present spatial concentrations of the highly specialized personnel who ensure oil's smooth transformation from an indistinct geological substance into a commodity. Near-identical work settings and sets of techniques are replicated from Mexico to the Gulf of Guinea, and are usually performed by cosmopolitan teams of workers using identical equipment (Appel 2011). The conditions of oil's physical abstraction thus result in specific types of spatial abstraction, which, according to Ferguson, allow forms of corporate profiteering that remain largely oblivious to whatever "curse" may have befallen its host nation.

This ties in with a third important element of oil's distributed materiality, the platform and the associated floating production, storage, and offloading vessel (FPSO). These oversized technologies of extraction have become iconic objectifications of detachment, gigantism, and destructiveness. The Deepwater Horizon going up in flames is but a terrible reminder of how oil's curse works. Michael Watts recounts the harrowing incident of eight Ilaje villagers from the Niger Delta clambering onto a platform operated by Chevron/Texaco in 1998 to discuss compensation for environmental destruction caused by the company. Their actions, intended to elicit rapid response from company executives, were met with violence: protestors were shot "by Nigerian government security forces and Chevron security personnel transported to the platform on Chevron-leased helicopters and paid for by the company" (Watts 2011:50). In 2008, however, members of the Nigerian rebel movement MEND managed to capture a Shell FPSO in Nigeria's Bonga field located seventy miles offshore, in a forceful demonstration of their might (Watts 2011:59). Their protest seems so spectacular precisely because it defies the infrastructure's apparent insurmountability. These are quite different enactments of oil's magic/materiality reliant on technocapitalist equipment that elicits some forms of protest and defeats others.

The pipeline is the sleeker assertion of extractive governance and oil's magic/materiality, snaking its way through deserts and rain forests as if pretending to be part of nature itself (Sawyer 2004). Invented to provide a less labor-intensive means of transportation of carbon energy sources over longer distances, the pipeline has nonetheless shown to be vulnerable to attacks and intrusions of those suffering under oil's curse (Mitchell 2009). Recent cases where the pipeline is among the key protagonists of resource-related protests throw into doubt the assertion that oil's lootability—its capacity to be picked up and put into one's pocket—is relatively low (cf. Le Billon 2005). As oil is being transported away from its original sites of extraction, it becomes more readily manipulated, and more readily an active object of contestation. The pipeline has become a formidable anchor for the grievances of an array of local and global actors. Contestations may involve local oil workers and indigenous people alongside cosmopolitan environmentalists who demand attention to the particular instance while seeking to generalize from it in order to claim human rights and entitlements. They question the collusions between the state's attempts to privatize and neoliberalize petroleum production and the ability of oil multinationals to implement their projects in a socially and environmentally responsible manner. In this context, pipelines readily generate their own politics and conflicts (Barry 2012; Reyna 2007; Valdivia 2008). Such contestations of oil's governance may even become a symptom of oil's curse itself (rather than simply a reaction to it). For example, so-called oil bunkering from both pipelines and ships provides a source of income for militarized insurgents asserting their entitlements through violence (Watts 2009, 2011). In this light, it seems unsurprising that the new Chad–Cameroon pipeline transporting oil from the contested Chadian oil fields to West African shores—which was supported with a pathbreaking World Bank program, heavily monitored by the international community, and aimed (but apparently failed) to counteract the resource curse—should have been removed from "illicit interference" simply by placing it underground (Ferguson 2006; Pegg 2005).

RETHINKING OIL'S MAGIC

I have argued that in order to grasp oil's magic/materiality it is necessary to account for the whole techno-socio-natural arrangement through which the substance we know as oil is spun. Magic, as anthropologists have shown, depends on the objects it commands—but the force of these objects is engendered in their practical association with other things, human and nonhuman. This is about both epistemology and ontology; how we know oil and what we know oil to be in all its versions, forms, and facets. My intention is not to invoke a new commodity fetishism but to suggest a richer approach to the problem of oil, one mindful of its material ramifications and implications. I hope to have shown the relational nature of oil's magical/material efficacy, involving pipelines, platforms, bureaucracies, workers' compounds, and other infrastructures typical of the industry that are partly, but never fully, to do with the specific aspects of the substance we call oil. Oil's magic cannot be reduced to any of the aspects discussed above. Quite the contrary. Its magic/materiality is not simply an essence, concentrated in its chemical and physical constitution, but is a potentiality that unfolds through a series of material processes of transformation, appropriation, and use. In this view, oil's magic is not an essence or "natural" quality. Nor is it the kind of generalizable abstraction or economic truth that the theory of the resource curse implies. It is an always historically determined material outcome of a complex set of relations between people, technologies, and things articulated in particular locales. This is the potential of oil's magic, not as a transcendent presence but as a latent absence.[14]

NOTES

1. The empirical research for this paper was funded by the British Academy and the John Fell OUP Research Fund, and would have been impossible without all the people in STP who shared their thoughts and insights with me.

2. http://omrpublic.iea.org/.

3. For a detailed discussion, see Seibert (2006, 2008).

4. The focus of this chapter differs from the interesting studies of discourses of the occult, specifically in the African context. There, it has become a truism of anthropological analysis that discourses and practices of magic and the occult often serve as a commentary on the social inequalities and exploitation attendant on capitalism in the contemporary world (e.g., Comaroff and Comaroff 2001; Meyer and Pels 2003).

5. My argument shares the concern with nonhuman "agency" often associated with actor-network theorists, including Michel Callon, Bruno Latour, John Law, and Annemarie Mol, but in retaining the focus on "magic" it aims to stay closer to the language in which oil's capacities or powers have been discussed and problematized.

6. More recently, anthropologists have begun to consider resource-related magic as an ontological problem, highlighting the imbrication of processes of resource extraction, local people's relationship to their biophysical or other-than-human environment, and their cosmological significance (e.g., Biersack 1999; de la Cadena 2010; Jorgensen 1998; Kirsch 2004).

7. The concept of "resource materiality" is developed in Richardson and Weszkalnys (m.s.).

8. Similar to Foucault's notion of power/knowledge, I conceive of magic/materiality as a dual concept to indicate the intimate involvement of one side of the pair (materiality) in the maintenance of the other.

9. A similar notion can already be found in the classic accounts of "sympathetic magic" by Frazer (1911) and Mauss (1973).

10. Environmental destruction has rarely been considered part of the "resource curse" phenomenon in political and economic analyses. Anthropologists have given the environmental impacts on local health and livelihood more attention (e.g., Kirsch 2004; Sawyer 2004).

11. The ambiguous capacity of natural resource extraction to hinder, rather than promote, economic growth was already noted by Adam Smith and, later, by development economists in the 1950s (Lederman and Maloney 2007:1).

12. Following test drills in 2006, Chevron Texaco has deemed existing reserves not to be economically viable. No official results have been reported for exploratory drills conducted by Addax and Sinopec in 2009 and 2010. However, the parties involved remain optimistic that hydrocarbon finds will be made in the future (see http://erhc.com/articles/october-2011-update/).

13. Conversely, we might consider modernity itself not just as the harbinger of rationality and reason but also as generative of magic (Pels 2003).

14. See Agamben (1999) for a related notion of potentiality as the presence of an absence.

REFERENCES

Agamben, Giorgio
1999 *Potentialities: Collected essays in philosophy*. Stanford, CA: Stanford University Press.

Appel, Hannah
2011 *Futures: Oil and the making of modularity in equatorial Guinea*. Ph.D. dissertation, Department of Anthropology, Stanford University, CA.

Auty, Richard
1993 *Sustaining development in mineral economies: The resource curse thesis*. London: Routledge.

Barry, Andrew
2006 Technological zones. *European Journal of Social Theory* 9(2): 239–253.
2012 Pipelines. In *Globalisation in practice*, edited by N. Thrift, A. Tickell, and S. Woolgar. Oxford: Oxford University Press.

Behrends, Andrea, Stephen Reyna, and Gunther Schlee, eds.
2011 *Crude domination: An anthropology of oil*. Oxford and New York: Berghahn.

Biersack, A.
1999 The Mount Kare python and his gold: Totemism and ecology in the Papua New Guinea Highlands. *American Anthropologist* 101(1): 68–87.

Clarke, Duncan
2010 *Crude continent: The struggle for Africa's oil prize*. London: Profile Books.

Collier, Paul and Anke Hoeffler
2000 Greed and grievance in civil war. In *World Bank policy research working paper* 2355.

Collier, Paul, Anke Hoeffler, and Dominic Rohner
2006 *Beyond greed and grievance: Feasibility and civil war*. Oxford: CSAE, University of Oxford.

Comaroff, Jean and John L. Comaroff, eds.
2001 *Millennial capitalism and the culture of neoliberalism*. Durham, NC: Duke University Press.

Coronil, Fernando
1997 *The magical state: Nature, money, and modernity in Venezuela*. Chicago: University of Chicago Press.

de la Cadena, Marisol
2010 Indigenous cosmopolitics in the Andes: Conceptual reflections beyond 'politics'. *Cultural Anthropology* 25(2): 334–370.

dos Prazeres, Luís
2008 *Dossier petróleo: Cronologia histórica 1876–2004*. S. Tomé: Banco Internacional de São Tomé e Príncipe.

Engelke, Matthew
2005 Sticky subjects and sticky objects: The substance of African Christian healing. In *Materiality*, edited by Daniel Miller, 118–139. Durham: Duke University Press.

Evans-Pritchard, E.E.
1950 *Witchcraft, oracles and magic among the Azande*. Oxford: Clarendon.

Favret-Saada, J.
1980 *Deadly words: Witchcraft in the Bocage*. Cambridge: Cambridge University Press.

Ferguson, James
2006 *Global shadows: Africa in the neo-liberal world order.* Durham, NC and London: Duke University Press.

Ferry, Elizabeth and Mandana E. Limbert, eds.
2008 *Timely assets*. Santa Fe, NM: School for Advanced Research Press.

Frazer, James G.
1911 *The golden bough: A study in magic and religion*. London: Macmillan.

Frynas, Jedrzej George and Manuel Paolo
2007 A new scramble for African oil? Historical, political, and business perspectives. *African Affairs* 106: 229–251.

Gelb, Alan
1988 *Oil windfalls: Blessing or curse?* New York: Oxford University Press.

Humphreys, Macartan, Jeffrey Sachs, and Joseph Stiglitz, eds.
2007 *Escaping the resource curse.* New York: Columbia University Press.

Jorgensen, Dan
1998 Whose nature? Invading bush spirits, travelling ancestors, and mining in Telefolmin. *Social Analysis* 42(3): 100–116.

Karl, Terry Lynn
1997 *The paradox of plenty: Oil booms and petro states.* Berkeley: University of California Press.

Keenan, Jeremy
2008 US militarization in Africa: What anthropologists should know about AFRICOM. *Anthropology Today* 24(5): 16–20.

Kirsch, Stuart
2004 Changing views of place and time along the Ok Tedi. In *Mining and Indigenous Lifeworlds in Australia and Papua New Guinea*, edited by A. Rumsey and J. Weiner, 182–207. Wantage, Australia: Sean Kingston.

Klare, Michael
2001 *Resource wars: The changing landscape of global conflict.* New York: Henry Holt.

Klare, Michael and Daniel Volman
2006 America, China and the scramble for Africa's oil. *Review of African Political Economy* 108: 297–309.

Latour, Bruno
1993 *We have never been modern*. London: Harvester Wheatsheaf.

Le Billon, Philippe
2005 *The geopolitics of resource wars: Resource dependence, governance and violence*. London and New York: Routledge.

Lederman, David and William F. Maloney
2007 Neither curse nor destiny: Introduction to natural resources and development. In *Natural resources: Neither curse nor destiny*, edited by D. Lederman and W. F. Maloney, 1–12. World Bank and Palo Alto, CA: Stanford University Press.

Maurer, Bill
2006 In the matter of Marxism. In *The handbook of material culture*, edited by Christopher Tilley, Webb Keane, Susanne Kuechler-Fogden, Mike Rowlands, and Patricia Spyer, 13–28. London: Sage Publications.

Mauss, Marcel
1975 *A general theory of magic*. New York: Norton.

Meyer, Birgit and Peter Pels
2003 *Magic and modernity: Interfaces of revelation and concealment*. Stanford, CA: Stanford University Press.

Mitchell, Timothy
2009 Carbon democracy. *Economy and Society* 38(3): 399–432.

Nash, June
1993 [1979] *We eat the mines and the mines eat us: Dependency and exploitation in Bolivian tin mines*. New York: Columbia University Press.

Obi, Cyril
2010 Oil as the 'curse' of conflict in Africa: Peering through the smoke and mirrors. *Review of African Political Economy* 37(126): 483–495.

Ochoa, Todd Ramón
2007 Versions of the dead: *Kalunga*, Cuban-Kongo materiality, and ethnography. *Cultural Anthropology* 22(4): 473–500.

Pegg, Scott
2005 Can policy intervention beat the resource curse? Evidence from the Chad-Cameroon pipeline project. *African Affairs* 105(418): 1–25.

Pels, Peter
2003 Introduction: Magic and modernity. In *Magic and modernity: Interfaces of revelation and concealment*, edited by Birgit Meyer and Peter Pels, 1–38. Stanford, CA: Stanford University Press.
2010 Magical things: On fetishes, commodities, and computers. In *The Oxford handbook of material culture*, edited by Dan Hicks and M. C. Beaudry, 613–633. Oxford: Oxford University Press.

Peluso, Nancy and Michael Watts
2001 Violent environments. In *Violent Environments*, edited by Nancy Lee Peluso and Michael Watts, 3–38. Ithaca and London: Cornell University Press.

Reyna, S.
2007 The traveling model that would not travel. *Social Analysis* 51(3): 78–102.

Richards, P.
1996 *Fighting for the rain forest: War, youth and resources in Sierra Leone*. Oxford: James Currey.

Richardson, Tanya and Gisa Weszkalnys
m.s. Resource Materialities.

Ross, Michael L.
1999 The political economy of the resource curse. *World Politics* 51(2): 297–322.
2012 *The oil curse: How petroleum wealth shapes the development of nations*. Princeton, NJ: Princeton University Press.

Sachs, Jeffrey D. and Andrew M. Warner
2001 The curse of natural resources. *European Economic Review* 45: 827–838.

Sawyer, Suzana
2004 Crude properties: The sublime and slime of oil operations in the Ecuadorian Amazon. In *Property in question: Value transformation in the global economy*, edited by C. Humphrey and K. Verdery, 85–111. Oxford and New York: Berg.

Shaw, Rosalind
2003 Robert Kaplan and "Juju Journalism" in Sierra Leone's rebel war: The primitivizing of an African conflict. In *Magic and modernity: Interfaces of revelation and concealment*, edited by Birgit Meyer and Peter Pels, 81–102. Stanford, CA: Stanford University Press.

Seibert, Gerhard
2006 *Comrades, clients and cousins: Colonialism, socialism and democratization in São Tomé e Príncipe*. Leiden and Boston: Brill.
2008 São Tomé and Príncipe: The troubles of oil in an aid-dependent micro-state. In *Extractive economies and conflicts in the global South: Multi-regional perspectives on rentier politics*, edited by Kenneth Omeje, 119–134. Aldershot, UK: Ashgate.

Selby, Jan
2005 Oil and water: The contrasting anatomies of resource conflicts. *Government and Opposition* 40(2): 200–224.

Soares de Oliveira, Ricardo
2007 *Oil and politics in the Gulf of Guinea*. New York: Columbia University Press.

Tambiah, Stanley J.
1968 The magical power of words. *Man* 3(2): 175–208.
1990 *Magic, science, religion, and the scope of rationality*. Cambridge: Cambridge University Press.

Taussig, Michael
1980 *The devil and commodity fetishism in South America*. Chapel Hill: The University of North Carolina Press.
2003 Viscerality, faith, and skepticism: Another theory of magic. In *Magic and modernity: Interfaces of revelation and concealment*, edited by B. Meyer and P. Pels, 272–306. Stanford, CA: Stanford University Press.

Taylor, Ian
2006 China's oil diplomacy in Africa. *International Affairs* 82(3): 937–959.

Valdivia, Gabriela
2008 Governing the relations between people and things: Citizenship, territory, and the political economy of petroleum in Ecuador. *Political Geography* 27: 456–477.

Watts, Michael

2004 Resource curse: Governmentality, oil, and power in the Niger Delta, Nigeria. *Geopolitics* 9: 50–80.

2009 Oil, development, and the politics of the bottom billion. *Macalester International* 24: 79–130.

2011 Blood oil: The anatomy of a petro-insurgency in the Niger Delta. In *Crude domination: An anthropology of oil*, edited by A. Behrends, S. Reyna, and G. Schlee, 49–80. Oxford and New York: Berghahn.

Weiner, Annette

1983 From words to objects to magic: Hard words and the boundaries of social interaction. *Man* 18(4): 690–709.

Weszkalnys, Gisa

2008 Hope and oil: Expectations in São Tomé e Príncipe. *Review of African Political Economy* 35(3): 473–482.

2011 Cursed resources, or articulations of economic theory in the Gulf of Guinea. *Economy and Society* 40(3): 345–372.

CHAPTER FIFTEEN

Energy Politics on the "Other" U.S.-Mexico Border

Lisa Breglia

Deep under the waters of the Gulf of Mexico, off the coasts of Texas and Tamaulipas, lies the "other" U.S.-Mexico border. This isn't the dusty border of cartel violence, "coyotes," and fences but rather a maritime border that few contemplate. Receiving little political or legal attention until the 1970s, this watery boundary is scarcely considered by the average citizen and hardly reaches the ideologically saturated symbolism of its land-based counterpart.

At least when viewed from the north. Since the discovery of valuable hydrocarbon resources in the deepwater region of the western Gulf of Mexico, the maritime border is starting to take on a new importance.[1] Especially for Mexicans, for whom oil is constitutionally protected as national property, the maritime border has quickly come to represent the protection of the nation's sovereign rights over its natural resources. In some contrast, for the United States the boundary is pragmatically economic: the maritime border delimits the area that might be leased to private companies for oil exploitation. However, for both nations, the deepwater area of the Gulf also represents a new frontier of possibility for increased self-sufficiency and energy independence—a new area of contestation between these "distant neighbors."

Over the past three decades, the Gulf has become increasingly important to the energy, and thus national security, of both the United States and Mexico. Both nations agree upon the centrality of Gulf oil resources to their own energy security agendas, both in the past and looking forward. The Gulf of Mexico is the heart of the U.S. petroleum industry and drilling there accounts for 30 percent of all U.S. crude production and more than 90 percent of all U.S. offshore production (Johnson 2011). Shallow-water drilling off the coasts of Louisiana and Texas started in the 1930s but is now in significant decline. Deepwater production is taking its place. The United States initiated a drive for deepwater expansion in the mid-1990s, when energy dependence was high and the intense rhetoric surrounding foreign (especially OPEC) imports spurred the now-familiar turn to domestic sources of crude.[2] Following a decade of intensified efforts (1995–2004), deepwater drilling yielded an astounding 535 percent rise in crude oil production (to 959,000 barrels per day).[3] The U.S. Minerals Management Service (a division of the Department of the Interior, now the Bureau of Oceans and Energy Management) reported in *Deepwater Gulf of Mexico: America's Expanding Frontier* that the deepwater program "succeeded probably beyond the most optimistic dreams of most of us and shows no sign of diminishment" (Richardson et al. 2004:xi).

Even though the Deepwater Horizon tragedy and the brief drilling moratorium caused a disruption in offshore exploration, hopes run high that deepwater areas of the Gulf will continue to be able to make up for shallow-water declines. The Energy Information Association's 2012 energy outlook projects that continued development of deepwater crude oil resources in the Gulf of Mexico will become an increasingly important component of U.S. domestic crude production (EIA 2012). Though currently onshore production outpaces offshore output, the EIA recently suggested that onshore sources, including shale production (so-called tight oil), will soon peak and decline while offshore supplies will continue to grow—especially since the lifting of the 2010 moratorium. The EIA predicts offshore crude oil production in the Gulf of Mexico trending upward over time, fluctuating between 1.4 and 2.0 million barrels per day, as new large development projects in deepwater come online.

Mexico's oil industry relies on the Gulf to an even greater extent than that of the United States. Since the initiation of offshore oil operations by Petróleos Mexicanos (Pemex, the country's national oil company) in the Campeche Sound in the 1979, crude has flowed abundantly from the Gulf. Over the past three decades, the Gulf has supplied up to 80 percent of the nation's petroleum. As oil is a state-owned industry in Mexico, income from Gulf crude has represented upwards of 40 percent of Mexico's federal budget. Gulf oil petroleum rents are especially critical to the fiscal health of coastal states such as Campeche, Tabasco, and Veracruz. These are the three Gulf coastal states recognized by Pemex as "priority states"—those that rank very high in both concentration of oil activity and in the national marginalization index (which measures income and access to infrastructure and services, including education and health care). Especially since the discovery of offshore oil riches in the 1970s, the distribution of resource wealth in Mexico demonstrates that "the centrality of these states to the oil industry, hence to Mexican public finances, is not necessarily reflected in better living conditions for the population" (Pirker et al. 2007:15). To address this issue, Pemex instituted its own Social Development agency in 1995 within the auspices of the parastatal in order to redistribute rents back to oil-affected regions at both the state and municipal level (Breglia 2013). The office of Social Development, through multiple funding mechanisms, focuses cash and in-kind (fuel, asphalt, etc.) support on oil-affected municipalities in priority states.

While some funds go toward social welfare programs, the aid is highly biased toward public works projects such as road and bridge construction.

Whereas the United States frames the dynamic of energy (in)dependence around issues of reliance on foreign imports to feed domestic demands, Mexico's oil dependence is couched in terms of the nation's reliance on oil rents at both the national and local level. The nation's resource security came under grave threat following the peak of the Gulf's supergiant oil field Cantarell in 2004—one of the largest known complexes in the world. Cantarell's production declined from an output of more than two million barrels per day in 2005 to less than 450,000 in 2011, reducing the country's total output by more than 20 percent (Pemex 2011, 2012). Whereas the country's goal was to increase production (based on export demands from the United States as well as rising domestic demand), at the close of 2011 Mexico was producing at levels equivalent to 1990. At the national level, hopes are that the Gulf will continue to provide Mexico with the riches of the past three decades—this time from deepwaters. Between 2007 and 2012, Mexico spent $3.6 billion in deepwater exploration, acquiring 3D seismic data on some ninety thousand square kilometers and drilling exploratory wells (McCulley 2012). According to Pemex estimates, deepwaters hold thirty billion barrels of recoverable oil, 58 percent of the nation's prospective reserves. These deepwater reservoirs are sought in order to rescue Pemex's declining production and, in turn, the federal budget. At the local level, coastal communities expect the Mexican state will act to protect the public interest. Historically, the state has acted as custodian of the resources emanating from the "patrimonial sea"—a marine resource claimed by both citizens and the state as both natural and cultural heritage. Even under multiple waves of neoliberal privatization, the Mexican state has maintained a constitutional mandate for the nationalization of "strategic" resources of the subsoil and seabed—most notably, oil.

This case study examines the spatial politics of not only the "other" U.S.-Mexican border but the "other" Gulf oil found in its crosshairs. Security, sovereignty, and nationalism are front-and-center issues when we confront the ongoing complexities of the U.S.-Mexico relationship, especially as manifest along the border. By turning to the often-overlooked U.S.-Mexico maritime border, I hope to illustrate the tensions that arise when two different property regimes for resource ownership come up against one another in the shared transnational space of the Gulf of Mexico. The insatiable quest for new reserves across the globe crosses political, cultural, and other boundaries as new frontiers and uncharted territories are brought within the purview of global market forces. With the "cheap and easy" oil gone, the new oil frontier of deepwater and ultra-deepwater hydrocarbons has proven more expensive and difficult to reach. Deepwater oil is also geopolitically contentious oil. Underwater oil and gas deposits lie in waters of the high seas previously within the global ocean commons. These waters have not been legally delimited, and sovereign rights over exploration and resource exploitation are often unclear and quickly become matters of dispute or arbitration. In some cases the stakes are raised, such as when hydrocarbons found in the western Gulf cross international maritime boundaries. Thus, since the discovery of valuable hydrocarbon resources in this region, the maritime border has begun to take on a new importance for both nations as the conflict in the territory and governance regimes over an increasingly valuable energy resource come into sharper focus. This is far from an abstract political issue in Mexico. In a country where *petroleo es nuestro*—where oil belongs to the people—the governance, custodianship, and exploitation (in a word, the "territory" of oil) is central to popular discourse and public debate. Over the past few years, as least when viewed from the south, the U.S.-Mexico maritime border is a zone of contestation between a culture of energy that has at its core a property-rights-based notion of oil as both national patrimony and part of the public good and another that values the territorializing practice of neoliberal privatization of energy resources. Which will prevail?

THE MEXICAN GULF, A PATRIMONIAL SEA

My interest in the cultural, political, and economic dynamics of the Gulf of Mexico stems from several years of conducting ethnographic research in the coastal communities of Mexico's Yucatán Peninsula. For centuries, Gulf resources have stood at the center of coastal communities' lives and livelihoods. My current research examines how Mexico's Gulf Coast communities are affected by the three decades of intensive offshore oil production. I am especially interested in the intersection between the peaks and declines of the fishing industry and oil production, both in the Campeche Sound (Breglia 2013). Since 2007, I have been engaged in a long-term ethnographic research project based in a coastal community in the Laguna de Términos region of Mexico's Yucatán Peninsula. The Laguna de Términos is a major Mesoamerican watershed system, featuring the lagoon and several rivers, with extensive wetlands and mangroves. The region is home to nearly three hundred species of birds, several endangered reptiles (including the hawksbill and white turtles), as well as marine mammals. Twelve to fourteen hundred bottle-nosed dolphins regularly feed and breed in the waters of the lagoon. Before falling into a serious decline concomitant with the rise of petroleum exploitation, the region was known to support abundant and varied fisheries, both artisanal and industrial. As an alternative to fishing, local communities are hoping that the region's status as a federally protected flora and fauna reserve (declared in 1994) will help ecotourism initiatives. The reserve encompasses both the watershed as well as the major urban center, Ciudad del Carmen—homebase for Mexico's offshore oil operations in the Gulf. Carmen—formerly known for little more than its robust shrimping industry—transformed rapidly from an overgrown fishing village into a multinational energy center following the 1970s discovery of Cantarell, a billion-barrel oilfield, just eighty kilometers off Carmen's coast.

For coastal communities, marine resources are shared patrimony under the stewardship of the nation. While all heritage resources have cultural as well political and economic dimensions, Gulf oil has a especially deep symbolic dimension for Mexicans. Nationalism, manifest as the desire to protect national resources from foreign plunder, plays a central role. Every March 18, Mexico celebrates Expropriation Day as a national holiday to commemorate the rescue of their oil fields from foreign owners in 1938—demonstrating the continued liveliness and power of the national heritage concept. The expropriation marked the state's reaffirmation of a much longer legacy of resource sovereignty owing not only to Mexico's Spanish colonial heritage but to the influence of Islamic jurisprudence in Spain. The twentieth-century act of expropriation was not an original act of subsoil nationalization but in fact an assertion of state control over a valuable national patrimony that had become alienated from the citizenry through private leasing, concessions, and outright sales to foreign companies during the dictatorship of Porfirio Díaz (1876–1910).

Following the letter of the 1917 constitution, hydrocarbons, as part of the subsoil, are constitutionally protected as property of the nation, and oil is exploited by a state-owned company, Pemex. However, under increasing waves of neoliberalism, the de jure nationalization of oil is looking, especially to critics of the past several administrations, like de facto moves toward privatization. According to Mexico's former secretary of energy, Georgina Kessel, 70 percent of the activities of exploration and production are carried out by private contractors (Gonzales Amador and Carrizales 2008). The number of workers on the offshore platforms reflects the penetration of the private sector as well: in the Campeche Sound, eighteen thousand Pemex employees work alongside fourteen thousand contract workers (Muñoz Rios 2007). Halliburton, one of the principal contractors with Pemex, relies especially on foreign rather than Mexican workers. Technical and professional positions on Gulf platforms once occupied by Mexican workers are now being replaced by foreigners.

Oil nationalism has been a political hot potato in Mexico for decades. Protests against the privatization of oil and Pemex reached a fever pitch in response to the energy reform efforts of President Felipe Calderón (2006–2012). Following in the incrementally liberalizing footsteps of previous administrations, in 2008 the Calderón administration made a move to regularize the de facto privatization of on- and offshore oil services through the mechanism of contracting. Although multinationals have always been involved in Mexico's oil industry to some extent, it was specifically through the "multiple service contract" (MSC) devised by the Fox administration (2000–2006) that multinational oil services providers began receiving millions of dollars annually in lucrative contracts while the oil industry remained under the de jure constitutional protection as a national industry.

The Gulf of Mexico, the nation's most strategically important production region, has thus been caught between a de facto privatization and a de jure maintenance of the commons. The public sense of the latter—the Gulf as public patrimony—carried the heavy weight of history. Indeed, Mexico had previously fought, and won, a border battle for the protection of patrimonial resources in the high seas of the Gulf. Protecting marine resources as part of Mexico's national patrimony has a history that predates the discovery of deepwater, or any offshore oil for that matter. The roots of legislating the Gulf to claim and protect—and, in turn, exploit—marine resources is rooted in the Gulf's valuable fisheries. Dating back to the 1940s, Mexico's hard-fought resource sovereignty issue was over not oil but shrimp. Mexico was concerned with extending its territorial waters to protect its valuable nationalized shrimp fishery from exploitation by U.S. boats from Texas. A decades-long (1940s–1970s) battle with U.S. fishermen over rights to exploit Gulf shrimp and Pacific tuna spurred Mexico to extend its territorial sovereignty from three to twelve nautical miles. The 1976 international law—United Nations Convention on the Law of the Sea (UNCLOS)—enabled the establishment of an exclusive economic zone (EEZ) of two hundred nautical miles seaward. Thus, Mexico follows international law in spirit in conceptualizing the sea and its resources as a part of the national ocean commons.

The United States established its EEZ as well, but not without the growing attentions of the private sector. By this time petroleum geology had come far enough to predict the occurrence of hydrocarbons in its EEZ, where a state has sovereign rights (but not sovereignty) over activities such as exploring, exploiting, conserving, and managing the natural resources on the surface and subsurface of the seabed. The EEZ is particularly important for the case of hydrocarbon resources because it also gives coastal states the right to conduct other activities with a view to exploring and extracting economic benefits from the zone.

Since the declaration of its EEZ in 1976, Mexico has converted its maritime jurisdiction in the Gulf of Mexico into more than a territorial sea or an exclusive economic zone. Because it contains resources vital to the nation's domestic security in terms of both economics and sociocultural identity, the Gulf of Mexico is also a patrimonial sea. The term *patrimonial sea* is specifically Latin American in origin and was first used as an antecedent to the concept of the EEZ. Edmundo Vargas coined the term in 1971 to describe "the maritime zone in which the coastal state has exclusive rights for the exploration, conservation, and exploitation of the natural resources of the waters adjacent to its coast, and of the sea-bed and the sub-soil thereof up to a limit determined by the said State in accordance with reasonable criteria and on the basis of its geographical, geological and biological characteristics" (Yturriaga 1997:26).

The concept reflected states' exercise of sovereignty over their waters specifically in the need to access fisheries and was codified in the 1972 Santo Domingo Declaration, one of the pillars of the 1982 United Nations Convention on the Law of the Sea. In the Vargas definition of patrimonial sea and the later definition of the EEZ, maritime zones extending two hundred nautical miles from a nation's coastlines are not territorial seas in which a nation has sovereignty. Rather, they are spaces in which nation-states enjoy exclusive economic rights. Even though the term has been used to define the economic jurisdiction for the territory we simply now refer to as the EEZ, a revival and slight redefinition of *patrimonial sea* can help us to better describe Mexico's interests in (and, in turn, America's problem with—or vice versa) securing energy resources in the Gulf of Mexico. All of the Mexican Gulf is a patrimonial sea for Mexicans because the resources it contains—from fisheries to hydrocarbons—are symbolically tied to the nation-state's project of providing for the national welfare.

Taking the concept of the patrimonial sea as offered by the Declaration of Santo Domingo and broadening its scope to include both practical and symbolic dimensions, we can include the quotidian meaning of "patrimony" for Mexican citizens. This allows us to account for how Mexicans understand the Gulf's oil resources. As valuable as oil is, the symbolic value of *petróleo* as *patrimonio* to Mexican citizens is priceless. By this I mean that the maritime jurisdictional redefinitions have produced a deep symbolic effect by underlining the importance of oil as Mexican national cultural patrimony. A proper analysis of past maritime boundary negotiation as well as the success of upcoming agreements over this "other" U.S.-Mexico border will depend a great deal upon maintaining an awareness of how the Gulf of Mexico functions, in legal and popular public discourse, as a patrimonial sea. The concept

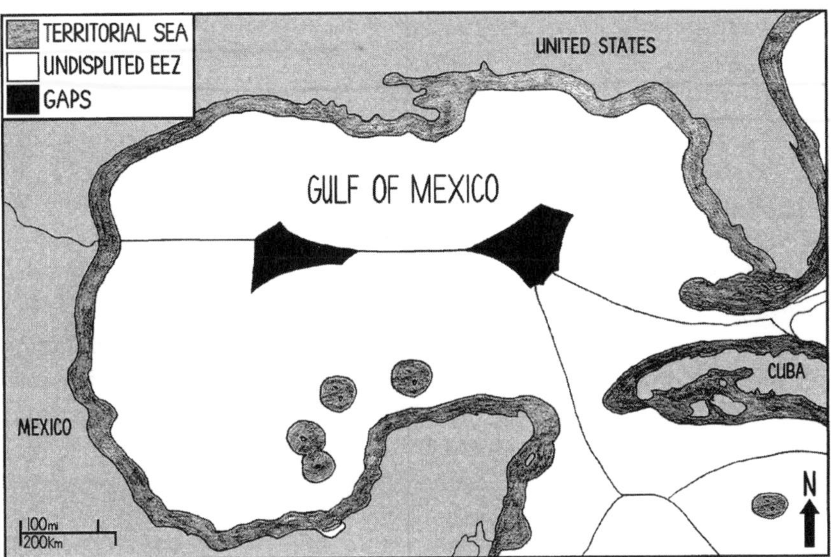

FIGURE 15.1. Gulf Of Mexico And U.S.-Mexico Maritime Boundaries

of the patrimonial sea is especially critical to bring to our understanding of Mexico's claim of sovereignty over hydrocarbon resources in the negotiation of transboundary reservoirs in the western Gulf along the maritime border with the United States.

The 1982 United Nations Convention on the Law of the Sea enabled states to assert an extensive territorialization of the Gulf, and the establishment of each state's EEZ captured 95 percent of its harvestable living resources and 98 percent of its petroleum resources within the two nations' exploitable zones.[4] Given that the EEZs cover nearly the whole of the Gulf and proscribe an exploitation regime for nearly all of the Gulf's valuable resources, the issues of sovereignty, territory, and hydrocarbon resource rights in the Gulf should be straightforward matters. But the creation of the EEZs, like all cartographic techniques, is a practice that is at once both technical and ideological. Both dimensions have proven deeply complicated. Technically, the drawing of each country's EEZ should have regularized the resource exploitation regimes in the Gulf. However, the extensions of each EEZ left open areas known as the western and eastern gaps—colloquially known as "doughnut holes." Until very recently, the western gap has commanded the most intense scrutiny and debate.[5] Estimates of the reserves in the 5,092-square-nautical-mile (4.5 million acre) western doughnut hole—territory belonging to neither the United States nor Mexico—consists of anywhere from 2.4 to 22 billion barrels of oil and a minimum of five to forty-four billion cubic feet of natural gas (Estrada 2009). The Clinton-Zedillo Treaty of 2000 splits the resource rights of the western gap between the United States and Mexico proportionally, based on an equidistant extension of each nation's EEZ.[6]

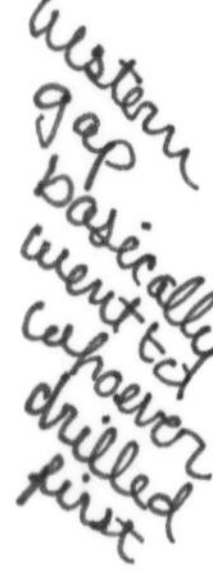

When the Clinton-Zedillo Treaty was negotiated, both parties believed that significant hydrocarbon resources existed in the deepwater of the doughnut hole, but nobody knew exactly where. Due to the seabed's geomorphology, the negotiators of the 2000 treaty wanted to preempt the anxieties arising (especially on the part of the Mexican committee) due to the strong likelihood of the presence of common petroleum deposits straddling the already-set maritime boundaries. Curiously enough, because hydrocarbons migrate, a shared reservoir may be drained by whoever drills it first, leading to a problem of potential reverse illegal migration across the U.S.-Mexican border popularly known as "drainage" or the "drinking straw" (*popote*) effect, meant to call to mind the sucking of oil from one side of the boundary to the other.

Thus, to avoid problems with the possible existence of such transboundary reservoirs, the 2000 treaty established "the area," a 1.4-nautical-mile buffer zone on each side of the maritime boundary (effectively creating a 2.8-nautical-mile strip).[7] The parties agreed to a ten-year moratorium on drilling in the area. After the expiration of the moratorium (originally January 2011), both parties to the treaty would be obliged to announce their decisions to lease, license, grant concessions, or otherwise make available portions of the buffer area for petroleum or natural gas exploration or development and to indicate when petroleum or natural gas resources would be put into production. In May 2010, at the height of the Deepwater Horizon disaster, Presidents Calderón and Obama negotiated an extension of the moratorium on drilling in the buffer zone.

The moratorium was, perhaps, beneficial for Pemex. After all, in 2000 the parastatal had been completely unprepared for, and perhaps uninterested in, (given the success of Cantarell) launching a deepwater drilling program on the maritime border. Yet north of the border, flush with private-sector capital incentivized by a generous leasing program managed by the agency formerly known as the Minerals Management Service, the story is very different. The United States began to push for deepwater production in the Gulf in the mid-1990s with the Outer Continental Shelf Deep Water Royalty Relief Act (1995)—essentially a subsidy to encourage deepwater drilling. By 1997, 70 percent of Gulf leases were in deep water (Kallaur 2001). Although this blanket act has now been suspended, royalty relief is still granted on a lease-by-lease basis. The worldwide dwindling of the flow of cheap and "easy" crude makes the exorbitant investment of time and capital into deep and ultradeep drilling look like a good option. Higher oil prices continue to make the investment required by the private companies worth the risk (Womack 2010).

To Mexico's consternation, many private companies—both multinationals and independents—now have drilling rights on dozens of parcels within ten miles of the Mexican maritime border. When Shell began drilling just eight miles north of the Mexican border at the 8,700-foot-deep Great White field in 2007, the company assured the public that "[n]one of these reservoirs extend beyond the U.S.-Mexican international boundary; all are completely within U.S. waters. So, there is no chance for cross-border drainage of oil" (Millard 2007). The Perdido spar, anchored in 7,816 feet (2,382 m), serves the Great White, Trident, Tobago, and Silver Tip fields and is the deepest oil development in the world. The venture, jointly owned by Shell (35 percent), Chevron (37.5 percent), and BP (27.5 percent), is expected to produce one hundred thousand barrels per day (*Rigzone* 2010).

Political grandstanding in Mexican energy reform debates of 2008 portrayed Mexico's oil patrimony as under imminent threat from this border activity. The Calderón administration's pro-reform ad "*El Tesoro: Fortalecimiento de Pemex*" suggests that deepwater oil is an undersea treasure (Pemex 2008). The video displays a map showing the Gulf maritime boundaries and, most significantly, the concentration of development on the United States side of the border. In order for Pemex to exploit the "undersea treasure," according to the rationale, the parastatal would have to rely more deeply upon the equipment, technology, and personnel of the private sector. Mixing nationalist rhetoric with a rationalization for private-sector partnerships, the campaign for the energy reform played on fears of border incursion by the United States. A key image in the campaign ad shows a map of the Gulf of Mexico divided into U.S., Mexican, and Cuban territory. Clusters of targets across the map indicate the oil and gas projects under development in the maritime border zone. The map graphic demonstrates the density of U.S. development right on the Mexican border. Meanwhile, no similar developments are portrayed on Mexico's side. Instead, only a swirling, golden glow appears in the Mexican patrimonial sea representing the "undersea treasure" of the oil and gas reservoirs. The dramatic image (intensified by accompanying music) plays on the power of national fears of border incursion. Whereas the United States has co-opted the land border as theirs to be violated by Mexicans, it seems that the reverse has been effected here. The state-produced video shows the threat of U.S. border incursion—not on Mexican territory per se but on Mexico's sovereign rights over its national patrimony. For a critically vociferous—though increasingly fractured—left, any Pemex development in the Gulf's deepwaters is a Potemkin village constructed by the right wing. At the same time, all the hype about moving production to the Gulf deepwaters is merely a thinly veiled excuse for privatizing Pemex.

Perspectives from abroad support the notion that Pemex lacks the technology and know-how to continue a productive drilling program, especially in deepwater. An analysis by the U.S. Energy Information Administration, the 2009 *International Energy Outlook*, pessimistic about Mexico's own deepwater capabilities and supportive of private-sector investment, unsurprisingly concurs with the energy reform policy of the Calderón administration. A 2010 article in the *New York Times* points out that the nation "probably still has plenty of oil, especially beneath the deep waters of the Gulf of Mexico, but Pemex lacks the technology and know-how to get it out." "Inviting foreign companies into the country," however, "is one of the touchiest propositions in Mexican politics" (Krauss and Malkin 2010). The *Times* article, published in anticipation of the 2010 Expropriation Day, called the continued oil nationalization a "straitjacket," the "outright reversal" of which was "unthinkable" (Krauss and Malkin 2010).

CONCLUSION

The Gulf of Mexico's rich deepwater hydrocarbon deposits are not only transboundary reservoirs but also contested spaces for national and transnational politics. Thus, like other peak resources, the underwater resources in the Gulf are caught in a fraught politics characteristic of what Michael Klare calls the "Era of Xtreme Energy," when the easily obtained conventional hydrocarbons, such as the crude from Cantarell, are gone. According to Klare, after the end of the Petroleum Age we've now entered a "no-man's land," the Era of Xtreme Energy, where "we will remain for years to come . . . until an age that will see the great flowering of renewable energy" (Klare 2009). The case of the Gulf of Mexico's deepwaters, especially in the western gap, quite literally produces a no-man's land, where exercise of resource sovereignty hangs in the balance. Here, the deterritorialized nature of offshore space puts the control over the planet's most valuable resource up for grabs—often to the highest bidder.

Maintaining the line between the commons and privatization will be more difficult than simply drawing a clear maritime border in the deepwater Gulf of Mexico. The highly politicized question of who will exploit hydrocarbons under what conditions is deepened by the ideological nature of the significance of the territory of the Gulf to stakeholder states. Because it contains resources vital to the nation's domestic security in terms of both economics and sociocultural identity, the Gulf of Mexico is also a patrimonial sea. The resources it contains—from fisheries to hydrocarbons—are symbolically tied to the state's project of providing for the national welfare.

Both the United States and Mexico will insist on the centrality of deep and ultradeep oil as key to their national energy security strategies. For those who wish to protect the environment and seek an energy future in alternatives to fossil fuels, the benefits to forging ahead on the deepwater frontier are uncertain at best. Meanwhile, the risks are multiple and extraordinarily high. What also hangs in the balance at the U.S.-Mexico maritime border (and, in the coming years, the U.S., Mexico, and Cuba border in the eastern Gulf) is the future of Mexico's resource sovereignty over hydrocarbons. The Gulf is a shared hydrological system but also a contested geopolitical zone. As long as multinational oil services providers have the upper hand in deciding the future of energy, the logic of drilling at any cost will continue to win out over the public interest.

NOTES

1. Projects in less than one-thousand-foot (305 m) water depths are considered to be shallow-water projects and those in greater than one thousand feet are considered to be deepwater projects.

2. Deepwater production began in 1979 with Shell's Cognac Field, but it took another five years before the next deepwater field (ExxonMobil's Lena Field) came on line. Both developments relied on extending the limits of platform technology used to develop the Gulf's shallow-water areas. Since then, deepwater exploration and production technology has advanced tremendously. In February 1997, there were seventeen producing deepwater projects, up from only six at the end of 1992 (Nixon et al. 2009).

3. For U.S. deepwater Gulf production statistics, see http://www.gomr.boemre.gov/homepg / offshore/deepwatr/summary.asp.

4. Even though the United States refused to become a signatory to the UNCLOS, the country still follows the spirit of the international law.

5. The Eastern Gulf of Mexico Planning Area remains under a congressional drilling moratorium (the Gulf of Mexico Energy Security Act of 2006) until 2022.

6. Treaty Between the Government of the United States of America and the Government of the United Mexican States on the Delimitation of the Continental Shelf in the Western Gulf of Mexico Beyond 200 Nautical Miles, June 9, 2000, U.S.-Mexico, S. TREATY DOC. No. 106–39.

7. A spokesperson for the U.S. Minerals Management Service called the buffer zone—which covers about 10 percent of the United States' part of the gap—an "unusual" arrangement made to allay Mexico's anxiety regarding the likelihood of companies working on the U.S. side much sooner than on the Mexican side and as a tool to help the treaty pass more quickly through the Mexican senate (*Oil and Gas Journal* 2000:30).

REFERENCES

Breglia, Lisa
2013 *Living with oil: Promises, peaks, and declines on Mexico's Gulf Coast*. Austin: University of Texas Press.

Energy Information Agency (EIA)
2012 Annual energy outlook early release. http://www.eia.gov/forecasts/aeo/er/early_production.cfm.

Gonzales Amador, Roberto and David Carrizales
2008 En manos privadas, 70% de la exploración y producción de crudo. *La Jornada*, November 11.

Estrada, Javier
2009 Reservoirs that cross country lines need special agreements. *Offshore Magazine*, July 1. http://www.offshore-mag.com/index/article-display/5400343081/s-articles/s-offshore/s-volume-69/s-issue-7/s-latin-america/s-reservoirs-that_cross.html.

Johnson, Toni
2011 U.S. deepwater drilling's future. Council on Foreign Relations report, January 11. http://www.cfr.org/united-states/us-deepwater-drillings-future/p22204.

Kallaur, Carolita
2001 The Deepwater Gulf of Mexico: Lessons learned. Proceedings: Institute of Petroleum's International Conference on Deepwater Exploration and Production in Association with OGP. London. http://www.gomr.mms.gov/homepg/offshore/deepwatr/lessons_learned.html.

Krauss, Charles and Elisabeth Malkin
2010 Mexico oil politics keeps riches just out of reach. *New York Times*, March 8. http://www.nytimes.com/2010/03/09/business/global/09pemex.html?pagewanted=all.

Klare, Michael
2009 The era of Xtreme energy: Life after the age of oil. TomsDispatch.com, September 22. http://www.tomdispatch.com/post/175127.

McCulley, Russell
2012 Pemex promises production boost. *Oil Online*, February 13. http://www.oilonline.com/default.asp?id=259&nid=37422&name=Pemex+promises+production+boost.

Millard, Peter
2007 As deepwater drilling booms, Mexico's oil could leak to U.S. *Dow Jones Newswires*. September 7. http://www.rigzone.com/news/article.asp?a_id=49955.

Muñoz Rios, Patricia
2007 Crece en Pemex el número de trabajadores subcontratados. *La Jornada*, December 27. http://www.jornada.unam.mx/2007/12/27/index.php?section=politica&article=012n2pol.

Nixon, Lesley, Nancy K. Shepard, Christy M. Bohannon, Tara M. Montgomery, Eric G. Kazanis, and Mike P. Gravois
2009 *Deepwater Gulf of Mexico 2009*. New Orleans: U.S. Department of the Interior, Mineral Management Service. http://www.gomr.boemre.gov/PDFs/2009/2009–016.pdf.

Oil and Gas Journal
2000 Gulf of Mexico western gap division agreed, exploration pending, July 10. http://www.ogj.com/articles/print/volume-98/issue-28/exploration-development/gulf-of-mexico-western-gap-division-agreed-exploration-pending.html.

Pemex
2012 Pemex fact sheet. http://www.ri.pemex.com/index.cfm?action=content§ionID=25&catID=12705.
2011 Anuario Estadístico 2011. http://www.ri.pemex.com/files/content/AE_E_Petroleos%20Mexicanos_ing_2011.pdf.
2008 *El Tesoro: Fortalecimiento de PEMEX*. http://www.youtube.com/watch?v=lgK_kotLHr4.

Pirker, Kristina, José Manuel Arias, and Hugo Ireta Guzmán
2007 *El acceso a la información para la contraloría social*. Mexico, DF: Fundar Centro de Análisis.

Rigzone
2010 Shell unlocks Perdido's riches. *Rigzone*, March 31. http://www.rigzone.com/news/article.asp?a_id=90358.

Richardson, G. Ed, Leanne S. French, Richie D. Baud, Robert H. Peterson, Carla D. Roark, Tara M. Montgomery, Eric G. Kazanis, G. Michael Conner, and Michael P. Gravois
2004 *Deepwater Gulf of Mexico: America's expanding frontier*. New Orleans: U.S. Department of the Interior, Mineral Management Service. http://www.gomr.boemre.gov/PDFs/2004/2004-021.pdf

Treaty on Maritime Boundaries Between the Government of the United States of America and the Government of the United Mexican States, May 4, 1978, U.S.-Mex., 17 I.L.M. 1073.

Womack, Jason
2010 Ultra-deepwater drilling is likely to remain profitable. *Rigzone*, February 26. http://www.rigzone.com/news/article.asp?a_id=88478.

Yturriaga, José Antonio
1997 *The international regime of fisheries: From* UNCLOS *1982 to the presential sea*. The Hague, the Netherlands: Martinus Nijoff Publishers.

CHAPTER SIXTEEN

Beyond the Horizon: Oil and Gas Along the Gulf of Mexico

Thomas McGuire and Diane Austin[1]

> In the distance, one could still see the vessels that had been working to insure that the wellhead remained capped. They seemed unreal—industrial apparitions, gray like haze. Kessler tossed a bucket into the sapphire-colored water, and said, "You get this blue, kind of Caribbean color." The sunlight was warm, and the breeze moved across the deck, and he added, "It's hard to believe that there was really nasty thick crude all over this just a few months ago."
>
> —Raffi Khatchadourian, *The New Yorker*

John Kessler, an oceanographer from Texas A&M University, was sampling the surface water of the Gulf in September of 2010, five months after the April 20 explosion of the *Deepwater Horizon*. He and his colleagues had already discovered the plume of dissolved hydrocarbons, a thousand meters deep. The writer for *The New Yorker*, on board the research vessel *Pisces*, reported on the results of one probe: "The word 'plume' was scientifically accurate, but the amounts of hydrocarbons in the samples were invisibly minute. In the syringes, the water was frigid and crystal clear" (Khatchadourian 2011:59). The scientists appeared to be working toward a consensus that microorganisms had degraded much of the oil, that the early and vociferous warnings about an ecological catastrophe were too shrill. Yet the social, economic, and emotional scars from the BP blowout will not heal anytime soon. The wounds are deep, because oil and gas is so tightly embedded in the history and culture of coastal Louisiana and Texas, because oil and gas—long before the blowout on the rig—have so profoundly altered the coastal landscape, and because the country watches the region through unsympathetic eyes. The *Deepwater Horizon* is not the first injury to be inflicted on this region by the petroleum industry and, for the complicated reasons an abused parent or spouse returns to and defends her abuser, so, too, have the people of this region gotten up, covered the cuts and bruises, and kept going. Occasionally, though, an event occurs that enables a victim to break the cycle.

Andrew Hoffman and P. Devereaux Jennings argue that the event in the Gulf has the making of a "cultural anomaly," a potential Kuhnian paradigm shifter in environmental management and fossil fuel production. Such anomalies focus "sustained public attention and invite the collective definition or redefinition of social problems and the actors themselves" (Hoffman and Jennings 2011:101). Yet the most sustained discourse coming out of the Gulf Coast in the wake of the spill surrounded the moratorium on offshore drilling. For much of 2010 and into 2011, activity in the Gulf was halted while new regulatory procedures and spill response measures were hammered out. The owner of a small oil company in Lafayette succinctly captured the local sentiment: "Idle rigs are on location, they're ready to work. Idle helicopters are ready to fly, idle boats are ready to sail. Mr. President, let us work" (Smith 2010). In this chapter, we explore the internal and external forces and tensions that influence the responses of those who live and work in the communities directly affected by offshore drilling and production.

THE OIL INDUSTRY AS A WAY OF LIFE

While the injury and loss of life resulting from the explosion had immediate and lasting effects on the people of the Gulf Coast—whether or not they worked offshore or knew anyone who was killed—so, too, did the assault on the offshore industry and, because of their participation in that industry, the associated attack on the people who live and work in the region. Two brief excerpts from field notes illustrate the centrality of the industry to the lives of individuals and communities along the Gulf Coast. A motel clerk in Lafourche Parish, Louisiana, a major staging area for offshore work, reported that Transocean, the drilling contractor that owned the *Deepwater Horizon*,

> had several rigs in the Gulf and all their crew change people stayed at the motel. The guys get to be like family. She said that the *Horizon* workers stayed at the motel when their rig had been out in that part of the Gulf. Many came back year after year. She got a bit teary at this point and told me that one of the guys who was killed in the explosion had stayed at the motel just a few weeks before the accident. (Diane Austin field notes, Lafourche Parish, Louisiana, October 29, 2010)

And an organizer of Morgan City's annual Shrimp and Petroleum Festival defended the affair even as oil from the BP spill continued to work its way through the Gulf's ecosystem:

> She said the attendance was good and, based on the number of vendors who want space for next year, the sales were good, too. A group of Vietnamese from the Catholic Church in Amelia had a booth, as they have for many years, and were selling Louisiana shrimp . . . [She] talked about having had some people from out of town come and question the festival, why they had it, whether it was going to be held this year because of the spill. She spoke passionately about how she defended the festival, saying that this was a festival celebrating the two major industries in the area and had been going on for a long time, telling others that oil was part of this community. (Diane Austin field notes, Morgan City, Louisiana, September 8, 2010)

Thus, even the seemingly distinct seafood sector in the Gulf, heavily impacted by the spill, is embedded in oil. This has historic roots. As the petroleum industry moved offshore, work patterns changed. Shifts on the rigs and platforms typically started as "7 and 7," seven days offshore, seven home. Then, as distances to offshore operation increased, these shifts were extended. Some have argued that such concentrated work schedules have allowed workers to continue as fishermen during their time back in their communities (cf. Gramling 1996). Oyster harvesters likewise have come to an uneasy accommodation with the oil industry: companies towing equipment across oyster reefs have long considered it simply a cost of doing a very lucrative business to cash out the damage claims of oyster-bed leaseholders (cf. McGuire 2008).

In many ways, the people of southern Louisiana know the offshore petroleum industry well and are entwined in it. They have lived through the proud moments when they were rewarded for contributing to the nation's wealth, energy independence, and position in the global oil industry, and also the devastation caused by industry downturns, restructuring, and environmental contamination. Yet the industry has changed in significant ways in the past two decades. The corporate leaders have become concentrated in urban hubs such as Houston. Offshore projects have gotten larger, the number of those projects fewer, and the people who work on them more scattered. Consequently, local familiarity with the specifics of what goes on offshore and among industry leaders has diminished. Nonetheless, both individual and collective understanding of what this industry is, what it does, and what it means to the region have developed over many decades and will not readily change.

From the explosion forward, the people of south Louisiana drew upon their experiences of "oil field culture" (see, e.g., Weaver 2010) to make sense of what was happening. Despite all the variation in those experiences, some commonalities emerged. No one knew what to expect from this event;

indeed, in the almost fifteen years that members of our research teams had been conducting research on the social effects of the Gulf of Mexico offshore petroleum industry, almost no one had ever even mentioned the possibility of a blowout of this magnitude and duration. Some tried to link it to prior experiences. While the oil was still gushing, for example, southern Louisiana natives talked of the "Grand Isle tie-dye," the black patterns formed when oil adhered to the swimsuits of beachgoers in the 1950s and 1960s. The following quote, taken from a 2001 interview with Charles Wallace, an oil industry veteran who began working for Pure Oil Company in 1947, is eerily familiar to anyone who listened to or read transcripts describing what happened on the *Deepwater Horizon*—with the added element of the 1957 Hurricane Audrey coming into the Gulf at the same time.

> I called the Lake Charles office at six o'clock . . . Charlie Small told me, he said, "Forget about the drilling report. You can call that report in in a few minutes. We got an announcement about a hurricane brewing out there in the Gulf." We all knew what we were supposed to do. But I was in trouble because during the night that night, we had run pipe and I told the engineers in the office that they wanted me to cement the well with cement with 10 percent gel. Well, 10 percent gel, you can't make it weigh enough to hold the formation back. And I told the engineers in the office that that's the way it was and they said, "Well, run it anyway." So that is what I did . . . So, I was going to take a nap. Except, on this occasion, I didn't take a nap because I told them that the cement wasn't heavy enough to hold the formation back. And so, what I did, I stayed up and I went up on the rig floor. . . . The well was coming in and it was blowing out . . . this was early in the morning, I'm talking about before daylight . . . The pressure kept building up and when it got to around 3,000 pounds, it ruptured that pipe. Well, the well was wild to the world and you could see how the water was boiling around it . . . And that's when they called and told me that the hurricane was coming. And there we were, I was standing in the middle of the ocean with the drill pipe and the derrick, the caisson blowing out, and the drilling tender was not on location and the hurricane is coming. (Charles Wallace, Larose, Louisiana, July 16, 2001)

Indeed, despite all the changes since the first successful well was drilled out of sight of land off the coast of Morgan City in 1947, some aspects of the industry persist—the danger, the uncertainty, the thrill of discovery, the need to think on your feet, the pressure to bring the oil in, the difference between being in the office and being out on the rig, and so forth. At the same time, as almost anyone in southern Louisiana will point out, thousands of wells have been drilled without incident. And when shallow-water blowouts spewed oil that coated beaches, the effects were much more localized. They did not continue for months, nor were they witnessed by millions of people around the world. The disaster on the *Deepwater Horizon*—another faulty cementing job—could have changed the game.

THE BLOWOUT AND THE MORATORIUM

BP's venture into the ultra-deepwaters of the Gulf of Mexico was problematic. Before 2005, when the company installed its production platform Thunder Horse over a twenty-five-well complex six thousand feet below the surface, it had spent a troubled decade attempting to remake its image and restructure its finances and organization. Under the leadership of CEO John Browne, the company cut costs largely by reducing its engineering staff and adding accountants to strictly monitor expenses and reward units within the vast operation that showed profits, acquired Amoco and Arco, other majors, and attempted to reinvent its image with a prolonged media campaign on "Beyond Petroleum." But with the decline of BP's major field in the North Sea—production peaked in the Forties field in 1979 and was barely producing a profit by 2003—it needed a new elephant field. With its discoveries 150 miles south of New Orleans in 1999, it thought it had achieved its goal.

Towed from its shipyard in South Korea, Thunder Horse was installed over the field early in 2005. The platform had to be abandoned as early-season Hurricane Dennis approached in March. Dennis gave the platform only a glancing blow, but returning workers found it listing on its port side, the edge of the top deck in the water. Underwater inspection revealed no hurricane damage; what nearly sank the platform were one-way valves designed to allow water to flow freely between ballast tanks to keep the platform level. They had been installed backwards (Steffy 2011:97–98). Once repaired in a Texas shipyard, Thunder Horse was repositioned. By then most of the welds on the network of piping on the ocean floor had failed, causing further delays. BP's partner in the project, Exxon, decried BP's lack of engineering capabilities and sent its own teams to oversee the repair work. Shortly after the Thunder Horse embarrassment, John Browne was replaced as CEO by another Englishman, the yachtsman Tony Hayward.

Like the Thunder Horse field, the Macondo well in Mississippi Canyon was drilled by a partnership of oil companies. Anadarko Petroleum was a major shareholder, but BP was the designated operator. As such, BP's exploration engineers in Houston and the "company men" on the rig were responsible for designing the well and overseeing the work of the contract drilling company. Transocean's rig, *Deepwater Horizon*, was capable of drilling in ten thousand feet of water and, until the blowout, had established a credible safety record—seven years with no "lost-time incidents" (Cavnar 2010:19–20). BP had the rig under contract for three years up to the event and was paying Transocean a day rate of $500,000 to drill in Mississippi Canyon. Also involved in the complex task were drilling and completion services companies, notably the world-renowned Halliburton, responsible for cementing the well. M-I Swaco provided the chemically complex drilling fluid, referred to in the industry as "mud." Beyond these key players was an array of other actors involved in finding the oil and gas, designing and building specialized components, transporting materials and workers to the drilling rig, and catering to those on board.

Mississippi Canyon Block 252, where the catastrophe occurred, was leased to BP by the Minerals Management Service (MMS), the bureau within the U.S. Department of the Interior responsible for managing the country's offshore oil and gas properties. Reorganized and renamed after the spill as the Bureau of Ocean Energy Management, Regulation and Enforcement (BOEMRE), and later reconfigured as the Bureau of Ocean Energy Management (BOEM) and the Bureau of Safety and Environmental Enforcement (BSEE), the bureau derives its authority from the Outer Continental Shelf Lands Act (OCSLA). It conducts periodic "lease sales" of blocks of ocean floor to oil companies for exploratory seismic work and exploratory drilling, then reviews and approves drilling plans, issues drilling permits, and receives revenues from the lease-sale bids and royalties from eventual hydrocarbon production from those leases. The OCSLA mandates that the agency review permit applications "within a reasonable time," a vague provision that would play heavily in litigation over Interior Secretary Ken Salazar's May 28, 2010, moratorium on new and already-permitted deepwater drilling.

The moratorium, requested by President Obama, placed a six-month suspension on drilling in waters over five hundred feet deep and required the industry to meet stringent engineering and safety measures before activity could resume. Blowout preventers (BOPs)—the *Deepwater Horizon*'s failed BOP was but one cause of the well's disaster—had to be improved, and companies had to demonstrate the ability to respond to spills the magnitude of Macondo's.

Along the Gulf, the reaction to the moratorium was swift. The well was still gushing—it would not be officially declared killed until September—when Louisiana's governor, Bobby Jindal, complained to the president and the interior secretary in early June that three to six thousand jobs would be lost in the coming weeks, and twenty thousand over the next year. He wrote: "During one of the most challenging economic periods in decades, the last thing we need is to enact public policies that will certainly destroy thousands of existing jobs while preventing the creation of thousands of jobs" (Vucci 2010). The Mid-Continent Oil and Gas Association made some quick calculations of its own: lost wages for workers on the thirty-three idled deepwater rigs could come to $330 million per month; $1 million a day would be lost in commerce from idled supply boats; $16.5 million a day in revenues from rig day rates would be foregone (Tilove 2010a).

In the communities already impacted by the closure of Gulf waters to fishing, shrimping, and oystering, the moratorium on deepwater drilling caused an immediate and negative reaction. Politicians showed no hesitation in using the opportunity to criticize President Obama and his administration. They assembled before a crowd of around fourteen thousand Gulf Coast residents in the Cajundome in Lafayette, Louisiana, in July. A project researcher captured the flavor of the rally in his field notes:

> The event had the feel of a large civic event, not a political/cultural/economic rally. This was reinforced by the "opening act," a regional country music artist who was actively "pumping up" the crowd before the rally officially started . . . [Governor] Jindal spoke somewhat briefly and somewhat generically—focusing on broad statements in support of oil exploration and oil work as a way of life—points that the crowd was sure to affirm through applause and cheers—and implored Obama to come down to Louisiana, talk to local residents, and learn how misguided his policies really were—that everyone just wanted to work, how this was stopping. (Ben McMahan field notes, Lafayette, Louisiana, July 2010; cf. Smith 2010)

The first legal challenge to the moratorium was launched by Hornbeck Offshore Services, owner and operator of a fleet of supply vessels for deepwater and ultra-deepwater exploration and production. Hornbeck was joined in the suit against Secretary Salazar by Bollinger, a shipyard conglomerate that builds and repairs offshore supply vessels, and another south Louisiana family-run entity, the extensive Edison Chouest Offshore, builders and operators of the most sophisticated fleet of support vessels. A dozen more service and supply companies came into the action to demand an immediate injunction on Salazar's moratorium. The government's attorneys were supported by lawyers from the Florida Wildlife Federation, Earthjustice, the Center for Biological Diversity, Natural Resources Defense Council, Sierra Club, and other environmental organizations. The oil companies stayed away from the suit. The properties they already had in production in deepwater were not affected by the moratorium and their contracts included legal language allowing them to stop paying the day-rate charges if drilling rigs were shut down.

On June 22, 2010, the federal district court in Louisiana ruled against the government. On appeal, the U.S. Court of Appeals for the Fifth Circuit agreed with the lower court: the secretary of the interior had failed to demonstrate that "offshore drilling of new deepwater wells poses an unacceptable threat of serious and irreparable harm to wildlife and the marine, coastal, and human environment" of the Gulf Coast. In his ruling of June 22, Judge Martin Feldman acknowledged that "[o]il and gas production is quite simply elemental to Gulf communities" and suspended the blanket moratorium of new wells in five hundred or more feet of water (Hornbeck v. Salazar 2010:6).

The Department of the Interior didn't rest, however. On July 12, it again ordered a suspension of activities, this one more specifically targeted to wells using subsea blowout preventers or surface BOPs on floating rigs. The specificity was an effort to address some of Judge Feldman's objections, that the original moratorium was too broad. Seemingly, the new moratorium would have allowed the shallow-water shelf activity to continue. However, a de facto moratorium ensnared shallow operations as well: the regulatory

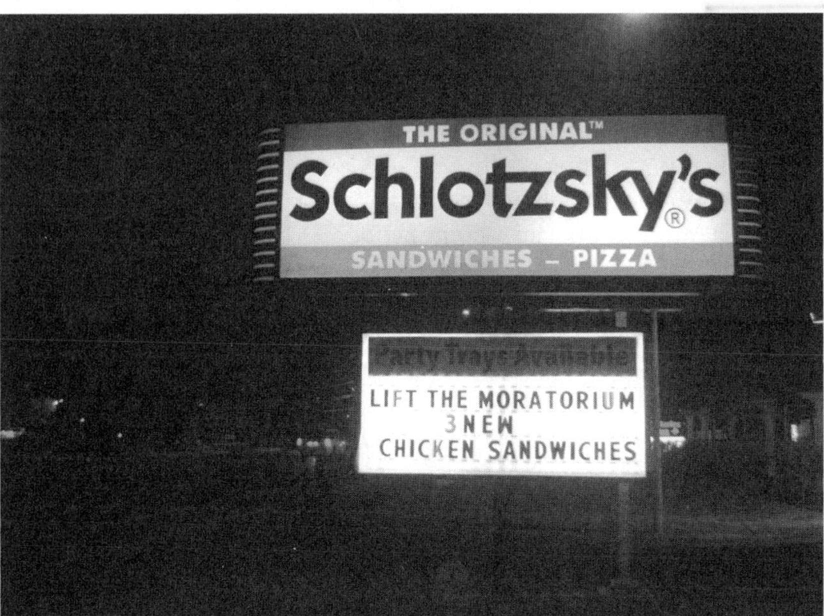

FIGURE 16.1. Sign in Lafayette, Louisiana. Photo taken September 2010.

agency, citing lack of personnel and resources to evaluate new permits under the heightened scrutiny demanded by the BP blowout and its ongoing gusher, slowed the issuance of shallow-water permits (Killalea 2010).

In an immediate response to the blowout, a new group, the Shallow Water Energy Security Coalition, formed to "enhance the understanding of shallow water operations as policymakers develop legislative and regulatory responses to recent events"—to lobby on behalf of some forty thousand workers directly involved in exploring, producing, and servicing oil and gas resources in waters less than one thousand feet deep. The coalition also claimed an additional 180,000 constituents, those indirectly employed by the industry: welders, equipment manufacturers, helicopter pilots, truckers, restaurant owners, supply boat captains, and deckhands (Shallow Water Energy Security Coalition n.d.).

A battle of numbers and scales ensued in September. The Inter-Agency Economic Report "Estimating the Economic Effects of the Deepwater Drilling Moratorium on the Gulf Coast Economy" was presented to the Senate Small Business Committee on September 13 (Economics and Statistics Administration 2010). Based on interviews with drilling companies and employment data for five of the most affected Louisiana parishes, the analysis concluded that eight to twelve thousand jobs would be temporarily lost due to the pause in activity, many fewer than the twenty-three thousand that BOEMRE's director, Michael Bromwich, had predicted in July. Moreover, the report suggested that at least some of this impact was mitigated by wages generated by spill cleanup activities and BP's Rig Worker Assistance Fund. Yet the latter, which was initially limited to people who worked on one of the thirty-three deepwater rigs operating in the Gulf of Mexico on May 6, 2010, was little used; fewer than eight hundred individuals applied and not even 350 received compensation during the initial time period.

The report was immediately attacked from several directions, though its economics were not challenged. Louisiana's Senator Mary Landrieu, chairwoman of the Small Business Committee, "expressed concern that the findings would generate headlines indicating that fears of the economic impact of the moratorium—which she and others had said could be worse than the spill itself—were overblown" (Tilove 2010b). A Louisiana State University economics professor, Joseph Mason, agreed generally with the job-loss calculations but not the framing: "The administration says 'only' 12,000 jobs will be lost . . . Many communities along the Gulf Coast are home to just a few thousand residents. This is almost as if they are telling a region that has been hit by disaster—both natural and man-made—time and again that we'll 'only' wipe out a handful of their communities" (Tilove 2010b). Jim Noe, executive director of the Shallow Water coalition, chided the report for only looking at the deepwater sector—the inter-agency committee's task—and not the de facto shallow-water pause and its impacts on the region and nationally, rather than just the five-parish area addressed in the report. Reminding others of the oil that had spewed into the Gulf, a spokesman for Environment America voiced a somewhat different perspective: "Despite what some might say, the sky is not falling because of the temporary drilling moratorium; but it has already fallen on tens of thousands of fishermen and tourism industry workers who are out of jobs because of the BP oil spill" (Tilove 2010b).

Senator Landrieu's fears came to fruition quickly. The *New York Times*'s headline on September 17 read: "Report Says Drilling Ban Had Little Effect" (Anon. 2010). A week earlier, the *Times* had editorialized for "A Necessary Moratorium" following a fire on a shallow-water production platform (not a drilling rig) off Louisiana, urging the administration to continue to resist pressures from the Gulf states' politicians and industry to lift the moratorium until new safety and engineering requirements were met (Broder 2010). The *Times*'s focus again was on deepwater drilling, not the reticulating net of businesses and workers across the industry.

By mid-October, the administration appeared ready to lift the ban on deepwater drilling. Michael Bromwich, the regulatory agency's director, announced that sufficient progress had been made in the previous months to enhance the safety of future drilling operations and to address some of the weaknesses in spill containment and spill response. Bromwich acknowledged, though, that the permit process would move more slowly than in the past, in part because of the post–*Deepwater Horizon* requirement for site-specific environmental impact assessments for new wells: "This is not going to be an agency that is going to be a permit-processing mill" (*Platts* 2010; Baker and Broder 2010).

Another suit against the Interior Department by Ensco, a deepwater driller that had five permits on hold during and after the moratorium, ended with an order from the U.S. District in Louisiana (the same judge that had ruled in favor of Hornbeck Offshore back in March 2010) directing the department to proceed in an "expeditious manner" in issuing permits. The first permit for the deepwater since the blowout was approved by BOEMRE on February 28, 2011. By the end of March, six additional permits were approved—but all for wells already in progress when the moratorium was put into effect. The anticipated exodus of rigs from the Gulf of Mexico did not materialize, perhaps largely due to the global financial crisis, but through 2011, activity was slow to pick up (Greenberg 2011:47) and concerns about lost jobs and economic continued (Greater New Orleans, Inc. 2012).

THE EVENT AS A CULTURAL ANOMALY?

Professors Hoffman and Jennings scrutinize the BP spill as a potential catalyzing event within the "third wave" of environmental management:

> The third wave began around 2008—with the growing recognition of the interconnected nature of the global ecosystem and a growing concern that humans are altering it on a global scale—and had not yet diminished at the time of the spill. (Hoffman and Jennings 2011:103)

Earlier such waves had their own catalyzing events: the publication of Rachel Carson's *Silent Spring* in 1962, the exposure of Love Canal near the "honeymoon capital of the world," the 1969 Santa Barbara oil spill, the 1989 *Exxon Valdez* spill in the pristine Prince William Sound. Each was a defining event, activating widespread media attention and public reaction, in part because of its context. The *Exxon Valdez* spill at the time was only the twenty-third largest spill in history. The next largest spill was from the Liberian tanker *Burmah Agate* off the Galveston beach in 1979, spilling six hundred thousand gallons more than the *Valdez*. With "[p]ictures of dying ducks and pristine Alaska coastline . . . splashed over the newsprint and TV media," the *Valdez* vastly exceeded the *Burmah Agate* mishap in national attention (Hoffman and Jennings 2011:102).

Yet for an event to become a game changer, a cultural anomaly, more than an attention-getting context is needed. *Identities* must be challenged—values, norms, systems of meaning: who are we, what are we, what do we do that makes us distinctive? As Hoffman and Jennings (2011:103) argue,

> If events raise these critical underlying questions, they can become cultural anomalies that alter the dominant institutions of society. Such challenging events in turn provide openings for high-profile 'cultural entrepreneurs' or networks of entrepreneurs to insert their interpretations of the event into the larger discourse, and to influence the rules of the game.

The BP spill contained many of the requisite factors for cultural anomalies:

> The context certainly led to a strong definition of the problem, and the problem challenged the identity of at least BP and the Gulf States. The conflict created by the event tapped all of the competing frameworks

> concerning the natural environment, but especially the belief in science and about the nature of risk . . . (Hoffman and Jennings 2011:108)

Yet the interim conclusion of these analysts is that the event will likely have only minor impacts on how the environment is managed, on how fossil fuels are extracted and utilized. They arrive at this conclusion based in part by a reading of the context. It occurred during one of the worst recessions in U.S. history: "the ongoing recession and the economic hazards of controls on offshore drilling overrode concerns for present and future environmental risks" (Hoffman and Jennings 2011:109).

But they also note the failure of the Gulf to generate its own cultural entrepreneurs:

> Oddly enough, no well-recognized spokesperson or active network arose as representative of the Gulf States community as a whole. In other words, unlike the *Valdez* and also Santa Barbara cases, no local coalition of representatives or specific NGOs was able to build directly off the event as the voice for change. (Hoffman and Jennings 2011:108)

There has been no shortage of NGOs in the region. Indeed, since Hurricane Katrina, there is a cottage industry of such groups. Rather, the issue lies in "community." It is at once meaningless in any anthropological sense to talk about a "Gulf States community," as Hoffman and Jennings do, and crucial to understand the depth and breadth of that community. As we have attempted to demonstrate in this chapter, residents of the Gulf grew up with an oil and gas industry, grew it out to deeper and deeper waters, and both prospered in good times and suffered together through the bad times. Those who worked and died on the *Deepwater Horizon* are part of a culture with deep historical roots. Those who escaped the tragedy helped to clean up the waters, beaches, and bayous. But they also pushed hard to get the industry moving again.

Critics of the petroleum industry are certainly present, but they recognize better than most the embeddedness of the industry. And, in the wake of the extensive and rapid coastal erosion occurring in the region, only some of which can be blamed on the canal dredging and widespread environmental change caused by the industry, they recognize, too, that solutions will be costly and that only the oil and gas industry will generate the revenue needed to address it. Ironically, then, the BP spill and the billions of dollars in fines that might be directed toward coastal restoration may actually prove to be a benefit.

NOTE

1. The authors would like to thank two social scientists, Harry Luton and Sidney Chaky (Bureau of Ocean Energy Management, Gulf of Mexico Region, U.S. Department of the Interior), for support of the work reported here.

REFERENCES

Baker, Peter and John M. Broder
2010 U.S. lifts the ban on deep drilling, with new rules. *New York Times*, October 13. http://www.nytimes.com (accessed November 27, 2011).

Broder, John M.
2010 Report says drilling ban had little effect. *New York Times*, September 17. http://www.nytimes.com (accessed November 27, 2011).

Cavnar, Bob
2010 *Disaster on the horizon: High stakes, high risk, and the story behind the Deepwater Well blowout*. White River Junction, VT: Chelsea Green Publishing Company.

Economics and Statistics Administration
2010 Estimating the economic effects of the Deepwater Drilling moratorium on the gulf coast economy. http://www.esa.doc.gov (accessed November 27, 2011).

Gramling, Robert
1996 *Oil on the edge: Offshore development, conflict, gridlock*. Albany: State University of New York Press.

Greater New Orleans, Inc.
2012 The impacts of decreased and delayed drilling permit approvals on Gulf of Mexico businesses. January 30. http://gnoinc.org/uploads/GNO_Inc_Permit_Slowdown_Impact_Survey_Results.pdf (accessed November 27, 2011).

Greenberg, Jerry
2011 Drilled again. *WorkBoat* 68(5): 46–50.

Hoffman, Andrew J. and P. Devereaux Jennings
2011 The BP oil spill as a cultural anomaly? Institutional context, conflict, and change. *Journal of Management Inquiry* 20(2): 100–112.

Hornbeck Offshore Services versus Kenneth Lee "Ken" Salazar et al.
2010 Order and reason. United States District Court, Eastern District of Louisiana.

Khatchadourian, Raffi
2011 The Gulf war. *The New Yorker* 87(4): 37–59.

Killalea, Mike
2010 Ruling halts Gulf of Mexico drilling moratorium, but Salazar working to reinstate it. *Drilling Contractor*, July 12. http://www.drillingcontractor.org/ruling-halts-gulf-of-mexico-drilling-moratorium-for-now-but-salazar-working-to-reinstate- it-628 (accessed November 27, 2011).

McGuire, Thomas R.
2008 Shell games on the water bottoms of Louisiana: Investigative journalism and anthropological inquiry. In *Against the grain: The Vayda tradition in human ecology and ecological anthropology*, edited by Bradley B. Walters, Bonnie J. McCay, Paige West, and Susan Lees, 117–134. Lanham, MD: AltaMira Press.

New York Times
2010 A necessary moratorium. *New York Times*, September 7. http://www.nytimes.com (accessed November 27, 2011).

Platts
2010 Salazar lifts drill ban 49 days early, but industry fears permitting delays." *Platts Inside Energy*, October 18. http://www.lexisnexis.com.ezproxy2.1ibrary.arizona.edu/lnacui2api/api/version1/getDocCui?lni=51GB-P4M1-DY6W-J4BY&csi=7989&hl=t&hv=t&hnsd=f&hns=t&hgn=t&oc=00240&perma=true (accessed November 27, 2011).

Shallow Water Energy Security Coalition
n.d. About the coalition. http://www.shallowwaterenergy.org/ (accessed November 27, 2011).

Smith, Sonia
2010 Gulf anti-moratorium rally draws big crowd. *Platts Oilgram News* 88(142): 9.

Steffy, Loren C.
2011 *Drowning in oil: BP and the reckless pursuit of profit*. New York: McGraw Hill.

Tilove, Jonathan
2010a Obama administration resists pressure to modify moratorium on new deepwater drilling in Gulf of Mexico. NOLA.com, June 3. http://www.nola.com/news/gulf-oil-spill/index.ssf/2010/06/obama_administration_resists_p.html (accessed November 27, 2011).
2010b Federal report downplaying drilling moratorium effects is disputed by Mary Landrieu, Vitter. NOLA.com, September 12. http://www.nola.com/news/gulf-oil-spill/index.ssf/2010/09/federal_report_that_drilling_m.html (accessed November 27, 2011).

Vucci, Evan
2010 Gov. Bobby Jindal urges Obama to gets deepwater drilling back quickly. NOLA.com, June 3. http://www.nola.com/news/gulf-oil spill/index.ssf/2010/06/gov_bobby_jindal_urges_obama_t.html (accessed November 27, 2011).

Weaver, Bobby D.
2010 *Oilfield trash: Life and labor in the oil patch*. College Station: Texas A&M University Press.

CONVERSATION 5

Energy Contested: Borders and Boundaries

Lisa Breglia, Thomas Love, Thomas McGuire, Gisa Weszkalnys

In this final conversation, contributors bring into focus energy as "power"—who is and isn't empowered as energy flows across human and natural boundaries. Yet, building from their case studies in the Gulf of Mexico and the Gulf of Guinea, they interrogate prevailing assumptions about the seemingly obvious "naturalness" of energy resources and their effects within existing political and other structures. Energy, they show, never just "is"; energy is always already connecting, integrating, and reinforcing people and nature while also dismantling, undermining, and reorganizing these relationships and boundaries. Why? Because the structured reality of the objective world is always affected by and reflected in the cultural filters people—particularly "local" people in resource-producing regions but also communities of engineers, managers, planners, oil workers, and wider publics—use to make sense of changes thrust upon them by processes emanating from the voracious appetite of the global growth system.

The three papers in this section are particularly concerned with oil—that magical black gold that is simultaneously a resource curse. As anthropologists, we are concerned to bring these "natural" qualities of oil back into social and cultural frameworks, for these processes and their complex local effects have histories as well as cultures, powerfully affecting yet being affected by what local people imagine about these newly valued resources that become "commodities" with "value" as they cross old and new boundaries.

Contributors in this conversation examined several questions, including: How are relations of power—between individuals and oil companies, between individuals and states, and between nations—reflected in and affected by conflict over energy resources? At the same time, though, they ask: Through what frameworks—technical, economic, social, magical, metaphorical—do people think about, apprehend, and experience oil? What role does oil play in people's lives, and how might this role be different depending on one's proximity to the source of oil?

Lisa Breglia, Thomas McGuire, and Gisa Weszkalnys pick up and develop several themes explored in earlier conversations: examining how energy often has its effects in its absence as well as more obviously its presence; the visibility of certain energy forms (particularly liquid—water, oil, ethanol) versus the invisibility of others (particularly electricity but also wind, solar, and undersea oil); precisely highlighting the way energy is inextricably bound up in the values and beliefs people share about it. The materiality of energy is not only about the potential of its chemical or kinetic energy but also about all that goes into its movement across boundaries—pipelines, platforms, and powerlines but also planning bureaus, maps, tables, government engineers, and reunions of experts. In the process new boundaries are created, powerful enough if not yet explicitly "political" or "legal." Such matters loop back and connect with earlier chapters in several ways, for example with Mason's earlier examination of the role of experts in creating shared expectations about energy futures.

GISA WESZKALNYS: The question of conflict and power relations involved in the generation of oil as a highly priced energy source is very close to the focus of my research. In my chapter, I try to develop an anthropological perspective on the much-talked-about phenomenon of the "resource curse," which has dominated scholarly accounts especially of developing oil states in Africa. I aim to highlight the material side of the curse, and how it may be understood as a peculiar capacity of oil. The idea of the curse, as it is formulated in scholarly and popular writing, has been centrally focused on conflicts of different types that emerge around resources—whether that's conflict between citizens and their governments, among members of political elites in oil-producing nations, or between inhabitants of extractive zones and those doing the extraction (that is, multinational corporations, state-run oil companies). Conflict is one hugely important mode of engagement that people have with this key energy source, alongside others such as management or extraction. Saying this also implies how conflict is constitutive of relations of power; it's not external to them. Anthropologists and human geographers have done a great deal in showing how conflicts around oil produce new subject positions for the residents of extractive sites—as citizens, as protestors, as rebels or insurgents. They forge new relations with the state, for example, through their contested relation with multinationals operating in such sites. If this citizen-subject position isn't entirely new, it is at least given an entirely new spin in these contexts. It provides what are often very marginalized populations a way to attempt to hold their government accountable, by assuming the position of a contesting citizenry, and in doing so to join forces with others, such as international environmental activists. This is power often heavily curtailed and occasionally violently suppressed, and yet it's not negligible as one of the remarkable effects of conflict as a mode of engagement with the resource.

LISA BREGLIA: One fascinating thing I learned from my case study on the politics of transboundary reservoirs on the U.S.-Mexico maritime border is just how problematic the migratory nature of oil can be—not only for legislating space and territorial jurisdiction under international law but also for claiming, owning, and possessing this valuable and increasing elusive substance. Indeed, the ability of oil, in its most material form, to disrespect borders and boundaries is one of the most magical qualities. Both states and multinationals imagine the subterranean world of hydrocarbon reservoirs perfectly mapped according to the logics of capital and strategic geopolitical interest and at the ready for extraction. However, the wily nature of hydrocarbons may thwart these logics and stymie contemporary modes of power.

Attempts to rein in the migration, flux, and flow of oil bump up against the boundaries and borders of cultures of energy created in local contexts of extraction: whether offshore enclaves or onshore communities affected by processes of extraction, production, and distribution. As a social scientist exploring these dynamics, I find that what we share is not only an abiding interest but a socially engaged concern for the politics of how oil flows through and against borders and boundaries—physical or political, economic, ecological, or ethical.

Especially in our ethnographic accounts of oil-affected communities, we have been more successful in describing the outright conflicts provoked by the presence of extractive industries (in scenarios characterized by the raw exercise of power over subjects) and less successful in giving texture to what you suggest as the more subtle modes of engagement engendered when local subjects engage in the already-existing relations of power surrounding resource extraction. Moving forward, I think we can begin to bring fresh eyes, new questions, and new tools for analysis to the ways in which power is operating in local contexts of extraction.

GISA WESZKALNYS: Incidentally, the drawing of maritime boundaries also played an important role in the context of São Tomé and Príncipe (STP), the small island state in the Gulf of Guinea where I carried out my research. Shortly after speculations about STP's potentially vast offshore oil resources began in the late 1990s, these boundaries became a matter of heated dispute. Nigeria claimed its own share of the prospective reserves. Out of this very brazen contest of STP's ownership of the as-yet-absent oil and its potential, a Joint Development Zone was forged in 2001, of which Nigeria has 60 percent and STP 40 percent. About one thousand times bigger than STP in terms of population, and with considerably greater experience in the oil business, Nigeria has thus been a constant and influential presence looming on the horizon of STP's oil economy. From the local perspective, this has also shaped the STP's specific oil materiality. Rumors abound that money from STP's emergent oil economy gets absorbed straight into Nigeria's own infamous oil-based economy of accumulation. This circumstance has had further implications, drawing boundaries between what is concealed and what can be out in the open, who can and can't be privy to certain types of information, and thus limiting, for example, the kinds of "ethical" oil cultures that can be set up around this supposedly joint oil in terms of transparency, accountability, and so on.

The resource curse here is not just about the "migration" of money in the form of revenues, kickbacks, and other sums that accompany the oil business; as Lisa notes, oil itself has this migratory nature irrespective of the boundaries we might draw around it. I'm also not sure, though, that states and corporations are looking at this situation through rose-tinted glasses seeing only well-mapped-out reservoirs. On the contrary. My sense is that discussions about how to navigate the complex and messy oil materiality loom large on the agenda of multinationals and government agencies, and that a great deal of effort goes into creating (rather than assuming) the more abstract realities that allow for what are seen as more ideal—though always vulnerable—conditions of extraction. In this sense, enclaves are but one technology of dealing with the mess. So in addition to the operations of power in and around extractive sites, I think the corporate headquarters and offices of national oil companies are a massive and relatively little-explored field of present-day energy cultures that would equally deserve our ethnographic attention.

THOMAS MCGUIRE: Lisa, you address several boundaries: the demarcation in the Gulf between Mexico and the U.S. (with Cuba looming larger as it begins to invite exploration), and a boundary of sorts within Mexico between the left and the right, between private investment and the national patrimony. You also touch on the chemical and physical materiality of oil: shared reservoirs, migrating oil.

While we don't really address either materiality or boundaries directly in our chapter, the *Deepwater Horizon* incident is about both. There are several boundaries by water depth, and the explosion is about working in the new ultradeep environment (the title of our chapter could perhaps be construed as moving beyond a comfort zone). This boundary was crossed but the "culture" of drilling in less risky environments was retained. There is a similar boundary in shallow-water, subsalt, deep gas zones. The physical materiality of both of these "frontier" plays is similar: high pressure. The distributed and relational materiality around the *Deepwater Horizon* is also richly textured: British-based operator, a foreign contract driller with a rig built in Korea and flagged in the Marianas, and a regulatory structure perhaps not up to the task of overseeing the frontier.

LISA BREGLIA: The *Deepwater Horizon* accident is a great case in point, Tom—illustrative of the kind of thinking that results in an instrumentalized (as you put it), colonized (as I would put it) culture of energy—one that is dictated by the needs of private capital over public good. The colonized culture of energy—where borders and boundaries are assumed to control and contain magical and migratory hydrocarbons—is where we might focus our concerns for imagining an alternative future.

THOMAS MCGUIRE: Yes, the tragic explosion on *Deepwater Horizon* provides space to interrogate many matters, for example the notion of "culture" in the exploration, production, and regulation of hydrocarbons. Postdisaster investigative commissions, internal company documents, and popular commentaries all highlighted corporate or organizational culture and, in particular, "safety culture" (or lack thereof) as a key ultimate cause for the failure of the well.

Setting to one side discussion of what the state, BP, and others mean by "culture"—a functionalist approach that mystifyies the fragmented, ambiguous, and contested nature of this "culture"—only about a dozen workers on the rig were employees of the operator, BP. The rest were contract workers for eight or ten firms specializing in numerous drilling and support operations, moving on and off the rig, taking orders and giving reports to superiors "on the beach" in offices spread across the Gulf. Had the "culture" of the rig foregrounded ambiguity, the signals from the negative pressure test might have been read properly.

One other point of useful contrast related to the *Deepwater* disaster is to recognize that Gisa's area is really constituted of enclaves, as she points out; why we make so much out of the moratorium in the Gulf of Mexico is that the region is decidedly not an enclave. Mexico perhaps is something of a different enclave: a state oil company that is now finding the need to transcend that historic status by seeking technical expertise elsewhere (as it contemplates crossing the deepwater boundary).

I'd like to close by adding something more or less politically incorrect but fitting for the way we're all implicated in cultures of energy: the *Deepwater Horizon* disaster was about private capital but occurred in the context of public demand for hydrocarbons. Until we "cross the energy divide," in the words of Robert and Edward Ayres, we are still utterly dependent on the stuff.

AFTERWORD

Maximizing Anthropology

Laura Nader

INTRODUCTION

Although anthropology, along with ecology, is one of the youngest sciences, from the start its goals were grand: to understand the human condition over millennia of existence. With the growth of sociocultural anthropology from the late nineteenth century the inspiration moved from early social evolutionary studies and the movement from savagery to barbarism to civilization, whether unilineal or multilineal, to more modest attainments by means of what we call ethnography—a type of description that makes the strange familiar and the familiar strange, according to some.

The forerunners of such work went to specific locales—islands in the Pacific, tribal groups in Africa, indigenous peoples of the New World—and in the process honed the contemporary study of peoples using techniques such as firsthand participant observation, long stays in the field, plus more sociological techniques such as interviews, life histories, archival work, demographic analysis, and more. While this work was highly successful and influential, the critiques began to pour in after the decay of Euro-American colonialism, sometimes followed by political if not economic independence.

By the late 1960s there were cries for "reinventing anthropology" (Hymes 1969), followed by various other trends such as interpretive anthropology, postcolonial anthropology, philosophical anthropologies—many of which actually fell into earlier pitfalls while also contributing new ideas. My own efforts were geared to maximizing anthropology by extending our subject matter to include industrial societies along with the so-called third and fourth worlds. "Up the Anthropologist" (Nader 1969), which appeared in *Reinventing Anthropology*, outlined a method for studying up, down, and sideways. At the same time critiques were articulated indicating what was hindering our maximizing potential—paradigmatic ideas such as objectivity or advocacy, exclusion of the powerful for study, problems of practicing "holism" when studying industrial societies. Outside of the discipline technological changes were varied and rapidly paced, technologies that had their own impacts on anthropological fieldwork.

CONTRIBUTIONS

The contributions at hand in this volume represent an advance from all the hand-wringing, finger-pointing, and self-critiques that followed in the latter part of the twentieth century. They cover the widest range of subject matters—household energy consumption; energy and history in centuries; energy as the sole domain of expert knowledge; conflicts over natural gas and renewable development projects in Alaska; coal, uranium, and wind energy cooperatives in Wyoming; the history of monopolies and collusions on price fixing among companies or by governments; lights for the first time in Peru and Zanzibar; continuities between Tanzania's socialist and neoliberal periods; processes maintaining "organized irresponsibility" in Brazil's booming biofuels industry; the power of local organizations to challenge public utilities in the United States; the life history of crude oil in the Gulf of Guinea from discovery to extraction to "bunkering" by rebels to consumption in the West; the different significance of the United States' and Mexico's interests in the Gulf of Mexico; national versus corporate ownership of energy; the Shrimp and Petroleum Festival in Louisiana. Amazing—*homo sapiens* writ large in the contemporary world—up, down and sideways.

A RUNAWAY WORLD

In 1968 the distinguished British anthropologist Sir Edmund Leach published the BBC Reith lectures as *A Runaway World?*, in which, in true contrarian style, Leach attacked the need for "objectivity" in science. He said words like *detachment*, *objectivity*, and *alienation* share the common element of separation: "We are accustomed to thinking of our human position as that of a passive spectator . . . But this detachment is an evasion of responsibility" (Leach 1968:2). Leach calls such demarcations as "an excuse for steering clear of politics . . . somewhere along the line, this kind of evasion has got to stop" (Leach 1968:6), and says "it has ceased to be true that nature is governed by immutable laws external to ourselves. We ourselves have become responsible" (Leach:15). Needless to say, Leach was appreciated by the media for meeting its needs and taken to task by journalists and journalist-academics for being confused and confusing. Members of the general public, however, who wrote Leach over five hundred letters, thanked him for his lucidity, whether or not they agreed with him.

How does this relate to the contents of this book? When we study in cultures from Peru or Zanzibar, as the authors do here, ranging from studying up, down, and sideways, the very process of adding it all up means we become implicated. We now know who the primary actors are, and the immediate consequences of their actions, and thus can no longer believe that humans are apart from nature, or that nature is predictable, given the right measures, or that scientists have all the answers. We lose our innocence. In fact the energy work is no longer about currents and flows but more like a series of jolts, disasters, traffic jams.

Indeed, we have been in transformation since the beginning of humankind, but the speed has changed, especially since the past two hundred years and the galloping speed of Western industrialization. It is this accelerating rate of change accompanied by Western types of industrialization that we have neglected as a key variable. Einstein warned that human technological and scientific discoveries were moving faster than humans have the mental capacity to absorb. We are confronted with a mismatch of great proportion, accompanied by scientific hubris or what Leach called the God complex. Who is more responsible for technological and social change, and how much change and with what amount of speed can humans cope? Marginal men are responsible, not *homo sapiens* writ large. It is and has been a few Western scientists, technologists, and their patrons who are changing the planet. How can anthropologists who write ethnographies about energy inspire a broader, more productive dialogue on the subject (Brooks 2012)?

When I was putting together *The Energy Reader* (Nader 2010) I included writings that were not primarily by anthropologists or social scientists. I primarily included non-anthropologists—physicists, engineers, economists, historians, chemists, lawyers, and writers—not because they were not social scientists but because we need to bond with specialist talents not found in anthropology, and who better than experts who share the anthropological perspective, one that connects dots forward and backwards, in awe and with humility, in thinking about the precarious nature of human survivability. Margaret Mead once said, "It only takes a few people to change the world," and it has been a few people most especially in the case of energy technologies.

One of the chapters in this volume mentioned the ways in which laypeople use the concepts of magic, science, and religion, but laypeople are not the only people who use such concepts. Scientists also practice magic, science, and religion, as Leach pointed out early on. The Trobrianders of the Western Pacific may have known the difference between the ability to predict Mother Nature when working in lagoons because that was where their knowledge lay. On the high seas, where they do not have enough experience, they use magic (Nader 1996:259–275). The same applies elsewhere in contemporary science. "Normal scientists" often do not differentiate between what they can experience and the unpredictable, as we are finding in all the talk about natural gas and fracking and from such nuclear catastrophes as Chernobyl, Three Mile

Island, and now Fukushima. And in spite of such catastrophes U.S. leadership is still saying, "It can't happen here." Or "we can deal with the problems," when there is little indication that they can.

TIME AND RATES OF CHANGE

Although the papers do not cover nuclear or energy conservation cultures, the heterogeneity reflected in the chapters do cover the major actors found in all energy scenarios—scientific expertise, producers, consumers, regulators, workers, consequence professionals, and externality benefactors. What might prove to be a path to the integrative power of anthropology is time dimension. The chapters document the uneven process of change in different settings. Minimally, this variation affords the reader the opportunity to assess directionality as related to development ideologies and the paradigm underlying development—progress, both technological and social. Thus the reader might assume a time dimension by lining up the chapters to indicate first-time electricity users, as in the Peruvian and Zanzibar cases, and oil and gas or coal in locales thought to be developed. However, such an exercise might not be theoretically enlightening given the differential distribution of social power in each locale, especially as to where social power is centralized or decentralized. What might prove a path to theoretical development is attention to the *longue durée*. It is possible to move from specific ethnographies to the evolution of forms over the millennia, over periods of time with different rates of change. The story of human evolution is intricately tied to energy use—whether human, animal, solar, wood, or, in the past two hundred years, coal, oil, nuclear, and ethanol.

Anthropologists who work with the *longue durée* can see more clearly perhaps that increasing energy does not solve the problem of finite material resources (or of happiness). Humans need to find ways of achieving steady state in lieu of growth in light of the observation that in the twentieth century alone humans used more natural resources (particularly the so-called developed societies) than in all prior history. The *longue durée* also teaches survival strategies—not all civilizations last.

Long-term perspectives remind us that solar systems have long durability, unlike changeability inherent in oil or coal. In other words, short-term thinking results in a naïve vision of life on this planet, a model of energy growth that can serve human beings everywhere, providing them with electricity, fuel, and whatever else they need to be "civilized" (another concept that needs dissection). For most of human history, sources of energy were part of the commons. What was aboveground and free is now underground, charged, and overcharged. Vast coordinate operating units use their power to benefit themselves. How did that happen?

Time in the modern sense has a static quality about it, often seen as universal, independent of particular cultures, directional, and associated with the idea of progress. Although we know that the notion of progress appeared with European industrialization, and that the Greeks for instance eschewed such a notion as problematic, the concept has spread worldwide. How did that happen? The Chinese can build dams without thinking about ancestors, integral to Chinese culture. It is not culture but centralized decision-making, the state and the corporation not Chinese people who are imagining the future of China. Concentrated power follows the notion of perpetual supply to meet perpetual demand, ideas coupled with growth models. The ideology of progress keeps this question mute. The future is taken for granted as an exponential improvement in people's lives. But recent environmental challenges render this assumption problematic, calling for inclusion of future generations in the democratic equation. What are the ethics of locking future generations into the choices their ancestors made, and does this vary cross-culturally?

Nuclear waste best epitomizes the quandary of future generations. Nuclear materials' persistence in time elicits the mind-boggling question of how to devise signs for people living thousands of years from now. The federal government in the United States may have buried the Yucca Mountain proposal for waste burial, but the whole range of waste issues relate to nuclear power life-cycle here and elsewhere. Anthropologists are particularly attentive to upstream stages in the nuclear production chain, such as uranium mining by Native Americans—an activity deleterious to health, the environment, and culture. Decades of nuclear testing in the United States, the former Soviet Union, and elsewhere also repeatedly affect indigenous peoples (Nader and Gusterson 2007). Our ancestors survived for 1.5 million years, a feat that industrial society has yet to prove possible. But, as I noted earlier, there has been a qualitative transformation: human technology is moving at breakneck speed relatively independent of consequential thinking. We all need a deeper sense of time, but an imperative need for a deeper sense of time applies first to those who operate with the price-earnings ratio, measuring growth by quarters instead of centuries.

Curt Stager, a paleoecologist, writes about the Age of Humans (Anthropocene). Human effects on the planet are so large that to speak of the natural world as if it were something separate from humanity is no longer possible. His book, *Deep Future: The Next 100,000 Years of Life on Earth* (2011), compares past climate changes to scenarios for future anthropogenic global warming. He takes Greenland as an example, linking long-term models to imagine how people will adapt to a world to come. Stager argues that it is false hope to think about returning to some prehuman state of nature.

Fortunately, Stager is not alone in writing about the planet in language laypeople can understand. A number of articles on the Anthropocene have been appearing in major newspapers and magazines hoping to enlarge the long-term perspective beyond specialized circles where in fact there might be less hope of change. The world has been knowingly changed, thus entailing that responsibility that Sir Edmund Leach spoke of when he was writing about Anthropocene in 1968!

BLACK-SWAN ANTHROPOLOGY

The original argument for black swans is recently attributed to Nasim Taleb (2007). Taleb's argument is a cautionary tale for anybody who pretends or thinks they can predict, in this case where energy cultures are going. There will always be events and ideas that turn things on their head, to wit black swans. Some thought coal was never-ending, but then came oil and those who thought that oil was never-ending, and then came nuclear, the technology that was going to solve all human energy problems—"too cheap to meter." And now shale gas with some combination of fracking. The "Big Boys" are all out there with their predictions, even in renewables—solar technologies, conservation technologies including garbage recycling, or all of the above. Most predictions are predicated on the continuation of corporate capitalism and corporate leaders looking for silver bullets. Challenging assumptions and thinking outside of acceptable assumptions is what anthropologists should be able to address, noticing the black swans before others come near to noticing, as did Edmund Leach. It is an important challenge for the skeptics in the discipline.

REFERENCES

Brooks, Andrew
2012 Radiating knowledge: The public anthropology of nuclear energy. *American Anthropologist* 114(1): 137–140.

Hymes, Dell, ed.
1969 *Reinventing Anthropology*. New York: Pantheon Books.

Leach, Edmund
1968 *A runaway world?* New York: Oxford University Press.

Nader, Laura
1969 Up the anthropologist: Perspectives gained from studying up. In *Reinventing Anthropology*, edited by Dell Hymes, 284–311. New York: Pantheon Books.
1996 *Naked science: Anthropological inquiry into boundaries, power, and knowledge*. London: Routledge.

Nader, Laura
2010 *The energy reader*. Oxford: Wiley-Blackwell.

Nader, Laura and Hugh Gusterson
2007 Nuclear legacies: Arrogance, secrecy, ignorance, lies, silence, suffering, action. In *Half lives and half truths: Confronting the radioactive legacies of the Cold War*, edited by Barbara Rose Johnston. Santa Fe, NM: School for Advanced Research.

Stager, Curt
2011 *Deep future: The next 100,000 years of life on Earth.* New York: St. Martin's Press.

Taleb, Nassim
2007 *The black swan: The impact of the highly improbable.* New York: Random House.

APPENDIX

Energy: Units and Concepts

I. ENERGY: The ability to do work, in a physical sense, is measured in **joules** in the International System of Units. A joule is defined as the energy expended (or work done) in applying a force of one newton through a distance of one meter (1 newton meter or N·m) or in passing an electric current of one ampere through a resistance of one ohm for one second.

There are several other units used to measure energy. A **BTU (British Thermal Unit)** is the energy required to raise the temperature of one pound of water by 1°F. A **therm** is 100,000 BTUs, while a **quad** is 10^{15} BTUs. There are 1,055 joules in a BTU. A **kilocalorie** is the amount of energy needed to raise one kilogram of water 1°C.

Another unit of energy especially relevant to anthropological discussions of fossil fueled–industrial modernity is **barrel of oil equivalent (BOE)**, the approximate energy released by burning one barrel (42 U.S. gallons, or 158.9873 liters) of crude oil. One BOE is equal to 6.1 gigajoules (or 5.8×10^6 BTU), though various grades of oil have slightly different heating values. One BOE is roughly equivalent to 5,800 cubic feet of natural gas. The BOE is used by oil and gas companies in their financial statements as a way of combining oil and natural gas reserves and production into a single measure. Other common multiples are the **BBOE** (billion barrel of oil equivalent), representing 10^9 barrels of oil, used to measure petroleum reserves, and **MMbd** (million barrels per day), used to measure daily production and consumption. Humanity currently consumes roughly 79 MMbd of crude oil.

II. POWER: The rate at which energy is converted (or, equivalently, at which work is performed), measured in **watts**. A watt is one joule per second. "When power from a potential energy source is flowing through and driving a useful process, we are accustomed to describing it as 'work.' Thus work is done when weights are lifted, when objects are arranged, or when anything useful happens" (Odum 1971:32).

Power is also measured in **kilowatt-hours (kWh)**, equal to 1,000 watt-hours or 3.6 megajoules, commonly known as a billing unit for energy delivered to consumers by electric utilities. Another measure of power is **horsepower**; one horsepower is equivalent to 33,000 foot-pounds per minute, or the power required to lift 550 pounds by one foot in one second, and is equivalent to about 746 watts.

III. NET ENERGY: The amount of energy left for all other uses after the energy needed to extract, process/convert, and deliver it. Difficult as they are to measure, this and the related concept of **energy return on energy invested (EROI** or **EROEI)** build crucially upon earlier insights by White and Steward and constitute crucial tools for assessing the actual amount of energy available for societal functioning (cf. Hall and Klitgaard 2011). This graph, for example, applies EROI analysis to U.S. energy use about 2005 (Hall et al. 2009), where the x-axis measures amount of energy used and y-axis measures energy return on energy invested to extract and process it. One readily sees that wind returns about 20 units of energy for every unit invested, but we don't use much wind energy. Natural gas has about the same return on energy investment, but we use a lot more of it. Coal provides a much better return and we use a lot more of it than either wind or gas. Some minimum energy return (5:1?) is needed to run a complex economy and society. The U.S. in 2005 consumed far more energy than the total amount of solar energy fixed on its surface area (but note that the 1930 number is for finding oil, while the others are for producing or obtaining it [C. Hall, personal correspondence]).

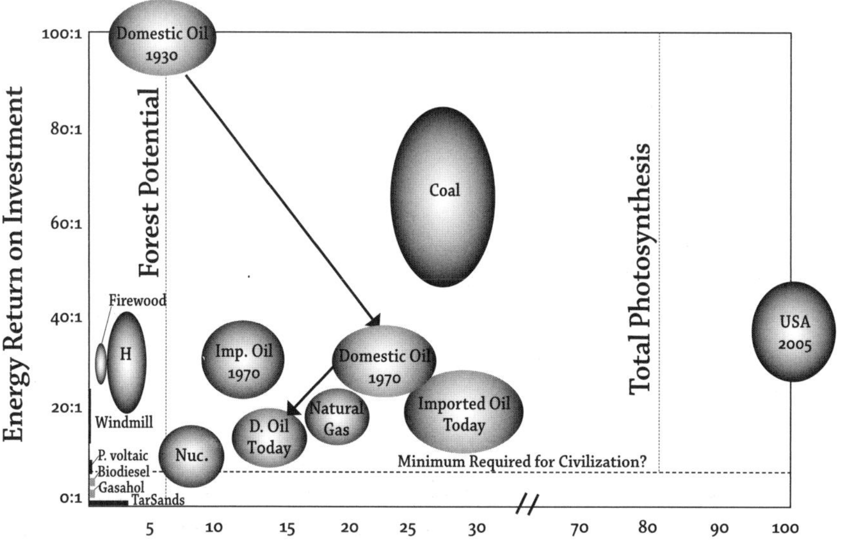

IV. ENERGY CONTENT OF DIFFERENT FUELS: As the chart demonstrates, fuels differ in their energy content, or density. Here is a useful list of the rough energy content of some common fuels.

Coal—12,000 BTU/lb
  Northern Appalachia coal—13,000 BTU/lb
  Powder River Basin coal—8,800 BTU/lb
Diesel—131,000 BTU/gal
Electricity—3,412 BTU/kWh
Ethanol—76,000 BTU/gal
Fuel oil #2—139,000 BTU/gal
Fuel oil #6—150,000 BTU/gal
Gasoline—125,000 BTU/gal
Heating oil—138,000 BTU/gal
Jet fuel—135,000 BTU/gal
Natural gas—1,000 BTU/cu ft
Petroleum—138,000 BTU/gal
Propane—91,600 BTU/gal
Wood pellets—7,000 BTU/lb

CURRENT ENERGY PRICES PER MILLION BTU (as of January 2010):

Coal
  Powder River Basin Coal—$0.56
  Northern Appalachia Coal—$2.08
Natural gas—$5.67
Ethanol subsidy—$5.92
Petroleum—$13.56
Propane—$13.92
#2 heating oil—$15.33
Jet fuel—$16.01
Diesel—$16.21
Gasoline—$18.16
Wood pellets—$18.57
Ethanol—$24.74
Electricity—$34.03

Source: Rapier 2010

SELECTED BIBLIOGRAPHY

Ackerman, Frank, Bruce Biewald, David White, Tim Woolf, and William Moomaw
1999 Grandfathering and coal plant emissions: The cost of cleaning up the Clean Air Act. *Energy Policy* 27(15): 929–931.

Adams, Richard Newbold
1975 *Energy and structure: A theory of social power.* Austin: University of Texas Press.
1988 *The eighth day: Social evolution as the self-organization of energy*. Austin: University of Texas Press.

Adas, Michael
1989 *Machines as the measure of men: Science, technology, and ideologies of western dominance.* Ithaca, NY: Cornell University Press.
2006 *Dominance by design: Technological imperatives and America's civilizing mission*. Cambridge, MA: Massachusetts Institute of Technology Press.

Adger, W. Neil.
2003 Social capital, collective action, and adaptation to climate change. *Economic Geography* 79(4): 387–404.

Akrich, Madeleine
1994 The description of technical objects.In *Shaping technology/building society: Studies in sociotechnical change*, edited by Wiebe E. Bijker and John Law, 205–224. Cambridge, MA: Massachusetts Institute of Technology Press.

Anderson, William P., Pavlos S. Kanaroglou, and Eric J. Miller
1996 Urban form, energy and the environment. *Urban Studies* 33: 7–35.

Andrews, Thomas
2008 *Killing for coal: America's deadliest labor war*. Cambridge, MA: Harvard University Press.

Appadurai, Arjun
1986 Introduction: Commodities and the politics of value. In *The social life of things: Commodities in a cultural perspective,* edited by Arjun Appadurai, 3–63. Cambridge: Cambridge University Press.
1990 Disjuncture and difference in the global cultural economy. *Public Culture* 2(2): 1–24.

Apter, Andrew
2005 *The Pan-African nation: Oil and the spectacle of culture in Nigeria*. Chicago and London: The University of Chicago Press.

Armstrong, Fraser and Katherine Blundell, eds.
2007 *Energy . . . beyond oil.* Oxford: Oxford University Press.

Atkinson, Adrian
2007a Cities after oil—1. Sustainable development and energy futures. *City* 11(2): 201–213.
2007b Cities after oil—2. *City* 11(3): 293–312.
2008 Cities after oil—3.*City* 12(1): 79–106.
2009 Cities after oil (one more time). *City* 13(4): 493–498.

Attali, Sophie and Harold Wilhite
2001 Assessing variables supporting and impeding the development of car sharing. *Proceedings of the ECEEE 2001 Summer Study*. Paris: European Council for an Energy Efficient Economy.

Auty, Richard

1993 *Sustaining development in mineral economies: The resource curse thesis.* London: Routledge.

Bakker, Karen and Gavin Bridge

2006 Material worlds? Resource geographies and the "matter of nature." *Progress in Human Geography* 30(1): 5–27.

Balzer, Marjorie Mandelstam

2006 The tension between might and rights: Siberians and energy developers in post-socialist binds. *Europe-Asia Studies* 58(4): 567–588.

Banerjee, Subhankar

2009 Terra incognita: Communities and resource wars. In *The Alaska Native reader: History, culture, politics,* edited by Maria Shaa Tláa Williams, 184–191. Durham, NC: Duke University Press.

Barnes, Peter

2003 *Who owns the sky? Our common assets and the future of capitalism.* Washington, DC: Island Press.

Barry, Andrew

2005 The British-Georgian case: The Baku-Tbilisi-Ceyhan pipeline. In *Territoires, environnement et nouveaux modes de gestion: La gouvernance en question,* edited by Bruno Latour and Christelle Gramaglia, 105–118. Paris: CNRS.

Batteau, Allen

2010 *Technology and culture.* Long Grove, IL: Waveland Press.

Barzelay, Michael

1986 *The politicized market economy: Alcohol in Brazil's energy strategy.* Berkeley: University of California Press.

Bebbington, Anthony, Leonith Hinojosa, Denise H. Bebbington,
Maria Luisa Burneo, and Ximena Warnaar

2008 Contention and ambiguity: Mining and the possibilities of development. *Development and Change* 39(6): 887–914.

Beck, Fred and Eric Martinot,

2004 Renewable energy policies and barriers. In *Encyclopedia of energy,* edited by Cutler J. Cleveland, 365–83. Waltham, MA: Academic Press.

Beck, Ulrich

2009 *World at risk.* Cambridge: Polity Press.

Behrends, Andrea

2008 Fighting for oil when there is no oil yet: The Darfur-Chad border. *Focaal* 52: 39–56.

Behrends, Andrea, Stephen Reyna, and Gunther Schlee, eds.

2011 *Crude domination: An anthropology of oil.* Oxford and New York: Berghahn.

Bell, Shannon Elizabeth and Yvonne A. Braun

2010 Coal, identity, and the gendering of environmental justice activism in central Appalachia. *Gender & Society* 24(6): 794–813.

Berger, Joel and Jon Beckmann

2010 Sexual predators, energy development, and conservation in greater Yellowstone. *Conservation Biology* 24(3): 891–896.

Besant-Jones, John E. and Bernard Tenenbaum

2001 Lessons from California's power crisis. *Finance and Development* 38(3): 24–28.

Bijker, Wiebe and John Law
1992 *Shaping technology, building society: Studies in sociotechnical change*. Cambridge, MA: Massachusetts Institute of Technology Press.

Bille, Mikkel and Tim Flohr Sørensen
2007 An anthropology of luminosity: The agency of light. *Journal of Material Culture* 12(3): 263–284.

Black, Brian
2000 *Petrolia: The landscape of America's first oil boom*. Baltimore: Johns Hopkins University Press.

Blottnitz, Harro von and Mary Ann Curran
2007 A review of assessments conducted on bio-ethanol as a transportation fuel from a net energy, greenhouse gas, and environmental life cycle perspective. *Journal of Cleaner Production* 15(7): 607–619.

Bolinger, M., R.Wiser, L. Milford, M. Stoddard, and K. Porter
2001 States emerge as clean energy investors: A review of state support for renewable energy. *The Electricity Journal* 14(9): 82–95.

Bohren, Lenora
2009 Car culture and decision-making: Choice and climate change. In *Anthropology and climate change*, edited by Susan Crate and Mark Nuttall, 370–379. Walnut Creek, CA: Left Coast Press.

Bowker, Geoffrey C.
1994 *Science on the run: Information management and industrial geophysics at Schlumberger, 1920–1940*. Cambridge, MA, and London: Massachusetts Institute of Technology Press.

Boyer, Dominic
2011 Energopolitics and the anthropology of energy. *Anthropology News* 52(5): 5–7.

Bridge, Gavin and Andrew Wood
2005 Geographies of knowledge, practices of globalization: Learning from the oil exploration and production industry. *Area* 37: 199–208.
2010 Less is more: Spectres of scarcity and the politics of resource access in the upstream oil sector. *Geoforum* 41: 565–576.

Brooks, Andrew
2012 Radiating knowledge: The public anthropology of nuclear energy. *American Anthropologist* 114(1): 137–140.

Bunker, Stephen G.
1985 *Underdeveloping the Amazon: Extraction, unequal exchange, and the failure of the modern state*. Chicago: University of Chicago Press.

Cheney, Richard and National Energy Policy Group
2001 *National energy policy: Report of the National Energy Policy Development Group*, May, 2001. Washington, DC: U.S. Government Printing Office. Available at http://www.ne.doe.gov/pdfFiles/nationalEnergyPolicy.pdf.

Castree, Noel
2001 Socializing nature: Theory, practice and politics. In *Social nature: Theory, practice and politics*, edited by Noel Castree and Bruce Braun, 1–21. Malden, MA: Blackwell Publishers.
2003 Environmental issues: Relational ontologies and hybrid politics. *Progress in Human Geography* 27: 203–211.

Cavner, Bob

2010 *Disaster on the horizon: High stakes, high risk, and the story behind the Deepwater Well blowout*. White River Junction, VT: Chelsea Green Publishing Company.

Choy, Timothy K.

2005 Articulated knowledges: Environmental forms after universality's demise. *American Anthropologist* 107(1): 5–18.

Clarke, Duncan

2010 *Africa—crude continent: The struggle for Africa's oil prize*. London: Profile Books.

Clemmer, Steven, Deborah Donovan, Alan Nogee, and Jeff Deyette

2001 *Clean energy blueprint: A smarter energy policy for today and the future*. Washington, DC: Union of Concerned Scientists. Available at http://www.ucsusa.org/clean_energy/solutions/big_picture_solutions/clean-energy- blueprint-a.html (accessed October 11, 2012).

Collins, Harry M. and Robert Evans

2002 The third wave of science studies: Studies of expertise and experience. *Social Studies of Science* 32(2): 235–296.

Connor, Linda, Sonia Freeman, and Nick Higginbotham

2009 Not just a coalmine: Shifting grounds of community opposition to coal mining in Southeastern Australia. *Ethnos* 74(4): 490–513.

Cooper, Gail

1998 *Air conditioning America: Engineers and the controlled environment, 1900–1960*. Baltimore: John Hopkins University Press.

Coronil, Fernando

1997 *The magical state: Nature, money, and modernity in Venezuela*. Chicago: University of Chicago Press.

Cowan, Ruth Schwartz

1985 How the refrigerator got its hum. In *The social shaping of technology: How the refrigerator got its hum*, edited by Donald MacKenzie and Judy Wajcman, 202–218. Milton Keynes, UK: Open University Press.

1999 The Industrial Revolution in the home. In *The social shaping of technology*, 2 edition, edited by Donald MacKenzie and Judy Wajcman, 281–300. Buckingham, UK: Open University Press.

Crate, Susan and Mark Nuttall, eds.

2009 *Anthropology and climate change*. Walnut Creek, CA: Left Coast Press.

Crosby, Alfred W.

2006 *Children of the sun: A history of humanity's unappeasable appetite for energy*. New York: Norton.

Daston, Lorraine

1995 The moral economy of science. *Osiris* 10: 3–24.

Debeir, Jean-Claude, Jean-Paul Deléage, and Daniel Hémery

1991 *In the servitude of power: Energy and civilization through the ages*. London: Zed Books.

Deffeyes, Kenneth

2005 *Beyond oil: The view from Hubbert's Peak*. New York: Hill and Wang.

Dethlefson, E. Stewart

1978 On redefining energy: A humanistic proposal. *Anthropology and Humanism Quarterly* 3(4): 21–5.

Diamond, Jared

2005 *Collapse: How societies choose to fail or succeed*. New York: Penguin.

Di John, Jonathan
2009 *From windfall to curse? Oil and industrialization in Venezuela, 1920 to present*. University Park: Penn State University Press.

Diesendorf, Mark
1979 *Energy and people: Social implications of different energy futures*. Canberra, Australia: Society for Social Responsibility in Science.

Djelic, Marie-Laure and Sigrid Quack
2010 *Transnational communities: Shaping global economic governance*. Cambridge: Cambridge University Press.

Douglas, Mary and Aaron Wildavsky
1983 *Risk and culture: An essay on the selection of technological dangers*. Berkeley: University of California Press.

Downey, Gary L.
1986 Risk in culture: The American conflict over nuclear power. *Cultural Anthropology* 1(4): 388–412.

Downey, Morgan
2009 *Oil 101*. New York: Wooden Table Press.

Drackle, Dorle and Werner Krauss
2011 Ethnographies of wind and power. *Anthropology News* (May): 9.

Dunn, Seth
2001 Routes to a hydrogen economy. *Renewable Energy World* (July-August): 19–29.

Energy Information Agency (EIA)
2008 International Energy Assessment, Chapter 1. Available at www.eia.doe.gov/oiaf/ieo/ (accessed October 11, 2012).

2012 Annual Energy Outlook Early Release. Available at www.eia.gov/forecasts/aeo/er/early_production.cfm (accessed October 11, 2012).

Ellis, Stephen
2003 Briefing: West Africa and its oil. *African Affairs* 102: 135–138.

Fahys-Smith, Virginia
2011 Migration of boom-town construction workers: The development of an analytic framework. *Environmental Geochemistry and Health* 5(4): 104–112.

Farhar, Barbara C., Partricia Weis, Charles T. Unseld, and Barbara A. Burns
1979 *Public opinion about energy: A literature review*. SERI/TR-53-155. Golden, CO: Solar Energy Research Institute.

Ferguson, James
2005 Seeing like an oil company: Space, security, and global capital in neoliberal Africa. *American Anthropologist* 107(3): 377–382.

Ferrer-Martí, Laia, Anna Garwood, José Chiroque, Rafael Escobar, Javier Coello, and Miguel Castro
2010 A community small-scale wind generation project in Peru. *Wind Engineering* 34(3): 277–288.

Ferry, Elizabeth and Mandana E. Limbert
2008 *Timely assets*. Santa Fe, NM: School for Advanced Research Press.

Fickett, Arnold, Clark Gellings, and Amory Lovins
1990 Energy end use technologies. *Scientific American* 263(3): 65–74.

Fishman, Robert
1987 *Bourgeois utopias: The rise and fall of suburbia*. New York: Basic Books.

Finn, Janet. 1998. Tracing the Veins: Of Copper, Culture, and Community from Butte to Chuquicamata. Berkeley: University of California Press.

Flavin, Christopher

2008 Low-carbon energy: A roadmap. *World Watch Report* 178: 52. Washington, DC: World Watch Institute. Available at http://www.worldwatch.org/node/5945.

Franklin, Sarah

1995 Science as culture, cultures of science. *Annual Review of Anthropology* 24: 163–84.

Freedman, David H.

2002 Fuel cells vs. the grid. *Technology Review* 40–47.

Friedel, Robert

2007 *A culture of improvement: Technology and the western millennium.* Cambridge, MA: Massachusetts Institute of Technology Press.

Frynas, Jedrzej George and Manuel Paolo

2007 A new scramble for African oil? Historical, political, and business perspectives. *African Affairs* 106: 229–251.

Fulkerson, William, Roddie R. Judkins, and Manoj K. Sanghvi

1990 Energy from fossil fuels. *Scientific American* 263(3): 128–135.

Gelb, Alan

1988 *Oil windfalls: Blessing or curse?* New York: Oxford University Press.

Georgescu-Roegen, Nicholas

1971 *The entropy law and the economic process.* Cambridge, MA: Harvard University Press.

Ghanadan, Rebecca H.

2008 *Public service or commodity goods? Electricity reforms, access and the politics of development in Tanzania.* PhD dissertation, Energy and Resources Group, University of California, Berkeley.

2009 Connected geographies and struggles over access: Electricity commercialisation in Tanzania. In *Electric Capitalism: Recolonising Africa on the Power Grid*, edited by David A. McDonald, 400–436. Cape Town, South Africa: HSRC Press.

Ghazvinian, John

2007 *Untapped: The scramble for Africa's oil.* New York: Harcourt.

Gillon, Sean

2010 Fields of dreams: Negotiating an ethanol agenda in the Midwest United States. *Journal of Peasant Studies* 37(4): 723–748.

Gilmore, John S.

1976 Boomtowns may hinder energy resource development. *Science* 191: 535–540.

Gold, Raymond L.

1974 *Social impacts of strip mining and other industrializations of coal resources.* Missoula: Institute for Social Science Research, University of Montana.

Goldberg, Marshall

2000 *Federal energy subsidies: Not all technologies are created equal. Renewable Energy Policy Project Research Report No. 11.* Washington, DC. Available at http://www.repp.org (accessed October 11, 2012).

Goldstein, Carolyn M.

1997 From service to sales: Home economics in light and power, 1920–1940. *Technology and Culture* 38(1): 121–152.

Goodell, Jeff

2006 *Big coal: The dirty secret behind America's energy future.* Boston: Houghton Mifflin.

Goodstein, David
2004 *Out of gas: The end of the age of oil.* New York: W. W. Norton.
Gore, Al
2006 *Earth in the Balance.* Emmaus, PA: Rodale Books.
Graeber, David
2006 Beyond power/knowledge: An exploration of the relation of power, ignorance, and stupidity. Malinowski Lecture, London School of Economics.
2011 *Debt: The first 5,000 years.* Brooklyn: Melville House.
Gramling, Robert
1996 *Oil on the edge: Offshore development, conflict, gridlock.* Albany: State University of New York Press.
Gratwick, Katharine Nawaal and Anton Eberhard
2008 Demise of the standard model for power sector reform and the emergence of hybrid power markets. *Energy Policy* 36(10): 3948–3960.
Gratwick, Katharine Nawaal, Rebecca Ghanadan, and Anton Eberhard
2006 Generating power and controversy: Understanding Tanzania's independent power projects. *Journal of Energy in Southern Africa* 17(4): 39–56.
Greenberg, Dolores
1990 Energy, power, and perceptions of social change in the early nineteenth century. *American Historical Review* 95(3): 693–714.
Gudeman, Stephen
1986 *Economics as culture: Models and metaphors of livelihood.* London: Routledge and Kegan Paul.
Gusterson, Hugh
1996 *Nuclear rites: A weapons laboratory at the end of the Cold War.* Berkeley: University of California Press.
2011 The lessons of Fukushima. *Bulletin of the Atomic Scientists*, March 16.
Hall, Charles A. S., Stephen Balogh ,and David Murphy
2009 What is the minimum EROI that a sustainable society must have? *Energies* 2: 25–47.
Hall, Charles A. S. and Kent Klitgaard
2011 *Energy and the wealth of nations: Understanding the biophysical economy.* New York: Springer.
Harrison, Conor and Jeff Popke
2011 "Because you got to have heat": The networked assemblage of energy poverty in eastern North Carolina. *Annals of the Association of American Geographers* 101(4): 949–961.
Hawken, Paul, Amory Lovins, and L. Hunter Lovins
2000 *Natural capitalism: Creating the next industrial revolution.* Boston: Back Bay Books.
Hayden, Corinne P.
2003 *When nature goes public: The making and unmaking of bioprospecting in Mexico.* Princeton, NJ: Princeton University Press.
Headrick, Daniel R.
2010 *Power over peoples: Technology, environments, and western imperialism, 1400 to the present.* Princeton, NJ: Princeton University Press.
Hecht, Gabrielle
1998 *The radiance of France: Nuclear power and national identity after World War II.* Cambridge, MA: Massachusetts Institute of Technology Press.

Heinberg, Richard
2003 *The party's over: Oil, war and the fate of industrial societies.* Gabriola Island, Canada: New Society Publishers.
2004 *Power down: Options and actions for a post-carbon world.* Gabriola Island, Canada: New Society Publishers.

Henning, Annette
2005 Climate change and energy use. *Anthropology Today* 21(3): 8–12.

Herring, Horace
1999 Does energy efficiency save energy? The debate and its consequences. *Applied Energy* 63: 209–226.

Hinrichs, Roger A. and Merlin H. Kleinbach
2005 *Energy: Its use and the environment.* Boston, MA: Brooks/Cole.

Hirsh, Richard
2001 *Power loss: The origins of deregulation and restructuring in the American electric utility system.* Cambridge, MA: Massachusetts Institute of Technology Press.

Hirsh, Richard F. and Adam H. Serchuk
1997 Power switch: Will the restructured electric utility system help the environment? *Environment* 41: 4–9, 32–39.

Hoffman, Andrew J. and P. Devereaux Jennings
2011 The BP oil spill as a cultural anomaly? Institutional context, conflict, and change. *Journal of Management Inquiry* 20(2): 100–112.

Hoffman, Peter
2001 *Tomorrow's energy: Hydrogen, fuel cells, and the prospects for a cleaner planet.* Cambridge, MA: Massachusetts Institute of Technology Press.

Holdren, John P.
1982 Energy hazards: What to measure, what to compare. *Technology Review* 85(3): 32–38, 74–75.
1990 Energy in transition. *Scientific American* 263(3): 157–163.

Holdren, John P. and Kirk R. Smith
2000 Energy, the environment, and health. In *World energy assessment: Energy and the challenge of sustainability,* UN Development Programme, UN Department of Economic and Social Affairs, New York and World Energy Council, 61–110.

Hollander, Gail
2010 Power is sweet: Sugarcane in the global ethanol assemblage. *Journal of Peasant Studies* 37(4): 699.

Holmes, Douglas and George E. Marcus
2005 Cultures of expertise and the management of globalization: Toward the re-functioning of ethnography. In *Global assemblages: Technology, politics, and ethics as anthropological problems*, edited by Aihwa Ong and Stephen J. Collier, 235–252. Malden, MA: Blackwell Publishers.

Holmes, John.
2002 End the moratorium: The Timor gap treaty as a model for the complete resolution of the western gap in the Gulf of Mexico. *Vanderbilt Journal of Transnational Law* 35: 925–952.

Horowitz, Daniel
2004 *Jimmy Carter and the energy crisis of the 1970s.* Boston: Bedford/St. Martin's.

Horowitz, Michael M.
1991 Victims upstream and down. *Journal of Refugee Studies* 4(2): 164–181.

Hornborg, Alf
1992 Machine fetishism, value, and the image of unlimited good: Toward a thermodynamics of imperialism. *Man* (n.s.) 27: 1–18.
2001 *The power of the machine: Global inequalities of economy, technology, and environment.* Walnut Creek, CA: AltaMira Press.
2011 *Global ecology and unequal exchange: Fetishism in a zero-sum world.* London: Routledge.

Hornborg, Alf, Brett Clark, and Kenneth Hermele, eds.
2012 *Ecology and power: Struggles over land and material resources in the past, present and future.* London: Routledge.

Hubbert, M. King
1956 Nuclear energy and the fossil fuels. Paper presented before the spring meeting of the Southern District Division of Production, American Petroleum Institute, San Antonio, Texas, March 8. Publication No. 95. Houston: Shell Development Company, Exploration and Production Research Division.

Huber, Matthew T.
2011 Enforcing scarcity: Oil, violence, and the making of the market. *Annals of the Association of American Geographers* 101(4): 816–826.

Hughes, Thomas
1983 *Networks of power: Electrification in western society, 1880–1930.* Baltimore, MD, and London: Johns Hopkins University Press.

Humphreys, Macartan, Jeffrey Sachs, and Joseph Stiglitz
2007 *Escaping the resource curse.* New York: Columbia University Press.

Illich, Ivan
1974 *Energy and equity.* New York: Harper and Row.

Inslee, Jay and Bracken Hendricks
2008 *Apollo's fire: Igniting America's clean energy economy.* Washington, DC: Island Press.

Irwin, Alan and Brian Wynne
2004 *Misunderstanding science? The public reconstruction of science and technology.* Cambridge: Cambridge University Press.

International Standardization Organization (ISO)
2004 Standards for a Sustainable Energy Future. Available at http://www.iso.org/iso/iso_focus_wec_special.pdf.

Jacobson, Mark Z. and Gilbert M. Masters
2001 Exploiting wind vs. coal. *Science* 293: 1438–1439.

Jasanoff, Sheila
1986 *Risk management and political culture: A comparative analysis of science in the policy context.* New York: Russell Sage Foundation.

Jasanoff, Sheila and Marybeth Long Martello
2004 *Earthly politics: Local and global in environmental governance.* Cambridge, MA: Massachusetts Institute of Technology Press.

Jochem, Eberhard
2000 Energy end-use efficiency. In *World energy assessment: Energy and the challenge of sustainability,* UN Development Programme, United Nations Department of Economic and Social Affairs and World Energy Council, 173–218. Available at http://www.undp.org/seed/eap/activities/wea/drafts- frame.html (accessed October 11, 2012).

Johnsen, Fred H.
1999 Burning with enthusiasm: Fuelwood scarcity in Tanzania in terms of severity, impacts and remedies. *Forum for Development Studies* 1(1): 107–131.

Johnson, Leigh
2010 The fearful symmetry of Arctic climate change: Accumulation by degradation. *Environment and Planning: Society and Space* 28(5): 828–47.

Johnson, Toni
2011 U.S. deepwater drilling's future. Council on Foreign Relations report. January 11. Available at http://www.cfr.org/united-states/us-deepwater-drillings-future/p22204 (accessed October 11, 2012).

Johnston, Barbara Rose
2011 In this nuclear world, what is the meaning of "safe"? *Bulletin of the Atomic Scientists*, March 18. Available at http://thebulletin.org/node/8641 (accessed April 6, 2012).

Karl, Terry Lynn
1997 *The paradox of plenty: Oil booms and petro states.* Berkeley: University of California Press.

Kashi, Ed and Michael Watts
2010 *Curse of the black gold: 50 years of oil in the Niger delta*. Brooklyn, NY: PowerHouse Books.

Kaup, Brent Z.
2008 Negotiating through nature: The resistant materiality and materiality of resistance in Bolivia's natural gas sector. *Geoforum* 39: 1734–1742.

Kelly, Ingrid
2008 *Energy in America: A tour of our fossil fuel culture and beyond*. Burlington: University of Vermont Press.

Kelsall, Tim
2002 Shop windows and smoke-filled rooms: Governance and the re-politicisation of Tanzania. *The Journal of Modern African Studies* 40(4): 597–619.

Kempton, Willet, James S. Boster, and Jennifer A. Hartley
1995 *Environmental values in American culture*. Cambridge, MA: Massachusetts Institute of Technology Press.

Kempton, Willett, Jeremy Firestone, Jonathan Lilley, Tracy Rouleau, and Phillip Whitaker
2005 The offshore wind power debate: Views from Cape Cod. *Coastal Management* 33:119–149.

Khan, Shahrukh Rafi, Zeb Rifaqat, and Sajid Kazmi
2007 *Harnessing and guiding social capital for rural development*. New York: Palgrave Macmillan.

Kirby, Peter Wynne
2011 *Troubled natures: Waste, environment, Japan*. Honolulu: University of Hawaii Press.

Klare, Michael
2001 *Resource wars: The changing landscape of global conflict*. New York: Henry Holt.
2012 *The race for what's left: The global scramble for the world's last resources*. New York: Metropolitan Books.

Klare, Michael and Daniel Volman
2006 America, China and the scramble for Africa's Oil. *Review of African Political Economy* 108: 297–309.

Klieman, Kairn
2008 Oil, politics, and development in the formation of a state: The Congolese petroleum wars, 1963–68. *International Journal of African Historical Studies* 41(2): 169–202.

Kolbert, Elizabeth
2007 *Field notes from a catastrophe: Man, nature, and climate change*. New York: Bloomsbury Publishing.

Konrad, John and Tom Shroder
2011 *Fire on the horizon: The untold story of the Gulf oil disaster*. New York: HarperCollins.

Koomey, Jonathan G., Chris Calwell, Skip Laitner, Jane Thornton, Richard E. Brown, Joseph H. Eto, Carrie Webber, and Cathy Cullicott
2002 Sorry, wrong number: The use and misuse of numerical facts in analysis and media reporting of energy issues. *Annual Review of Energy and the Environment* 27: 119–158.

Koselleck, Reinhart
2004 *Futures past*. New York: Columbia University Press.

Kunstler, James Howard
2005 *The long emergency: Surviving the end of oil, climate change, and other converging catastrophes of the twenty-first century*. New York: Grove Press.

Lahiri-Dutt, Kuntala
2011 The shifting gender of coal: Feminist musings on women's work in Indian collieries. *South Asia: Journal of South Asian Studies* 35(2): 456–476.

Latour, Bruno
1992 Where are the missing masses? The sociology of a few mundane artifacts. In *Shaping technology/building society: Studies in Sociotechnical Change*, edited by Wiebke Bijker and John Law, 225–258. Cambridge, MA: Massachusetts Institute of Technology Press.
1996 *Aramis: Or the love of technology*. Cambridge, MA: Harvard University Press.

Le Billon, Philippe
2005 *The geopolitics of resource wars: Resource dependence, governance and violence*. London and New York: Routledge.

Lie, Marianne and Knut H. Sørensen
1996 *Making technology our own? Domesticating technology into everyday life*. Oslo, Norway: Scandinavian University Press.

Limerick, Patricia Nelson, Claudia Puska, Andrwe Hildner, and Eric Skovsted
2003 *What every westerner should know about energy*. Boulder: Center of the American West, University of Colorado at Boulder.

LiPuma, Edward and Benjamin Lee
2004 *Financial derivatives and the globalization of risk*. Durham, NC: Duke University Press.

Lloyd, Alan C.
1999 The power plant in your basement. *Scientific American* 281(1): 64–69.

Logan, Jeffrey and Jiqiang Zhang
1998 Powering non-nuclear growth in China with natural gas and renewable energy technologies. *China Environment* (Series 2): 12–19.Washington, DC: Woodrow Wilson Center.

Lonergan, Stephen C.
1988 Theory and measurement of unequal exchange: A comparison between a Marxist approach and an energy theory of value. *Ecological Modeling* 41: 127–145.

Lopreato, Sally C. and Marian W. Meriwether
1976 *Energy attitudinal surveys: Summary, annotations, research recommendations*. Austin: University of Texas Press.

Love, Thomas
2008 Anthropology and the fossil fuel era. *Anthropology Today* 24(2): 3–4.

Love, Thomas and Anna Garwood
2011 Wind, sun and water: Complexities of alternative energy development in rural northern Peru. *Rural Society* 20: 294–307.

Lovins, Amory
1977 *Soft energy paths: Toward a durable peace*. New York: Penguin.

Lovins, Amory and E. Kyle Datta
2004 *Winning the oil end game*. Snowmass, CO: Rocky Mountain Institute.

Lutzenhiser, Loren
1993 Social and behavioral aspects of energy use. *Annual Review of Energy and Environment* 18: 247–289.

Mann, Charles C.
2002 Getting over oil. *Technology Review* (January/February): 32–38.

Marsden, Ben and Crosbie Smith
2005 *Engineering empires: A cultural history of technology in nineteenth-century Britain*. Houndmills: Palgrave Macmillan.

Martin, Stephanie, Mary Killorin, and Steve Colt
2008 *Fuel costs, migration, and community viability*. Final report for Denali Commission. Institute of Social and Economic Research, University of Alaska Anchorage.

Martinelli, Luiz A. and Solange Filoso
2008 Expansion of sugarcane ethanol production in Brazil: Environmental and social challenges. *Ecological Applications* 18(4): 885–898.

Martines-Filho, Joao, Heloisa L. Burnquist, and Carlos E. F. Vian
2006 Bioenergy and the rise of sugarcane-based ethanol in Brazil. *Choices* 21(2): 91–96.

Martinez-Alier, Juan
1987 *Ecological economics: Energy, environment and society*. Oxford: Blackwell.

Martinot, Eric, Akanksha Chaurey, Jose Moreira, Debra Lew, and Njeri Wamukonya
2002 Renewable energy markets in developing countries. *Annual Review of Energy and the Environment* 27: 309–348.

Mason, Arthur
2006 Images of the energy future. *Environmental Research Letters* 1(1): 20–25.
2007 The rise of consultant forecasting in liberalized natural gas markets. *Public Culture* 19(2): 367–379.
2010 New research on Russian arctic gas development. *International Arctic Social Sciences* 34: 17–21.

Massey, Garth
1980 Critical dimensions in urban life: Energy extraction and community collapse in Wyoming. *Urban Life* 9(2): 187–199.

McBeath, Jerry, Mathew Berman, Jonathon Rosenberg, and Mary F. Ehrlander
2008 *The political economy of oil in Alaska: Multinationals vs. the state*. Boulder, CO: Lynne Rienner Publishers, Inc.

McNeil, Bryan
2011 *Combating mountaintop removal: New directions in the fight against Big Coal*. Champaign: University of Illinois Press.

McNeish, John-Andrew and Owen Logan
2010 *Flammable societies: Studies on the socio-economics of oil and gas*. London: Pluto Press.

Miller, Daniel
1998 Why some things matter. In *Material cultures: Why some things matter*, edited by Daniel Miller, 3–21. Chicago: The University of Chicago Press.

Miller, Jon D., K. Prewitt, and R. Pearson
1980 *The attitudes of the U.S. public toward science and technology*. A final report to the National Science Foundation, under NSF Grant 8105662. DeKalb, IL: Public Opinion Laboratory.

Miller, Jon D., Robert W. Suchner, and Alan M. Voelker
1980 *Citizenship in an age of science*. New York: Pergamon.

Milstein, J. S.
1978 How consumers feel about energy: Attitudes and behaviors during winter and spring 1976–77. In *Energy policy in the United States,* edited by Seymour Warkov, 79–90. New York: Praeger Publishers.

Mitchell, Timothy
2009 Carbon democracy. *Economy and Society* 38(3): 399–432.
2011 *Carbon democracy: Political power in the age of oil.* New York: Verso.

Moen, Elizabeth
1981 Women in energy boom-towns. *Psychology of Women Quarterly* 6(1): 99–112.

Mouhot, Jean-François
2011 Past connections and present similarities in slave ownership and fossil fuel usage. *Climatic Change* 105: 329–355.

Murray, J. R., M. J. Minor, R. F. Cotterman, and N. M. Bradburn
1974 *The Impact of the 1973–74 oil embargo on the American household*. National Opinion Research Center: University of Chicago.

Murray, James and David King
2012 Climate policy: Oil's tipping point has passed. *Nature* 481 (January 26): 433–435.

Mwampamba, Tuyeni Heita
2007 Has the woodfuel crisis returned? Urban charcoal consumption in Tanzania and its implications to present and future forest availability. *Energy Policy* 35(8): 4221–4234.

Nader, Laura
1980 *Energy choices in a democratic society*. Supporting Paper No. 7, Study of Nuclear and Alternative Energy Systems. Washington, DC: National Research Council.
1996a *Naked science: Anthropological inquiry into boundaries, power, and knowledge*. London: Routledge.
1996b Anthropological inquiry into boundaries, power, and knowledge. In *Naked science: Anthropological inquiry into boundaries, power, and knowledge*. New York and London: Routledge.
2010 *The energy reader*. Oxford: Wiley-Blackwell.

Nader, Laura and Hugh Gusterson
2007 Nuclear legacies: Arrogance, secrecy, ignorance, lies, silence, suffering, action. In *Half lives and half truths: Confronting the radioactive legacies of the Cold War*, edited by Barbara Rose Johnston. Santa Fe, NM: School for Advanced Research.

Nash, June
1993 [1979] *We eat the mines and the mines eat us: Dependency and exploitation in Bolivian tin mines*. New York: Columbia University Press.

Nass, Luciano, Pedro Pereira, and David Ellis
2007 Biofuels in Brazil: An overview. *Crop Science* 47(6): 2228–2237.

National Commission on the BP Deepwater Horizon Oil Spill and Offshore Drilling
2011 Deepwater: The Gulf oil disaster and the future of offshore drilling. Report to the President.

Nealey, Stanley M. and William L. Rankin.
1978 *Nuclear knowledge and nuclear attitudes: Is ignorance bliss?* Technical Report B-HARC-411–02. Seattle: Battelle Human Affairs Research Center.

NEETF/Roper
2002 *Americans' low "energy IQ": A risk to our energy future. Why America needs a refresher course on energy.* Washington, DC: National Environmental Educational Training Foundation and Roper ASW. Available at www.neefusa.org/pdf/roper/Roper2002.pdf (accessed December 15, 2011).

Nelkin, Dorothy
1975 The political impact of technical expertise. *Social Studies of Science* 5(1): 35–54.

National Research Council
1984 *Energy use: The human dimension.* Edited by P. Stern and E. Aronson. New York: Freeman and Company.

Nye, David E.
1990 *Electrifying America: Social meanings of a new technology, 1880–1940.* Cambridge, MA: Massachusetts Institute of Technology Press.
1998 *Consuming power: A social history of American energies.* Cambridge, MA: Massachusetts Institute of Technology Press.

Obi, Cyril
2010 Oil as the "curse" of conflict in Africa: Peering through the smoke and mirrors. *Review of African Political Economy* 37(126): 483–495.

Obi, Cyril and Siri Aas Rustad
2011 *Oil and insurgency in the Niger Delta: Managing the complex politics of petro-violence.* London: Zed Books.

Odum, Howard T.
1971 *Environment, power and society.* New York: Wiley.
1996 *Environmental accounting: Emergy and environmental decision making.* New York: John Wiley & Sons.

Odum, Howard T. and Elisabeth C. Odum
2001 *A prosperous way down: Principles and policies.* Boulder: University Press of Colorado.

Pantzar, Mika
1997 Domestication of everyday life technology: Dynamic views on the social histories of artifacts. *Design Issues* 13(3): 52–65.

Pegg, Scott
2005 Can policy intervention beat the resource curse? Evidence from the Chad-Cameroon pipeline project. *African Affairs* 105(418): 1–25.

Peluso, Nancy and Michael Watts
2001 Violent environments. In *Violent environments*, edited by Nancy Lee Peluso and Michael Watts, 3–38. Ithaca and London: Cornell University Press.

Perlman, Robert and Roland Leslie Warren
1977 *Families in the energy crisis: Impacts and implications for theory and policy.* Cambridge: Ballinger.

Phillips, E. Barbara
2009 *City lights: Urban-suburban life in the global society*. Oxford: Oxford University Press.

Pimentel, David, Alison Marklein, Megan Toth, Marissa Karpoff, Gillian Paul, Robert McCormack, Joanna Kyriazis, and Tim Krueger
2009 Food versus biofuels: Environmental and economic costs. *Human Ecology* 37(1): 1–12.

Pimentel, David and Tad W. Patzek
2005 Ethanol production using corn, switchgrass, and wood; Biodiesel production using soybean and sunflower. *National Resources Research* 14(1): 65–76.

Pomeranz, Kenneth
2000 *The great divergence: China, Europe, and the making of the modern world economy*. Princeton, NJ: Princeton University Press.

Porter, Theodore M.
1996 *Trust in Numbers*. Princeton, NJ: Princeton University Press.

Power, Michael
1999 *The audit society: Rituals of verification*. Oxford: Oxford University Press.

Prindle, David F.
1981 *Petroleum politics and the Texas railroad commission*. Austin: University of Texas Press.

Pyne, Stephen
1997 *Fire in America: A cultural history of wildland and rural fire*. Seattle: University of Washington Press.

Rabinbach, Anson
1990 *The human motor: Energy, fatigue, and the origins of modernity*. New York: Basic Books.

Rapier, Robert
2010 *Prices of Various Energy Sources*. http://www.consumerenergyreport.com/2010/01/19/prices-of-various-energy-sources/ (accessed August 20, 2012)

Rappaport, Roy
1971 The flow of energy in an agricultural society. *Scientific American* 225(3): 117–132.

Reckwitz, Andreas
2002 Toward a theory of social practices: A development of culturist theorizing. *European Journal of Social Theory* 5: 243–263.

Reno, Joshua
2011 Beyond risk: Emplacement and the production of environmental evidence. *American Ethnologist* 38(3): 516–530.

Rolston, Jessica Smith
2010 Risky business: Neoliberalism and workplace safety in Wyoming coal mines. *Human Organization* 45(1): 43–52.

Rome, Adam
2001 *The bulldozer in the countryside: Suburban sprawl and the rise of American environmentalism*. Cambridge: Cambridge University Press.

Romm, Joseph
2007 *Hell and high water: Global warming—the solution and the politics—and what we should do*. New York: HarperCollins/William Morrow.

Røpke, Inge
2009 Theories of practice—New inspiration for ecological economic studies on consumption. *Ecological Economics* 68(10): 2490–2497.

Ross, Michael L.
1999 The political economy of the resource curse. *World Politics* 51(2): 297–322.
2012 *The oil curse: How petroleum wealth shapes the development of nations*. Princeton, NJ: Princeton University Press.

Ruddell, Rick
2011 Boomtown policing: Responding to the dark side of resource development. *Policing: A Journal of Policy and Practice* 5(4): 328–342.

Rupp, Stephanie
Forthcoming Powerplay: Ghana, China, and the politics of energy. In *African Studies Review*, special issue "Africa and China: Ethnographic and historical perspectives," edited by Jamie Monson and Stephanie Rupp.

Sachs, Jeffrey D. and Andrew M. Warner
2001 The curse of natural resources. *European Economic Review* 45: 827–838.

Sampson, Anthony
1975 *The seven sisters*. New York: Bantam Press.

Sànchez, Teodoro
2010 *The hidden energy crisis; How policies are failing the world's poor*. Bourton on Dunsmore, Rugby, Warwickshire, UK: Practical Action Publishing.

Sawyer, Suzana
2004 *Crude chronicles: Indigenous politics, multinational oil, and neoliberalism in Ecuador*. Durham, NC: Duke University Press.
2009 Human energy. *Dialectical Anthropology* 34: 67–75.

Sayre, Kenneth M.
1977 *Values in the electric power industry*. Notre Dame, IN: University of Notre Dame Press.

Scheer, Hermann
1999 *The solar economy: Renewable energy for a sustainable future.* Sterling, VA: Earthscan.

Schipper, Lee and International Energy Agency
1997 *Indicators of energy use and efficiency: Understanding the link between energy and human activity,* Paris: OECD and IEA.

Schwegler, Tara and Michael G. Powell
2008 Unruly experts: Methods and forms of collaboration in the anthropology of public policy. *Anthropology in Action* 15: 1–9.

Scientific American
2006 *Energy's future beyond carbon: How to power the economy and still fight global warming*. Special Issue (September).

Scott, James C.
1998 *Seeing like a state: How certain schemes to improve the human condition have failed.* New Haven, CT: Yale University Press.

Selby, Jan
2005 Oil and water: The contrasting anatomies of resource conflicts. *Government and Opposition* 40(2): 200–224.

Serchuk, Adam
2000 The environmental imperative for renewable energy: Executive summary, 3–9. *Renewable Energy Policy Project Special Report*, Special Earth Day Report, April.

Shove, Elizabeth
2003 *Comfort, cleanliness, and convenience: The social organization of normality*. Oxford and New York: Berg.
2010 Social theory and climate change: Questions often, sometimes, and not yet asked. *Theory, Culture, and Society* 27(3): 277–288.

Sieferle, Rolf Peter
1990 The energy system: A basic concept of environmental history. In *The silent countdown: Essays in European environmental history*, edited by Peter Brimblecombe and Christian Pfister, 9–20. Germany: Springer-Verlag.

Smil, Vaclav
1992 Elusive links: Energy, value, economic growth, and quality of life. *OPEC Review* (Spring): 1–21.
1994 *Energy in world history*. Boulder, CO: Westview Press.
2000 Energy in the twentieth century: Resources, conversions, costs, uses, and consequences. *Annual Review of Energy and the Environment* 25: 21–51.
2003 *Energy at the crossroads: Global perspectives and uncertainties*. Cambridge, MA: Massachusetts Institute of Technology Press.
2008 *Energy in nature and society: General energetics of complex systems*. Cambridge, MA: Massachusetts Institute of Technology Press.

Smith, Crosbie
1998 *The science of energy: A cultural history of energy physics in Victorian Britain*. Chicago: University of Chicago Press.

Soares de Oliveira, Ricardo
2007 *Oil and politics in the Gulf of Guinea*. New York: Columbia University Press.

Spreng, Daniel T.
1988 *Net-energy analysis and the energy requirements of energy systems*. Westport, CT: Praeger.

Stammler, Florian and Emma Wilson
2006 Dialogue for development: An exploration of relations between oil and gas companies, communities, and the state. *Sibirica* 5(2): 1–42.

Steffy, Loren C.
2011 *Drowning in oil: BP and the reckless pursuit of profit*. New York: McGraw Hill.

Strauss, Sarah and Ben Orlove, eds.
2003 *Weather, climate, culture*. Oxford: Berg Publishers.

Taleb, Nassim
2007 *The black swan: The impact of the highly improbable*. New York: Random House.

Tamminen, Terry
2006 *Lives per gallon: The true cost of our oil addiction government*. Washington, DC: Island Press.

Taubes, Gary
2002 Whose nuclear waste? *Technology Review* 105(1): 60–67.

Tauxe, Caroline
1993 *Farms, mines and main streets: Uneven development in a Dakota County*. Philadelphia: Temple University Press.

Taylor, Ian
2006 China's oil diplomacy in Africa. *International Affairs* 82(3): 937–959.

Temples, James R.
1980 The politics of nuclear power. *Political Science Quarterly* 95(2): 239–260.

Topçu, Sezin
2008 Confronting nuclear risks: Counter-expertise as politics within the French nuclear industry debate. *Nature and Culture* 3(2): 225–45.

Tugwell, Franklin
1980 Energy and political economy. *Comparative Politics* 13: 103–118.

Tussing, Arlon R. and Bob Tippee
1995 *The natural gas industry: Evolution, structure, and economics.* Tulsa, OK: PennWell Books.

Urquhart, Ian
2010 Between the sands and a hard place? Aboriginal peoples and the oil sands. Buffett Center for International and Comparative Studies Working Paper: Energy Series.

Valdivia, Gabriela
2008 Governing the relations between people and things: Citizenship, territory, and the political economy of petroleum in Ecuador. *Political Geography* 27: 456-477.

Victor, David, Amy M. Jaffe, and Mark H. Hayes
2006 *Natural gas and geopolitics: From 1970 to 2040.* Cambridge: Cambridge University Press.

Wackernagel, Mathis and William E. Rees
1996 *Our ecological footprint: Reducing human impact on the Earth.* Gabriola Island, Canada: New Society Publishers.

Walsh, Anna C. and Jeanne Simonelli
1986 Migrant women in the oil field: The functions of social networks. *Human Organization* 45(1): 43–52.

Wamukonya, Njeri
2003 African power sector reforms: Some emerging lessons. *Energy* 2(1): 7–15.

Wang, Michael
2004 Resource curse: Governmentality, oil, and power in the Niger Delta, Nigeria. *Geopolitics* 9(1): 50–80.
2011 Blood oil: The anatomy of a petro-insurgency in the Niger Delta. In *Crude domination: An anthropology of oil*, edited by Andrea Behrends, Stephen P. Reyna, and Gunther Schlee, 49–80. Oxford and New York: Berghahn.

Warkov, Seymour, ed.
1978 *Energy policy in the United States: Social and behavioral dimensions.* New York: Praeger Publishers.

Watts, Michael
2004 Violent environments: Petroleum conflict and the political ecology of rule in the Niger Delta, Nigeria. In *Liberation ecologies*, edited by Richard Peet and Michael Watts. London: Routledge.

Watts, Michael, with Ed Kashi (photographer)
2008 *The curse of the black gold.* New York: Powerhouse Press.

Weaver, Bobby D.
2010 *Oilfield trash: Life and labor in the oil patch.* College Station: Texas A&M University Press.

Wernham, Aaron
2007 Inupiat health and proposed Alaskan oil development: Results of the first integrated health impact assessment/environmental impact statement for proposed oil development on Alaska's North slope. *EcoHealth* 4: 500–513.

Weszkalnys, Gisa
2008 Hope and oil: Expectations in São Tomé e Príncipe. *Review of African Political Economy* 35(3): 473–482.
2010 Re-conceiving the resource curse and the role of anthropology. *Suomen Antropologi* 35(1): 87–90.
2011 Cursed resources, or articulations of economic theory in the Gulf of Guinea. *Economy and Society* 40(3): 345–372.

White, Leslie
1943 Energy and the evolution of culture. *American Anthropologist* 45(3): 335–356.
1949 *The science of culture: A study of man and civilization*. New York: Farrar, Straus and Giroux.
1959 *The evolution of culture: The development of civilization to the fall of Rome*. New York: McGraw-Hill.

White, Richard
1995 *The organic machine*. New York: Hill and Wang
2007 *The Evolution of Culture*. Walnut Creek, CA: Left Coast Press..

Whitehead, John
1987 The partition of energy by social systems: A possible anthropological tool. *American Anthropologist* 89: 686–700.

Wilhite, Harold
2008a *Consumption and the transformation of everyday life: A view from South India*. Basingstoke and New York: Palgrave Macmillan.
2008b New thinking on the agentive relationship between end-use technologies and energy-using practices. *Journal of Energy Efficiency* 1(2): 121–130.
2009 The conditioning of comfort. *Building Research & Information* 37(1): 84–88.

Wilhite, Harold, Hidetoshi Nakagami, Takashi Masuda, Yukiko Yamaga, and Hiroshi Haneda
1996 A cross-cultural analysis of household energy-use behaviour in Japan and Norway." *Energy Policy* 24(9): 795–803.

Wilhite, Harold and Jorgen Norgaard
2004 Equating efficiency with reduction: A self-deception in energy policy. *Energy and Environment* 15(6): 991–1009.

Wilhite, Harold, Elizabeth Shove, Loren Lutzenhiser, and Willett Kempton
2000 The legacy of twenty years of demand side management: We know more about individual behavior but next to nothing about demand. In *Society, behaviour and climate change,* edited by Eberhard Jochem, Jayant A. Stathaye, and Daniel Bouille, 109–126. Dordrecht, the Netherlands: Kluwer Academic Press.

Wilk, Rick
2002 Culture and energy consumption. In *Energy: Science, policy and the pursuit of sustain ability,* edited by Robert Bent, Lloyd Orr, and Randall Baker, 109–130. Washington, DC: Island Press.
2009 Consuming ourselves to death: The anthropology of consumer culture and climate change. In *Anthropology and climate change: From encounters to actions,* edited by Susan Crate and Mark Nuttall, 265–276. Walnut Creek, CA: Left Coast Press.

Wilk, Rick and Harold Wilhite
1985 Why don't people weatherize their homes? An ethnographic solution. *Energy: The International Journal* 10(5): 621–630.

Wilkinson, Kenneth P., James G. Thompson, Robert R. Reynolds, and Lawrence M. Ostresh
1982 Local social disruption and Western energy development. *Pacific Sociological Review* 25(3): 275–296.

Wilson, Ernest J.
1987 World politics and international energy markets. *International Organization* 41(1): 125–149.

Winther, Tanja
2008 *The impact of electricity: Development, desires and dilemmas.* Oxford: Berghahn.

Wolfson, Richard
2008 *Energy, environment and climate.* New York: W. W. Norton & Company.

Yanagisako, Sylvia and Carol Delaney
1994 Naturalizing Power. In *Naturalizing power: Essays in feminist cultural analysis*, edited by Sylvia J. Yanagisako and Carol Delaney, 1–22. London: Taylor and Francis.

Yergin, Daniel
1991 *The prize: The epic quest for oil, money and power.* New York: Simon & Schuster.

Zalik, Anna
2011 Protest as violence in oilfields: The contested representation of profiteering in two extractive sites. In *Accumulating insecurity,* edited by Shelley Feldman, Charles Geisler, and Gayatri A. Menon, 261–284. Athens: University of Georgia Press.

CONTRIBUTORS

DIANE AUSTIN is associate research professor in the University of Arizona's Bureau of Applied Research in Anthropology. She studies community dynamics amid large-scale industrial activity, such as offshore petroleum, and community-based collaborative research and outreach programs. She has developed and maintained multisectoral partnerships with Native American communities, U.S. and Mexican border communities, and Gulf Coastal communities since the mid-1990s.

LISA BREGLIA is an interdisciplinary scholar specializing in Latin America. She has more than fifteen years of experience conducting research in Mexico and is the author of *Monumental Ambivalence: The Politics of Heritage* (2006) and *Living with Oil: Promises, Peaks, and Decline on Mexico's Gulf Coast* (2013). She teaches global affairs at George Mason University and lives in Washington, DC.

ELIZABETH CARTWRIGHT is professor of anthropology and nursing at Idaho State University; she works in Latin America and in the United States. Publications include *Women's Health: New Frontiers in Advocacy & Social Justice Research* (2007) and *Spaces of Illness and Curing: The Amuzgos of Oaxaca between the Sierra Sur and the Agricultural Camps of Sonora* (2003) (in Spanish). She was a guest editor for a special double issue of the journal *Medical Anthropology* focusing on structural vulnerability (2011). Her work focuses on health, environment, and community justice issues.

CHELSEA CHAPMAN is a doctoral candidate in the department of anthropology at the University of Wisconsin–Madison. Her present research involves hydrocarbon and renewable energy in central Alaska, where she focuses on regional power production projects; histories of ecological, economic and political change; and cultural conceptions of energy and land.

MICHAEL DEGANI is a PhD candidate in anthropology at Yale University. He is currently conducting dissertation research on electricity, informal economies, and the formation of urban infrastructures in Dar es Salaam, Tanzania.

ANNA GARWOOD is the executive director of Green Empowerment, an international NGO that works with partner organizations in Latin America, Africa, and Southeast Asia to provide access to renewable energy, clean water, and sustainable solutions. With a degree in anthropology from the University of Virginia, graduate certificate in sustainability from Portland State

University, and twelve years of experience in Peru, Nicaragua, and other countries, she specializes in the socioeconomic and cross-cultural issues of rural electrification and community development.

ALF HORNBORG is an anthropologist and professor of human ecology at Lund University, Sweden. He is the author of *The Power of the Machine* (2001) and *Global Ecology and Unequal Exchange* (2011) and lead editor of *The World System and the Earth System* (2007), *Rethinking Environmental History* (2007), *International Trade and Environmental Justice* (2010), *Ethnicity in Ancient Amazonia* (2011), and *Ecology and Power* (2012).

THOMAS LOVE is professor of anthropology at Linfield College. Working in the central Andes and the Pacific Northwest, he has written on energy and human ecology issues in *Anthropology Today*, *American Ethnologist*, *Ambio*, and *Journal of Sustainable Forestry*. He coedited *State, Capital and Rural Society: Anthropological Perspectives on Political Economy in Mexico and the Andes* (1989) with B. Orlove and M. Foley, and is completing a monograph, *The Independent Republic of Arequipa*, with University of Texas Press.

ARTHUR MASON is visiting assistant professor at the energy and resources group, University of California, Berkeley. He holds a PhD in cultural anthropology from UC Berkeley and is director of StudioPolar, a National Science Foundation initiative that examines the work of consultant expertise in stabilizing perspectives on arctic natural gas development.

THOMAS MCGUIRE is research anthropologist in the Bureau of Applied Research in Anthropology and professor at the School of Anthropology, University of Arizona. His interests focus on natural resource regimes: fisheries, oil and gas, water. He is a student of the Gulfs of Mexico and California and of the greater Southwest.

LAURA NADER is professor of anthropology at the University of California, Berkeley. Her current work focuses on how central dogmas are made and how they work. *Energy Choices in a Democratic Society* (1980) is a multidisciplinary collaborative effort of the National Academy of Sciences. In 2010 *The Energy Reader* was published. Nader is a member of the Academy of Arts and Sciences.

DEREK NEWBERRY is a PhD candidate in the Department of Anthropology at the University of Pennsylvania. His research explores the intersection of corporate ethics, rural development, and international environmental standards in shaping the Brazilian biofuel commodity chain. Prior to graduate school, he was an analyst of sustainable development trends in Latin America with experience in the government and NGO sectors.

DEVON REESER is a master's degree candidate in the Department of Global and Area Studies, with a minor in environment and natural resources, at the University of Wyoming. Her research focuses on adaptation to climate change within vulnerable communities. She is currently volunteering and researching in Paraguay with the U.S. Peace Corps.

JESSICA SMITH ROLSTON is the Hennebach Assistant Professor of Energy and Society in the Division of Liberal Arts and International Studies at the Colorado School of Mines. With the support of a National Endowment for the Humanities Fellowship, she is completing the manuscript for *Mining Coal and Undermining Gender: Rhythms of Work and Family in the American West*.

STEPHANIE RUPP is assistant professor in the anthropology department at Lehman College, City University of New York. She is the author of *Forests of Belonging: Identities, Ethnicities, and Stereotypes in the Congo River Basin* (2011). Her research areas include cultures of energy, African forest communities, and engagements between African and Asian nations.

SARAH STRAUSS is associate professor of anthropology at the University of Wyoming. In addition to the University of Wyoming, she has been a visiting scientist at the National Center for Atmospheric Research (NCAR) in Boulder, Colorado, and visiting professor at the Department of Geosciences of the University of Fribourg, Switzerland. Her books include *Positioning Yoga* (2004) and *Weather, Climate, Culture* (2003, edited with Ben Orlove). She works in India, Switzerland, and North America on questions of climate change, energy, and health.

SCOTT VANDEHEY earned his PhD in cultural anthropology from the University of California, San Diego in 2009. His research focuses on suburban citizenship in the United States and he has recently tuned his attention to suburban sustainability issues. He teaches at Willamette University and Linfield College and resides outside of Portland, Oregon.

GISA WESZKALNYS is lecturer in anthropology at the London School of Economics. She has conducted research on natural resource development, interdisciplinary research practices, and urban planning, and is author of

Berlin, Alexanderplatz: Transforming place in a unified Germany (2010), and *Elusive Promises: Planning in the Contemporary World* (2013, coedited with Simone Abram). Her current work examines the temporality and materiality of oil exploitation in the Gulf of Guinea.

HAROLD WILHITE is a social anthropologist and research director at the University of Oslo's Centre for Development and Environment. He has published widely on consumption, energy, and development, based on research in several regions of the world. His 2008 book *Consumption and the Transformation of Everyday Life: A View from South India* examines how consumption is changing in the encounter between transnational capitalism and deeply anchored cultural practices.

TANJA WINTHER is senior researcher at the Centre for Development and Environment, University of Oslo. With her background in social anthropology (PhD) and power engineering (MSc), her research centers on the social, cultural, and political dimensions of electricity from production to end use. Winther is author of *The Impact of Electricity: Development, Desires and Dilemmas* (2008), which was based on long-term fieldwork in Zanzibar, Tanzania. More recently she has focused on energy's social dimensions in India, Kenya, and Norway.

ACKNOWLEDGMENTS

As editors we have thoroughly enjoyed and benefited from the three years of collaboration involved in this project. From start to finish we have prodded, encouraged, and critiqued each others' work and perspectives. Likewise, we have enjoyed engaging with colleagues around the world who are hard at work on urgent energy issues and are committed to anthropological engagement with energy in both practice and analysis.

We are grateful to Jennifer Collier, our editor at Left Coast Press, who has enthusiastically supported this project since its conception. She steadfastly held the larger project in perspective (and on schedule), even as she welcomed the diversity of materials and theories that an anthropological overview of energy issues necessarily entails. Rachel Fudge came to our rescue at the eleventh hour, as did Scott Smiley; they were both integral to the production process.

We also extend our deep thanks to Jin Choi and Thomas Shine for allowing us to use an image of their spectacular power pylons—an award-winning architectural design that humanizes the electrical grid—for the cover of this volume.

We are all grateful to our departments and colleagues for their support and forbearance during the long process of bringing this project to fruition. Sarah Strauss thanks her husband, Carrick Eggleston, fellow energy researcher and off-grid solar installation expert, for support and insights from the world of physical sciences, as well as her children, Rory and Lia, for their patience through this long process. Stephanie Rupp is grateful for the joint-juggling provided by her husband, Ju-Hon, and for the budding interest that Kai-Lin, Kai-Shan, and Kai-Jin have developed for everything related to energy. Tom Love thanks Linfield College for release time and stipend in the latter stages of this project by being awarded the Kelley Fellowship, and his wife Penny for her support and encouragement throughout.

INDEX

Note: page numbers in *italics* refer to figures.